Methods in Molecular Biology™

AF768709

Series Editor
John M. Walker
School of Life Sciences
University of Hertfordshire
Hatfield, Hertfordshire, AL10 9AB, UK

For other titles published in this series, go to
www.springer.com/series/7651

High-Throughput Next Generation Sequencing

Methods and Applications

Edited by

Young Min Kwon

Department of Poultry Science, and Cell and Molecular Biology Program, University of Arkansas, Fayetteville, AR, USA

Steven C. Ricke

Center for Food Safety & Department of Food Science and the Cell and Molecular Biology Program, University of Arkansas, Fayetteville, AR, USA

Humana Press

Editors
Young Min Kwon, Ph.D.
Department of Poultry Science
and Cell and Molecular Biology Program
University of Arkansas
Fayetteville, AR
USA
ykwon@uark.edu

Steven C. Ricke, Ph.D.
Center for Food Safety
& Department of Food Science
and the Cell and Molecular
Biology Program
University of Arkansas
Fayetteville, AR
USA
sricke@uark.edu

ISSN 1064-3745 e-ISSN 1940-6029
ISBN 978-1-4939-6164-1 ISBN 978-1-61779-089-8 (eBook)
DOI 10.1007/978-1-61779-089-8
Springer New York Dordrecht Heidelberg London

Printed on acid-free paper

Humana Press is part of Springer Science+Business Media (www.springer.com)

Preface

The increasing demand for more cost-effective high-throughput DNA sequencing in this postgenome era has triggered the advent of the "next generation sequencing" methods. Due to their novel concepts and extraordinary high-throughput sequencing capacity, these methods allow researchers to grasp system-wide landscapes of the complex molecular events taking place in various biological systems, including microorganisms and microbial communities. These methods are now being recognized as an essential tool for more comprehensive and deeper understanding of the mechanisms underlying many biological processes. With realistic expectation that these methods will continue to improve at a rapid pace, biological scientists are excited about the growing possibilities for new research approaches that can be offered by these technologies. In *High-Throughput Next Generation Sequencing: Methods and Applications*, expert researchers explore the most recent advances in the applications of next generation sequencing technologies with emphasis on microorganisms and their community. However, the methods described in this book will also find general applications on the study of any living organisms. As part of the highly successful *Methods in Molecular Biology*™ series, the chapters compile step-by-step readily reproducible laboratory protocols, lists of the necessary materials and reagents, and tips on troubleshooting and avoiding known pitfalls.

Comprehensive and cutting-edge, *High-Throughput Next Generation Sequencing: Methods and Applications* is an excellent collection of chapters to aid all scientists who wish to apply this innovative research tools to enhance their own pursuits in microbiology and also biology in general.

Fayetteville, AR

Young Min Kwon
Steven C. Ricke

Contents

Contributors

BRIAN J. AKERLEY • *Department of Molecular Genetics and Microbiology, University of Massachusetts Medical School, Worcester, MA, USA*
JONATHAN BADALAMENTI • *Center for Environmental Biotechnology, Biodesign Institute, Arizona State University, Tempe, AZ, USA*
ANDREW CAMILLI • *Howard Hughes Medical Institute and Tufts University School of Medicine, Boston, MA, USA*
NICHOLAS CARUCCIO • *Epicentre Biotechnologies, Madison, WI, USA*
MARIE CAUSEY • *Helicos BioSciences Corporation, Cambridge, MA, USA*
ZHOUTAO CHEN • *Life Technologies, Carlsbad, CA, USA*
SCOT E. DOWD • *Research and Testing Laboratory, Lubbock, TX, USA*
XIAOPING DUAN • *Life Technologies, Carlsbad, CA, USA*
TIM J. DUMONCEAUX • *Agriculture and Agri-Food Canada Saskatoon Research Centre, Saskatoon, SK, Canada*
SUHUA FENG • *Department of Molecular, Cell and Developmental Biology, University of California, Los Angeles, CA, USA*
JEFFREY D. GAWRONSKI • *Department of Molecular Genetics and Microbiology, University of Massachusetts Medical School, Worcester, MA, USA*
JACK A. GILBERT • *Plymouth Marine Laboratory, The Hoe, Plymouth, UK*
IRENE B. HANNING • *University of Tennessee, Department of Food Science and Technology, Knoxville, TN, USA*
CHRISTOPHER E. HART • *Helicos BioSciences Corporation, Cambridge, MA, USA*
SEAN M. HEMMINGSEN • *National Research Council Plant Biotechnology Institute, Saskatoon, SK, Canada*
JANET E. HILL • *Department of Veterinary Microbiology, University of Saskatchewan, Saskatoon, SK, Canada*
YUICHI HONGOH • *Department of Biological Sciences, Graduate School of Bioscience and Biotechnology, Tokyo Institute of Technology, Tokyo, Japan*
MARGARET HUGHES • *School of Biological Sciences, University of Liverpool, Liverpool, UK*
STEVEN E. JACOBSEN • *Department of Molecular, Cell and Developmental Biology, University of California, Los Angeles, CA, USA*
DANIEL R. JONES • *Helicos BioSciences Corporation, Cambridge, MA, USA*
ALIX KIEU • *Helicos BioSciences Corporation, Cambridge, MA, USA*
TAEJOONG KIM • *Department of Population Health, College of Veterinary Medicine, University of Georgia, Athens, GA, USA*
BYUNG-WHI KONG • *Department of Poultry Science, and Cell and Molecular Biology Program, University of Arkansas, Fayetteville, AR, USA*
IWANKA KOZAREWA • *The Wellcome Trust Sanger Institute, Hinxton, Cambridge, UK*
ROSA KRAJMALNIK-BROWN • *Center for Environmental Biotechnology,*

Biodesign Institute, Arizona State University, Tempe, AZ, USA
YOUNG MIN KWON • *Department of Poultry Science, and Cell and Molecular Biology Program, University of Arkansas, Fayetteville, AR, USA*
DAVID LAPOINTE • *Information Services, University of Massachusetts Medical School, Worcester, MA, USA*
BONNIE LAVEROCK • *Plymouth Marine Laboratory, The Hoe, Plymouth, UK*
STAN LETOVSKY • *Helicos BioSciences Corporation, Cambridge, MA, USA*
MATTHEW G. LINKS • *Agriculture and Agri-Food Canada Saskatoon Research Centre, Saskatoon, SK, Canada; Department of Veterinary Microbiology, University of Saskatchewan, Saskatoon, SK, Canada*
DORON LIPSON • *Helicos BioSciences Corporation, Cambridge, MA, USA*
JANE M. LIU • *Department of Chemistry, Drew University, Madison, NJ, USA*
PATRICE M. MILOS • *Helicos BioSciences Corporation, Cambridge, MA, USA*
MARTIN MUHLING • *TU Bergakademie Freiberg, IÖZ – Interdisciplinary Centre for Ecology, Freiberg, Germany*
EGBERT MUNDT • *Department of Population Health, College of Veterinary Medicine, University of Georgia, Athens, GA, USA*
RAJESH NAYAK • *U.S. Food and Drug Administration, National Center for Toxicological Research, Jefferson, AR, USA*
FATIH OZSOLAK • *Helicos BioSciences Corporation, Cambridge, MA, USA*
PRATHAP PARAMESWARAN • *Center for Environmental Biotechnology, Biodesign Institute, Arizona State University, Tempe, AZ, USA*
MATTEO PELLEGRINI • *Department of Molecular, Cell and Developmental Biology, University of California, Los Angeles, CA, USA*
GEOFFREY A. PETERS • *National Microbiology Laboratory, Public Health Agency of Canada, Winnipeg, MB, Canada*
TAL RAZ • *Helicos BioSciences Corporation, Cambridge, MA, USA*
STEVEN C. RICKE • *Center for Food Safety & Department of Food Science and the Cell and Molecular Biology Program, University of Arkansas, Fayetteville, AR, USA*
BRUCE E. RITTMANN • *Center for Environmental Biotechnology, Biodesign Institute, Arizona State University, Tempe, AZ, USA*
LIUDMILLA RUBBI • *Department of Molecular, Cell and Developmental Biology, University of California, Los Angeles, CA, USA*
JOHN SCHELLENBERG • *Department of Medical Microbiology, University of Manitoba, Winnipeg, MB, Canada*
KATHLEEN E. STEINMANN • *Helicos BioSciences Corporation, Cambridge, MA, USA*
PALLAVI SINGH • *Cell and Molecular Biology Program, University of Arkansas, Fayetteville, AR, USA*
YAN SUN • *Research and Testing Laboratory, Lubbock, TX, USA*
BEN TEMPERTON • *Plymouth Marine Laboratory, The Hoe, Plymouth, UK*
EDWARD THAYER • *Helicos BioSciences Corporation, Cambridge, MA, USA*
SIMON THOMAS • *Plymouth Marine Laboratory, The Hoe, Plymouth, UK*
JOHN F. THOMPSON • *Helicos BioSciences Corporation, Cambridge, MA, USA*
ATSUSHI TOYODA • *Comparative Genomics laboratory, National Institute of Genetics,*

Shizuoka, Japan
DANIEL J. TURNER • *The Wellcome Trust Sanger Institute, Hinxton, Cambridge, UK*
RANDALL D. WOLCOTT • *Southwest Regional Wound Care Center, Lubbock, TX, USA*
SANDY M.S. WONG • *Department of Molecular Genetics and Microbiology, University of Massachusetts Medical School, Worcester, MA, USA*
HUSEN ZHANG • *Department of Civil, Environmental, and Construction Engineering, University of Central Florida, Orlando, FL, USA*

Part I

Genome Sequencing

Chapter 1

Helicos Single-Molecule Sequencing of Bacterial Genomes

Kathleen E. Steinmann, Christopher E. Hart, John F. Thompson, and Patrice M. Milos

Abstract

With the advent of high-throughput sequencing technologies, multiple bacterial genomes can be sequenced in days. While the ultimate goal of de novo assembly of bacterial genomes is progressing, changes in the genomic sequence of closely related bacterial strains and isolates are now easily monitored by comparison of their sequences to those of a reference genome. Such studies can be applied to the fields of bacterial evolution, epidemiology, and diagnostics. We present a protocol for single-molecule sequencing of bacterial DNA whose end result is the identification of single nucleotide variants, and various size insertions and deletions relative to a reference genome. The protocol is characterized by the simplicity of sample preparation and the lack of amplification-related sequencing bias.

Key words: Bacterial genome, Single-molecule sequencing, Sequencing bias, SNV, PolyA tail, HeliScope™ sequencer

1. Introduction

Sequencing of multiple bacterial genomes at a depth sufficient to allow their alignment to a reference genome can now be achieved in a single channel of a Heliscope™ Genetic Analysis System (1) and other high-throughput sequencing platforms (2). The ability to determine differences between closely associated bacterial strains and isolates through alignment to a reference genome has contributed greatly to an understanding of bacterial evolution (3, 4). This method of bacterial genome resequencing can be used to track the acquisition of virulence factors and antibiotic resistance through sequencing of multiple isolates (5–7). The genetic changes underlying mutant phenotypes can now be found more easily and at less cost by sequencing than by traditional genetic mapping (8).

Young Min Kwon and Steven C. Ricke (eds.), *High-Throughput Next Generation Sequencing: Methods and Applications*, Methods in Molecular Biology, vol. 733, DOI 10.1007/978-1-61779-089-8_1, © Springer Science+Business Media, LLC 2011

The DNA sample preparation technique used for single-molecule sequencing that is described below is unique in that no amplification or ligation is required. The DNA is fragmented. A polyA tail, added with terminal transferase and dATP, allows hybridization of the DNA to an oligodT-coated flow cell. Minimal sample manipulation results in the lack of GC bias associated with amplification-based technologies (9, 10). Helicos BioSciences demonstrated this lack of bias by showing even coverage for three bacterial genomes with differing GC contents (1). The mean sequencing coverage within 200-bp sliding windows across the *Escherichia coli* K12 MG1655 (50.8% GC), *Staphylococcus aureus* USA 3000 (37.7% GC), and *Rhodobacter sphaeroides* 2.4.1 (68.8% GC) genomes was plotted against the GC content in the same window (Fig. 1). These data demonstrated flat coverage and sequencing accuracy of sequence contexts ranging from 20 to 80% GC. The overall sequence coverage required for accurate SNP calling is related to the evenness of coverage (11). The lack of bias observed with single-molecule sequencing on the HeliScope™ Genetic Analysis System points to its use as a cost-effective and accurate method for bacterial genome resequencing. At present, a single HeliScope™ Sequencer channel provides over 80-fold coverage for a single bacterial genome, far exceeding the 20× to 25× coverage needed for variant detection. When amplification-free barcoding methods are employed, three or more bacterial genomes (depending on the genome size) can be resequenced per lane while retaining the lack of amplification bias.

2. Materials

2.1. DNA Isolation and Ultrasonic Shearing of DNA

1. Qiagen DNA purification Kit (Qiagen, Valencia, CA) (see Note 1).
2. S2 instrument (Covaris, Inc., Woburn, MA) (see Note 2).
3. Preparation station (Covaris, Inc., Woburn, MA).
4. MicroTube holder (single tube). (Covaris, Inc., Woburn, MA).
5. Snap-Cap microTube with AFA fiber and Pre-split Teflon/silicone/Teflon septa (Covaris, Inc., Woburn, MA).
6. Distilled water (Invitrogen, Carlsbad, CA).
7. 10× TE, pH 8.0 (Invitrogen, Carlsbad, CA).
8. 1.5 mL MAXYMum recovery tubes (Axygen Scientific, Union City, CA) (see Note 3).

2.2. Size Selection Using Solid Phase Reversible Immobilization

1. Agencourt®AMPure® XP Kit (Agencourt Bioscience Corp., Beverly, MA).
2. 100% Ethanol (Sigma, St Louis, MO).

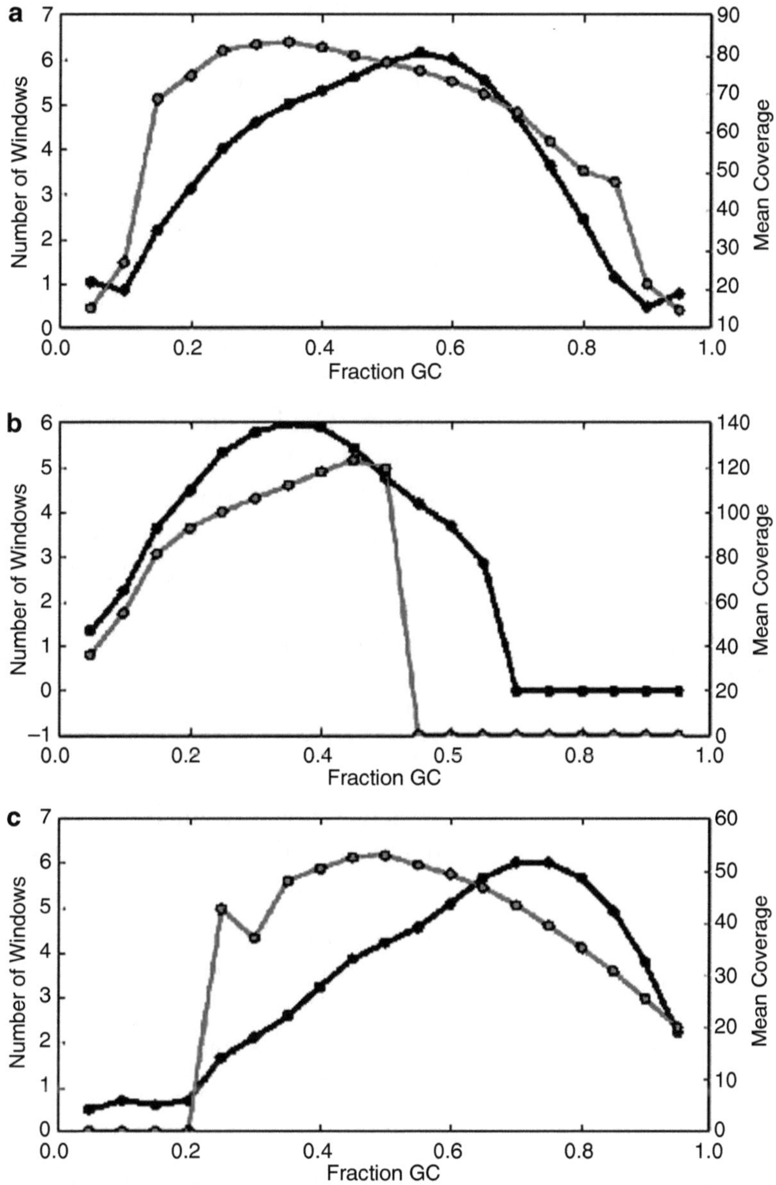

Fig. 1. Single-molecule DNA sequencing provides minimal sequence bias across diverse genomic content. The local GC content and observed mean sequencing coverage were tabulated using a 200-bp sliding window. Windows were then aggregated into GC-content bins ranging from 0 to 1 with a step size of 0.1. Plotted is the mean coverage (GRAY; Right Y-axis) for each window within each of the aggregated GC-content bins (BLACK; Left Y-axis, Log scale). (**a**) *E. coli*. (**b**) *Staphylococcus aureus*. (**c**) *Rhodobacter sphaeroides*. (Reproduced from ref. 1 with permission from Helicos BioSciences Corporation.).

3. Distilled water (Invitrogen, Carlsbad, CA).
4. Dynal® Magnet: DynaMag®-2 Magnet (Invitrogen, Carlsbad, CA) or similar.
5. Heatblock equipped with block milled for 1.5 mL tubes (VWR, Batavia, IL).

2.3. Calculating the Approximate Concentration of 3′ Ends

1. 4–20% TBE gel, 1.0 mM, 12 well or similar (Invitrogen, Carlsbad, CA) (see Note 4).
2. Ultrapure 10× TBE buffer (Invitrogen, Carlsbad, CA).
3. Para-film.
4. 10× BlueJuice™ gel loading buffer (Invitrogen, Carlsbad, CA).
5. 25 bp DNA ladder (Invitrogen, Carlsbad, CA).
6. 1 kB DNA ladder (Invitrogen, Carlsbad, CA).
7. SYBR® Gold nucleic acid gel stain (Invitrogen, Carlsbad, CA).
8. Photodocumentation system compatible with a SYBR® Gold photographic filter.
9. SYBR® Gold Nucleic photographic filter (Invitrogen, Carlsbad, CA).
10. XCell *Surelock*™ Mini-cell (Invitrogen, Carlsbad, CA).
11. Nanodrop™ 1000, 2000, 2000c, or 8000 spectrophotometer (Thermo Fisher Scientific, Waltham, MA).

2.4. PolyA Tailing Reaction

1. Terminal transferase kit: includes terminal transferase enzyme (20,000 U/mL), $CoCl_2$ at 2.5 mM, 10× terminal transferase buffer (New England BioLabs, Ipswich, MA).
2. Helicos™ DNA sample preparation reagents kit: includes Helicos™ PolyA tailing control oligonucleotide TR and Helicos™ PolyA tailing dATP. (Helicos BioSciences Corporation, Cambridge, MA). Store Kit at −80°C (see Note 5).
3. Distilled water (Invitrogen, Carlsbad, CA).
4. 0.2 mL MAXYMum recovery thin wall PCR tubes and 1.5 mL MAXYMum recovery tubes (Axygen Scientific, Union City, CA) (see Note 3).
5. Aluminum block milled for 0.2 mL tubes (VWR, Batavia, IL).
6. DNA engine thermal cycler (BioRad Laboratories, Hercules, CA).

2.5. Determining the Success of Tailing Reaction

1. 4–20% TBE gel, 1.0 mM, 12 well or similar (Invitrogen, Carlsbad, CA) (see Note 4).
2. Ultrapure 10× TBE buffer (Invitrogen, Carlsbad, CA).
3. 10× BlueJuice™ gel loading buffer (Invitrogen, Carlsbad, CA).
4. 100 bp DNA ladder (Invitrogen, Carlsbad, CA).
5. SYBR® Gold nucleic acid gel stain (Invitrogen, Carlsbad, CA).
6. Photodocumentation system compatible with a SYBR® Gold photographic filter.
7. SYBR® Gold nucleic photographic filter (Invitrogen, Carlsbad, CA).
8. XCell *Surelock*™ Mini-cell (Invitrogen, Carlsbad, CA).

2.6. Short Tail Correction

1. Terminal transferase enzyme (20,000 U/mL) (New England BioLabs, Ipswich, MA).
2. Helicos™ PolyA tailing dATP (Helicos BioSciences Corporation, Cambridge, MA).
3. Aluminum block milled for 0.2 mL tubes (VWR, Batavia, IL).
4. DNA engine thermal cycler (BioRad Laboratories, Hercules, CA).

2.7. 3′ Blocking Reaction

1. Terminal transferase enzyme (20,000 U/mL) (New England BioLabs, Ipswich, MA).
2. 1 mM Biotin-11-ddATP (PerkinElmer, Waltham, MA).
3. Aluminum block milled for 0.2 mL tubes (VWR, Batavia, IL).
4. DNA engine thermal cycler (BioRad Laboratories, Hercules, CA).

3. Methods

The major steps in preparing bacterial DNA for single-molecule sequencing on the HeliScope™ Sequencer consist of shearing the DNA to 200–300 bp, adding a polyA tail to allow hybridization to the oligonucleotide-coated Helicos™ Flow Cell and blocking the 3′ end of the DNA with ddATP to prevent that end of the DNA from acting as a substrate for the sequencing-by-synthesis reaction. A bead-based size-selection step after the shearing step removes salts and small nucleic acids that would be tailed, but are not sufficiently long to yield meaningful sequence information. A quality control step ensures that samples are sheared to the appropriate size. A second quality control step uses a control reaction to monitor the tail length of a PolyA tailing control oligonucleotide TR spike to insure sample polyA tail lengths of between 90 and 200 dA. Sequencing of the resultant sample and subsequent data analysis using Helicos variant detection software results in the enumeration of single nucleotide variants (SNVs), and insertions and deletions (indels) of up to 4 bp in length between the sequenced bacteria and a reference using alignment-based methods. Alternative approaches for detecting much longer variants using assembly-based methods are under development.

3.1. DNA Isolation and Ultrasonic Shearing of DNA

The method used for bacterial DNA isolation is dependent upon the bacterial source. The Qiagen DNA Purification Kit has been used successfully for bacterial DNA isolation following manufacturer's instructions (see Note 1).

1. Prepare the Covaris S2 instrument for ultrasonic shearing of DNA by filling the tank on the Covaris S2 instrument with

deionized water to level 12 on the fill line label. The water should cover the visible parts of the microTube when it is in the microTube holder (i.e., to the bottom of the snap cap).

2. Set the chiller to 4°C and turn it on.
3. Turn on the S2 unit by depressing the red switch located at the upper right corner of the instrument.
4. After the instrument is on, open the software. Click the ON button on the control panel under the word DEGAS to begin the degassing procedure. The instrument is ready to use when the water has been degassed for 30 min and the temperature software display is between 6°C and 8°C.
5. Prepare 500 ng to 3 μg of DNA in 120 μl of TE, pH 8.0. If the DNA is not in 120 μL of TE, add the appropriate amount 10× TE, pH 8.0 to make the overall concentration of TE in the solution 1× (see Note 6).
6. Place an unfilled Covaris microTube into the preparation station holder.
7. Keeping the cap on the tube, use a p200 pipette and 200-μL aerosol-free tip to transfer the 120 μL of DNA sample by inserting the tip through the pre-split septa. Place the tip along the interior wall of the tube. Slowly discharge the fluid into the tube, moving the pipette tip up along the interior wall as the tube fills. Be careful not to introduce a bubble into the bottom of the tube. If a bubble appears, remove the bubble by briefly (1–2 s) centrifuging the tube in a low-speed tabletop centrifuge equipped with appropriate adaptors.
8. Slide the tube into the microTube holder while keeping the tube vertical. Make sure the tube is centered in the holder. Carefully insert the holder into the machine. Take care not to introduce bubbles into the bottom of the tube during this process.
9. Click on Configure. On the Method Configuration Screen, set the Mode to Frequency Sweeping and the Bath Temperature Limit to 20°C. In the Treatment 1 box, set the Duty Cycle to 10%, the Intensity to 5 and the Cycles/Burst to 200. Set the time to 60 s and the Number of Cycles to 3. Click on Return to Main Panel. Click Start and Start again when the second screen appears.
10. After shearing is complete, remove the tube from the S2 holder and place it into the preparation station. Remove the snap cap with the tool supplied with the preparation station. Use a p200 pipette to transfer the sheared DNA to a new, clean 1.5-mL tube. A brief centrifugation may be used to collect any DNA remaining in the microTube.
11. Samples may be stored at –20°C after this step.

12. When the shearing is completed, click the OFF button under DEGAS, empty the water tank, turn off the chiller, close the software, and power down the instrument.

3.2. Size Selection Using Solid Phase Reversible Immobilization

1. Warm the AMPure XP bead solution to room temperature and vortex thoroughly to resuspend all beads.
2. Prepare 70% ethanol. Prepare fresh by diluting 7 mL of absolute ethanol into 3 mL of distilled water. Do not use a stock 70% ethanol.
3. Vortex the AMPure XP beads and add 360 μL of the AMPure XP bead slurry to each tube of sheared DNA. Pipette up and down ten times to mix.
4. Incubate the sample slurry for 5–10 min at room temperature.
5. Capture the AMPure XP beads by placing the tube(s) on the Dynal™ magnet until the beads are separated from the solution (approximately 5 min).
6. Carefully aspirate the supernatant keeping the tube(s) on the magnet. Do not disturb the beads adhering to the side of the tube. Take care not to remove any AMPure XP beads (see Note 7).
7. Add 700 ml of 70% EtOH to each tube on the Dynal™ magnet. Wait for 30 s.
8. Keep the tubes on the magnet and carefully aspirate the supernatant (see Note 7).
9. Repeat steps 7 and 8.
10. Briefly centrifuge the tubes to collect any remaining 70% EtOH to the bottom of the tube. Place the tubes back on the magnet and remove the last drops of 70% EtOH with a p10 pipette.
11. Dry the pellet at 37°C in a heat block milled for 1.5 mL tubes. Pellets should be dried until cracks appear in them (approximately 1–5 min). Take care not to over dry the pellets as they will be difficult to resuspend. This step can be performed at room temperature with the drying time being extended to a minimum of 10 min before cracks appear.
12. Elute the sheared DNA sample from the AMPure beads by adding 20 μL of distilled water to each tube. A brief (1–2 s) centrifugation may be necessary to collect all the beads at the bottom of the tube.
13. Pipette the entire volume of each tube up and down 20 times so that the beads are completely resuspended.
14. Place the tube back on the magnet. After the beads are separated from the solution, collect the 20 μL of solution and

place it into a new 1.5-mL tube. This supernatant contains the sheared, size-selected DNA (see Note 8).

15. Add another 20 μL of water to the tube. Repeat steps 13 and 14, this time adding the supernatant to the first elute. The final sheared, size-selected DNA volume should be 40 μL.
16. The DNA can be stored at −20°C after this step.

3.3. Calculating the Approximate Concentration of 3′ Ends

1. These instructions assume the use of an XCell *SureLock*™ Mini-cell and Invitrogen 4–20% gradient gels. An equivalent electrophoresis apparatus compatible with 10×10 cm gel cassettes can also be used. Non-gradient gels are not recommended.
2. Remove the 4–20% TBE gel from its storage pouch. Remove the comb, rinse the wells with water two to three times and remove the plastic strip at the bottom of the gel. Complete the assembly of the gel unit.
3. Prepare 1× TBE running buffer by diluting 100 mL of 10× TBE with 900 mL of water in a graduated cylinder. Cover with para-film and invert to mix. Running buffer may be stored at room temperature.
4. Add running buffer to the center reservoir of the gel apparatus. Check for leaks and reassemble if necessary. Add 1–2 in. of buffer to the bottom reservoir.
5. Load 2 μl aliquots of the samples in 1× BlueJuice™ buffer in a total volume of 10 μl. Make a 1:10 dilution of the 1 kB and 25-bp ladders in distilled water. Load 1 μl of the diluted markers in 1× BlueJuice™ buffer in a total volume of 10 μl.
6. Run the gel at 180 V for 45 minutes.
7. After removing the gel from the cassette, using the tool provided with the XCell *SureLock*™ Mini-cell, stain the gel for 10 min in freshly prepared SYBR® Gold nucleic acid gel stain diluted 1:100,000 in water (see Note 9).
8. Destain the gel in water for 10–15 min, changing the water every 2 min.
9. Image with a photodocumentation system compatible with a SYBR® Gold photographic filter.
10. Determine the average size of your sample by comparing the size of the middle of the sample smear to the size standards (see Note 10).
11. Determine the double-stranded DNA concentration in ng DNA/μL at this step using a Nanodrop™ 1000, 2000, 2000c, or 8000 spectrophotometer.

12. Calculate the pmoles of ends in the sample using the following formula:

    pmoles ends/μL = (X ng DNA/μL) × (1,000 pg/ng) × (pmole/660 pg) × (1/average # bp as determined from the gel photo) × 2 ends/molecule.

3.4. PolyA Tailing Reaction

1. Based on the calculations in step 12 of Subheading 3.3, prepare a Sample DNA tube for each DNA to be tailed by determining the volume of DNA that would give 3 pmoles of ends. Put that volume of DNA into a 0.2-mL PCR tube along with distilled water to bring the final volume to 26 μL.
2. Prepare a Control DNA tube for each DNA to be tailed by determining the volume of DNA that would give 0.8 pmoles of ends. Put that volume of DNA into a 0.2-mL PCR tube along with 1 μL of Helicos™ PolyA Tailing Control Oligo TR and distilled water to bring the final volume to 26 μL.
3. Prepare a separate Oligo TR Control DNA tube (without DNA sample) by putting 4 μL of Helicos™ PolyA Tailing Control Oligo TR in a tube containing 22 μL of distilled water.
4. Prepare a sample master mix by adding 4.4 μL of 10× terminal transferase buffer, 4.4 μL of $CoCl_2$ (2.5 mM), 4.2 μL of Helicos™ PolyA Tailing dATP, and 2.2 μL terminal transferase enzyme (20 U/μL) per sample (see Note 11). The master mix volume includes a 10% scale-up. Mix thoroughly by pipetting the entire mix up and down several times (see Note 12). Keep on ice.
5. Prepare a control master mix by adding 4.4 μL of 10× terminal transferase buffer, 4.4 μL of $CoCl_2$ (2.5 mM), 3 μL of distilled water, 1.4 μL of Helicos™ PolyA Tailing dATP, and 2.2 μL terminal transferase enzyme (20 U/μL) per control reaction (see Note 13). The master mix includes a 10% scale-up. Mix thoroughly by pipetting the entire mix up and down several times (see Note 12). Keep on ice.
6. Heat the Sample and Control Tubes DNA tubes to 95°C for 5 min in a thermocycler. Immediately remove the DNA tubes from the thermocycler and snap cool for a minimum of 2 min by placing the tubes in an aluminum block milled for 0.2 μL tubes that has been prechilled in ice water (see Note 14).
7. Add 14 μL of sample master mix to the sample DNA tubes and 14 μL of control master mix to the control DNA tubes. Mix thoroughly by pipetting up and down ten times (see Note 12).
8. Collect the contents of the tubes into the bottom by briefly centrifuging.

9. Place the tubes in the thermocycler and incubate at 37°C for 60 min, 70°C for 10 min followed by a 4°C hold.
10. The tailed DNA can be stored at −20°C after this step.

3.5. Determining the Success of the Tailing Reaction

1. Gel type and running instructions are identical to those in Subheading 3.3. Instructions here will be limited to how to prepare samples for loading and how to interpret results.
2. Load 20 μl aliquots of the control reactions in 1× BlueJuice™ buffer (18 μL of the control reaction and 2 μL of 10× BlueJuice™).
3. Make a 1:10 dilution of the 100-bp ladder in distilled water. Load 1 μl of the diluted markers in 1× BlueJuice™ buffer in a total volume of 20 μl.
4. The sample itself is difficult to visualize. The band corresponding to the TR oligo spike is visible in the control lanes and monitors the tail length of the sample. All control reactions should migrate at the size of the Oligo TR Control Sample. A longer polyA tail may be indicative of a sample with a reduced number of strands ending in a 3′OH. Only strands having a 3′OH can be tailed. Tailed oligos with 90–200 dA are expected to migrate below the 600-bp band to midway between the 200- and 300-bp bands on the 100-bp ladder (see Note 15 and Fig. 2). If the TR oligo band in the Control reaction lane migrates anywhere between 250 bp and 600 bp, you may proceed to steps in Subheading 3.7.
5. In the rare instances where the band in the Control reaction lane migrates below 250 bp, the sample has a polyA tail shorter than 90 nucleotides. Proceed to steps in Subheading 3.6.
6. If the band in the Control reaction lane migrates above 600 bp, the polyA tail contains more than 200 dA. In this case, the sample could be run on the Helicos™ Genetic Analysis System. However, if sample is not limiting, we recommend repeating the PolyA tailing reaction on another sheared DNA aliquot using twice the amount of input DNA.

3.6. Short Tail Correction

1. Both the control and sample reactions undergo the correction. The denaturation step and thermocycler incubation conditions are as for the PolyA tailing reactions in Subheading 3.4.
2. After snap cooling the tubes, the following reagents are added. For the 3 pmole sample reactions, add 3.9 μL of dATP and 2 μL of terminal transferase. For the 0.8 pmole control reactions, prepare a 1:2 dilution of the dATP stock in water. Add 1.3 μL of diluted dATP and 1 μL of terminal transferase. Mix by pipetting up and down thoroughly ten times.

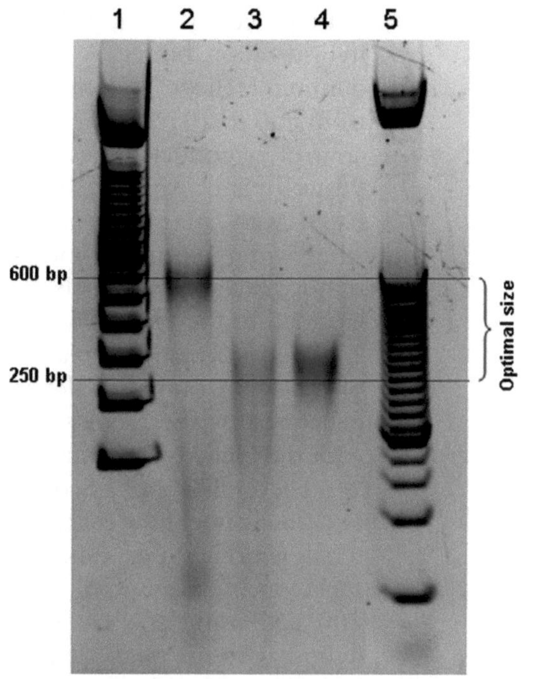

Fig. 2. Migration pattern of tailed TR oligonucleotide in control reactions with optimal dA tail lengths. Tailed TR oligos with 90–200 dA are expected to migrate below the 600-bp and to midway between the 200- and 300-bp bands on the 100-bp ladder. PolyA tailed samples do not migrate normally in the gel. Lane 1: 100-bp DNA ladder. Lane 2: A control reaction with 200 dA. Lane 3: A control reaction with 90 dA. Lane 4: Tailing Control Oligo TR with 90 dA. Lane 5: 25-bp DNA ladder.

3. Steps in Subheading 3.5 maybe performed on the control reaction to determine if the polyA tail length is greater than 90 dA.

3.7. 3′ Blocking Reaction

1. The 3′ Blocking reaction is performed only on the Sample reactions. The denaturation step and thermocycler incubation conditions are as for the PolyA tailing reactions in Subheading 3.4. The reagents to be added to the tubes are outlined below.
2. Dilute the 1-mM Biotin-11-ddATP 1:6 in water. After snap cooling the sample tubes, add 1 μL of the diluted Biotin-11-ddATP and 2 μL of terminal transferase to each tube.
3. After the blocking reaction is completed, add 1 μL of 500 mM EDTA to the samples. Samples should be stored at −20°C.

3.8. Sample Quantification and Sequencing

Sample quantification and sample loading are typically performed by the operator of the Heliscope™ Single Molecule Sequencer, and will therefore be outlined only briefly here.

Sample concentration is determined using the Helicos™ Optihyb™ assay. For the assay, a 1:50 dilution of a DNA sample is prepared by adding 2 μL of a DNA sample to 98 μL of hybridization buffer. A standard curve is generated by serially diluting Helicos™ Control Oligonucleotides at concentrations from 500 to 20 pM. The PolyA tailed and biotin-ddATP blocked control oligonucleotides and DNA samples are captured in a 96-well plate. After washing and blocking, the plates are incubated with an HRP (*h*orse*r*adish *p*eroxidase) conjugate. The streptavidin–biotin complexes that are formed are washed to remove excess HRP. TMB, a chromogenic substrate for the HRP, is added to the plate. The reaction is stopped by the addition of 1 N HCl and read on a plate reader at 450 nm. The concentration of the DNA sample is determined by comparison of the signal generated by the DNA sample to the standard curve generated with the Helicos™ Control Oligonucleotides.

The HeliScope™ Sequencer obtains sequence from one or two 25 channel flow cells, making it possible to sequence 50 bacterial genomes per run. DNA Samples are loaded onto the Heliscope™ Flow Cell at a recommended concentration of 200 pM using the Heliscope™ Sample Loader. The samples hybridize to the oligo dT primers on the flow cell surface and are locked into place by a procedure that ensures that sequencing-by-synthesis starts immediately after the first nonA base on the DNA samples. 120 nucleotide cycle additions are performed in an 8-day run, though the timing is adjustable based on user throughput needs. The HeliScope™ analysis engine on the instrument creates .srf files that correlate template position images with nucleotide addition images to generate sequence information. After the run is completed, .srf files are converted to .sms files which are used for subsequent data analysis.

3.9. Data Analysis

HeliScope™ data analysis can be done using a Unix system with at least 5 GB per CPU core. The HeliSphere™ data analysis pipeline is an open-source software written in the Python programming language. It is available for download at: http://open.helicosbio.com/mwiki/index.php/Releases. Download both the HeliSphere™ package and the examples. Follow the HeliSphere™ User's Guide documentation available at http://open.helicosbio.com/helisphere_user_guide/index.html after installation.

3.10. Running the Resequencing Pipeline

The resequencing pipline uses the raw sequence input file (.sms) to generate reads aligned to a reference genome and to report SNVs and short insertions and deletions (indels) between the sequenced material and the reference. The current version of the resequencing pipeline analyzes data from a single channel. Reliable SNV calling requires roughly 20× depth. With current machine performance of roughly 12–16 M reads per channel and an average read length of 35 bp, we would then recommend this single channel pipeline for resequencing applications where the genome is less than 25 Mb.

In order to run the pipeline, you must specify certain analysis parameters. A detailed description of all analysis parameters and how to apply them under more complex situations (e.g., for a barcoded sample) can be found in the Helisphere™ Users Guide. Default values for most analysis parameters are appropriate for running the pipeline on a single bacterial genome in a single channel. The following example outlines what must be specified for each analysis, how to set up the corresponding run configuration file, and how to launch the analysis.

1. Determine the input file directory and the .sms file name of the file to be processed. For example: /ifs/bioinf/workspace/bacreseq/smsDirectory/file.sms
2. Determine the flow cell and channel to be processed. For example: the flow cell is 1 and the channel is 5.
3. Chose an output directory. For the example: /ifs/bioinf/workspace/bacreseq/outputDir.
4. A reference fasta file has to be chosen and placed in the reference data directory. This directory is defined as the referenceDir variable inside the file (helicos installation root)/pypeline/config/pypeline-site.conf. If an indexDPgenomic database does not exist for this reference it needs to be created with preprocessDB and placed in the same directory. In the example, the reference is human.fasta and the indexDPgenomic data base prefix is human.seed18
5. For accurate mutation detection using SNPSniffer, consideration should be given only to mutations for which the allele with maximum p-Value is 1e-10 or less. Such data can be generated automatically by running the tool with the flag --pvalue_threshold 1e-10.
6. The above information is incorporated into a config file for the run. For example:

   [Global]

   channels = 1:5

   input = /ifs/bioinf/workspace/bacreseq/smsDirectory/file.sms

   outdir = /ifs/bioinf/workspace/bacreseq/outputDir

   referenceName = human

   [snpSniffer]

   p-valueThreshold = 1e-10

   It is saved in a .conf file. For example, this file is named run.resequencing.conf.
7. To incorporate the parameters used in the example, the pipeline should be launched with the following command:

   pypeline -p resequencing -c run.resequencing.conf

3.11. Resequencing Summary Report

The resequencing pipeline generates various reports. The Resequencing summary report (reseq.summary.txt) can be used to assess the quality of a run. The table generated contains the following information:

1. *Group*: flow cell and channel of the processed data.
2. *Reference*: reference used to align the data.
3. *Raw*: number of unfiltered reads in the input file.
4. *Filtered*: number of reads remaining after a filtering step for read length and some sequence contexts. To accurately detect indels of longer lengths (up to 4) during bacterial genome resequencing, filtering out reads shorter than 25 is recommended.
5. *Aligned*: number of reads aligning to the given reference at a minAlignScore threshold of 1.3 or higher. This score takes into account length of the read, the number of matched nucleotides, and penalties for misalignments using a scoring scheme unique to the resequencing pipeline.
6. *%filtered*: filtered reads/raw reads.
7. *%aligned*: aligned reads/filtered reads.
8. *MeanLen*: mean length of aligned read. Mean length is around 35.
9. *Del*, *Ins*, *Sub*, *Error*: assessed per-base deletion, insertion, substitution, and total error rates, respectively based on alignments reference.
10. *NumSNPs*: number of sequence variants detected by SNPSniffer.

3.12. The Mutation Analysis Report

SNPSniffer generates a table of SNPs named: <out_prefix>_SNP.txt. The Mutation Analysis report has the following fields (see Table 1):

1. *Num*: mutation number
2. *RefName*: reference name
3. *Start*: start position of mutation
4. *End*: end position of mutation
5. *Ref*: reference nucleotide/s
6. *Type*: type of mutation SUB (substitution), INS (insertion), DEL (deletion) or LEN (homopolymer deletion)
7. *ModelScore*: defined only for LEN mutation. Deletions in homopolymers are indicated by the LEN mutation type. This type of mutation is determined by matching the observed distribution of HP (homopolymer) length in the reads to a set of possible models. Two alleles will be present in the output table when the best matching model is the mixture of the

Table 1
Mutation Table resulting from an alignment of the sequence generated from *Escherichia coli K12 MG1655* to the *E. coli EDL933* reference sequence

Part A

Num	RefName	Start	End	Ref	Type	Model score	Shift	Allele 1	Allele2	Depth	Count 1	Count 2
14722	gi|56384585|gb|AE005174.2|	1671091	1671091	A	SUB		0	G		252	233	
14718	gi|56384585|gb|AE005174.2|	1670442	1670442	C	SUB		0	G		232	225	
14719	gi|56384585|gb|AE005174.2|	1670606	1670606	T	SUB		0	C		230	228	
14720	gi|56384585|gb|AE005174.2|	1670736	1670736	G	SUB		0	T		223	218	
14717	gi|56384585|gb|AE005174.2|	1670224	1670224	T	SUB		0	C		179	170	
14721	gi|56384585|gb|AE005174.2|	1670843	1670843	A	SUB		0	G		173	168	
29003	gi|56384585|gb|AE005174.2|	3109283	3109283	G	SUB		0	A	G	172	56	112
14723	gi|56384585|gb|AE005174.2|	1671310	1671310	C	SUB		0	T		164	162	
4359	gi|56384585|gb|AE005174.2|	406325	406325	C	SUB		0	T	C	157	41	114
29002	gi|56384585|gb|AE005174.2|	3109275	3109275	T	SUB		0	C	T	151	70	78
9367	gi|56384585|gb|AE005174.2|	867170	867170	A	SUB		0	G		122	120	
53715	gi|56384585|gb|AE005174.2|	5210721	5210721	G	SUB		0	T	G	111	41	66
51180	gi|56384585|gb|AE005174.2|	5013503	5013503	T	SUB		0	G	T	109	28	79
40900	gi|56384585|gb|AE005174.2|	4076594	4076594	T	SUB		0	C	T	108	41	67
51432	gi|56384585|gb|AE005174.2|	5029061	5029061	T	SUB		0	C	T	101	42	59
42574	gi|56384585|gb|AE005174.2|	4225504	4225504	N	SUB		0	C		100	98	
34786	gi|56384585|gb|AE005174.2|	3572816	3572816	A	SUB		0	A	C	98	41	55

(continued)

Table 1 (continued)

Part B

Num	Freq1	Freq2	*p*-Value1	*p*-Value2	Freq A	Freq C	Freq G	Freq T	Freq -
14722	0.924603		0		0.0238095	0	0.924603	0	0.0515873
14718	0.969828		0		0.0043103	0	0.969828	0	0.0258621
14719	0.991304		0		0	0.991304	0	0	0.0086957
14720	0.977578		0		0	0	0	0.977578	0.0224215
14717	0.949721		0		0.0055866	0.949721	0	0	0.0446927
14721	0.971098		0		0	0	0.971098	0	0.0289017
29003	0.325581	0.651163	7.48E-90	1.79E-223	0.325581	0	0.651163	0	0.0232558
14723	0.987805		0		0.0060976	0	0	0.987805	0.0060976
4359	0.261146	0.726115	1.18E-61	1.71E-236	0	0.726115	0	0.261146	0.0127389
29002	0.463576	0.516556	2.29E-125	1.74E-144	0	0.463576	0	0.516556	0.0198676
9367	0.983607		6.87E-286		0.0081967	0	0.983607	0	0.0081967
53715	0.369369	0.594595	5.97E-69	2.47E-128	0	0.009009	0.594595	0.369369	0.027027
51180	0.256881	0.724771	2.19E-42	2.87E-164	0.0091743	0	0.256881	0.724771	0.0091743
40900	0.37963	0.62037	1.49E-69	3.92E-132	0	0.37963	0	0.62037	0
51432	0.415842	0.584158	2.67E-73	3.23E-114	0	0.415842	0	0.584158	0
42574	0.98		4.49E-233		0	0.98	0	0	0.02
34786	0.418367	0.561224	9.92E-72	3.52E-105	0.418367	0.561224	0	0.0102041	0.0102041

Part C

Num	*p*-Value A	*p*-Value C	*p*-Value G	p-Value T	*p*-Value -	Flanking
14722	0.00052		0		0.000336	GCGACAGCAGTAAGACTTCCTTCCTAGTATTGCTTACGCCAGAG AAATAAC
14718	0.59641		0		0.186581	TTTCACTGTTGAAGCCGCCGGTAGTCACCGCCCAGTGCAGTGCC TCACGAT
14719		0			0.893853	TGTGCCCGTTTCGATGGCGGTACAGTAGGTTTTCGCTCAAGCAAC AGCGCA
14720				0	0.307612	CCCATACCCGACGATAACCATACGTGGGCAGCTCTCCGATAACAT GGTGTA
14717	0.50345	0			0.010203	GTCACGCTTTATCGTTTTCACGAAGTTCTCTGCTATTCCGTTACT CTCCGG
14721			0		0.15894	TCATCGGTTCGTCTGAGAATGACGTACAACTGCGCACGCGACAC CCGGAGA
29003	7.48E-90		1.79E-223		0.315601	ACGCCGCATCCGACATCTAACGCCCGAGCCGGTTGCCTGATGCG ACGCTGG
14723	0.47345			0	0.934513	AAGACTATCACTTATTTAAGTGATACTGGTTGTCTGGAGATTCAGG GGGCC
4359		1.71E-236		1.18E-61	0.73285	ACCGATGCCTGATGCGCCGCTGACGCGACTTATCAGGCCTACGG GGTGAAC
29002		2.29E-125		1.74E-144	0.454427	AAGCGGTCACGCCGCATCCGACATCTAACGCCCGAGCCGGTTGC CTGATGC
9367	0.37944		6.87E-286		0.868376	TTGCGTCAGCAACGGCCCGTAGGGCAAGCGAAGCGAGTCATCCT GCACGAC
53715		0.352165	2.47E-128	5.97E-69	0.276962	GTAAACGCCTTATCCGGCCTACGGAGGGTGCGGGAATTTGTAGG CCTGATA

(continued)

Table 1 (continued)

Part C

Num	*p*-Value A	*p*-Value C	*p*-Value G	p-Value T	*p*-Value -	Flanking
51180	0.34708		2.19E-42	2.87E-164	0.836628	CGCAAATTCAATATATTGCAGAGATTGCGTAGGCCTGATAAGCGT AGCGCA
40900		1.49E-69		3.92E-132		ATAAGCCGCTTTCTTTTGGGTATAGTGTCGTGGACAGTCATTCAT CTTTCT
51432		2.67E-73		3.23E-114		TTGCGGCACTGGAGTTTGGCAACAGTGCCGGATGCGGCGCAAG CGCCTTAT
42574		4.49E-233			0.492269	TGTNTGGCAGTTTATGGCGGGCGTCNTGCCCGCCACCCTCCGG GCCGTTGC
34786	9.92E-72	3.52E-105		0.318375	0.803853	GGTAACCCTGAGCACGCAGTTCTTCAGTCAGGCGTGGTGCACC GTAACGCT

The complete table was sorted for descending depth and then for ascending *p*-values for allele 1. Only the first 17 lines of the table are presented. It is divided into three parts for ease of viewing distributions corresponding to the two lengths. The score with which the best matching model matches the data is the model score. The closer it is to one the better the match. A score above .98 is a good score. A score of .99 and above is an excellent score. A score below .95 is not particularly good. In order to be confident of a length mutation, it is necessary to both have a good model score and a low *p*-value.

8. *Shift:* 0 for substitutions and deletions negative for insertions. A shift of −1 indicates that the insertion is immediately to the left of the position indicated. A shift of −2 indicates that the insertion is to the left of the first insertion at −1.
9. *Allele1/Allele2*: this indicates the nucleotide composition of each allele.
10. *Depth*: the number of reads that span the mutation. A minimum depth of 20 is acceptable for mutation calling for any diploid organism when there are no mixtures of strains. The table can be filtered to exclude SNPS in locations with depths less than 20.
11. *Count1/Count2*: the number of reads that have each one of the alleles.
12. *Freq1/Freq2*: frequency of each allele that is Count *i*/Freq *i* where $i = 1$ or 2.
13. *p-Value1/p-Value2*: each variation has a *p*-value associated with it. This is the probability that the mutation observed was generated by chance due to sequencing errors. It is recommended to consider only mutations for which the allele with maximum *p*-value is 1e-10 or less. Such data is generated automatically by including *p*-valueThreshold = 1e-10 in the config file.
14. *Freq A/C/G/T/-*: counts/depth for each nucleotide type.
15. *p-Value A/C/G/T/-*: *p*-values for each of the nucleotides.
16. *Flanking*: flanking region for mutation with an additional 25 nucleotides on each side.

4. Notes

1. The method used for bacterial DNA isolation is dependent upon the bacterial source. The Qiagen kit has been used successfully for bacterial DNA isolation.

   It can be used for Gram-negative and some Gram-positive bacteria. Various phenol–chloroform extraction and isopropanol precipitation methods have also successfully been used to purify bacterial DNA for sequencing. We discourage the use of any protocol that involves bead-beating, as it has been shown to reduce the yield of DNA in the correct size range for sequencing.

2. For higher throughput applications, the Covaris E210 Instrument uses the same shearing parameters as the S2 instrument, but is capable of shearing up to 96 samples unattended. Higher throughput enzymatic methods for obtaining fragment sizes between 200 and 300 bp (New England Biolabs, Ipswich, MA; Epicentre® Biotechnologies, Madison, WI) that are compatible with this protocol can also be used.
3. Axygen MAXYMum Recovery tubes have been shown to increase sample yield and reduce variability in the polyA tailing reaction. MAXYMum Recovery tubes should be used throughout the sample preparation process and for sample storage.
4. 4–20% Gradient gels should be used in these steps. They optimize both the ability to determine that there is no detectable DNA less than 50 bp after size selection and the ability to visualize the TR oligo in the control tailing reaction.
5. The TR oligo and dATP should be purchased from Helicos BioSciences Corporation. The reagents have been optimized to produce correct tail lengths and stabilized to permit long-term storage.
6. Smaller quantities of DNA can be sheared. If the sample reaction is modified to use 0.5–1 pmole of DNA (see Note 11) and the control reaction is modified to use 0.5 pmoles of DNA (see Note 13), as little as 100 ng of high quality DNA can be used at this step.
7. If you notice you are removing beads during aspiration, do not attempt to remove all the beads with a p1000 pipette. Rather, remove the last 20–50 μL with a p200 pipette.
8. If 100 ng of DNA has been sheared, the elution volumes should be reduced to 10 μL in both steps 12 and 15. In all cases, care should be taken to avoid getting beads in the supernatant. This can be achieved more easily by using a p10 pipette to aspirate the supernatant and by leaving the last microliter behind.
9. Care should be taken to avoid touching the gel. Clean containers and clean gloves should be used.
10. This gel step is largely a QC step. If you are processing many similar samples and find through initial gel studies that they shear to approximately the same size, this gel step can be omitted and an average size be used in subsequent calculations. Portions of the sheared samples may be retained to run on a gel for troubleshooting purposes if the tailing reaction does not succeed.

11. If sample is limiting, as little as 1 pmole of DNA can be put in a 40-μL sample reaction. For quantities of DNA less than 3 pmoles, the amount of dATP added must be scaled down in proportion to the amount of DNA added (e.g., for 2 pmoles of DNA, add 2.8 μL of dATP to the sample master mix; for 1 pmole, add 1.4 μL dATP). For quantities less than 1 pmole, the volume of the entire reaction should be scaled down (e.g., for 0.5 pmoles of DNA, the DNA should be in a final volume of 13 μL. The sample master mix should contain 2.2 μL of 10× terminal transferase buffer, 2.2 μL of $CoCl_2$ (2.5 mM), 1.4 μL of a 1:2 dilution of Helicos™ PolyA Tailing dATP in water, 0.8 μL distilled water and 1.1 μL terminal transferase enzyme (20 U/μL) per sample. Seven microliters of sample master mix should be added to each 0.5 pmole DNA tube).
12. The mixing step is crucial to the success of the tailing reaction.
13. If sample is limiting, the control reaction can be scaled down to use 0.4 pmoles of DNA. The DNA should be in a final volume of 12 μL and 1 μL of a 1:2 dilution of the Helicos™ PolyA Tailing Control Oligo TR in water should be added. Two microliter of TR oligo should be used in the Oligo TR Control. The control master mix should contain 2.2 μL of 10× terminal transferase buffer, 2.2 μL of $CoCl_2$ (2.5 mM), 1.4 μL of a 1:2 dilution of Helicos™ PolyA Tailing dATP in water, 0.8 μL of distilled water and 1.1 μL terminal transferase enzyme (20 U/μL) per sample. Seven microliters of control master mix should be added to each tube. Scaling this reaction down does preclude verifying the outcome of a short tail correction reaction (steps in Subheading 3.6).
14. It is essential to chill the block to 0°C in an ice and water slurry, and cool the DNA as quickly as possible to 0°C to prevent re-annealing of the denatured, single-stranded DNA products.
15. The size of the single-stranded polyA tailed samples cannot be determined by direct comparison to the double-stranded DNA ladders. The migration patterns of TR oligos containing dA90 and dA200 were determined experimentally.

Acknowledgments

The authors would like to acknowledge the valuable contributions of Katica Ilic, Kristen Kerouac, and Eldar Giladi.

References

1. Eliminating amplification bias from genome analysis. Helicos BioSciences Corporation Tech Note, Available for download at: http://helicosbio.com/HeliSphereCenter/PublicationsLibrary/HelicosMarketingCollateral/tabid/168/Default.aspx.
2. MacLean, D., Jones, J. D. G., and Studholme, D. J. (2009) Application of 'next-generation' sequencing technologies to microbial genetics. *Nat Rev Microbiol* 7, 287–296.
3. Holt, K. E., Parkhill, J., Mazzoni, C. J., Roumagnac, P., Weill, F. -X., Goodhead, I., Rance, R., Baker, S., Maskell, D. J., Wain, J., Dolecek, C., Achtman, M., and Dougan, G. (2008) High-throughput sequencing provides insights into genome variation and evolution in *Salmonella* Typhi. *Nat Genet* **40**, 987–993.
4. Harris, S. R., Feil, E.J., Holden, M. T. G., Quail, M. A., Nickerson, E. K., Chantratita, N., Gardete, S., Tavares, A., Day, N., Lindsay, J. A., Edgeworth, J. D., de Lencastre, H., Parkhill, J., Peacock, S. J., and Bentley, S. D. (2010) Evolution of MRSA during hospital and intercontinental spread. *Science* **327**, 469–474.
5. Smith, E. E., Buckley, D. G., Wu, Z., Saenphimmachak, C., Hoffman, L. R., D'Argenio, D. A., Miller, S. I., Ramsey, B. W., Speert, D. P., Moskowitz, S. M., Burns, J. L., Kaul, R. and Olson, M. V. (2006) Genetic adaptation by *Pseudomonas aeruginosa* to the airways of cystic fibrosis patients. *Proc Natl Acad Sci USA* **103**, 8487–8492.
6. Mwangi, M. M., Wu, S. W., Zhou, Y., Sieradzki, K. de Lencastre, H., Richardson, P., Bruce, D., Rubin, E., Myers, E., Siggia, E. D., and Tomasz, A. (2007) Tracking the *in vivo* evolution of multidrug resistance in *Staphylococcus aureus* by whole-genome sequencing. *Proc Natl Acad Sci USA* **104**, 9451–9456.
7. La Scola, B., Elkarkouri, K., Li, W., Wahab, T., Fournous, G., Rolain, J. -M., Biswas, S., Drancourt, M., Robert, C., Audic, S., Lofdahl, S., and Raoult, D. (2010) Rapid comparative genomic analysis for clinical microbiology: the *Francisella tularensis* paradigm. *Genome Res* **18**, 742–750.
8. Srivatsan, A., Han, Y., Peng, J., Tehranchi, A. K., Gibbs, R. Wang, J. D. and Chen, R. (2008) High-precision, whole-genome sequencing of laboratory strains facilitates genetic studies. *PLoS Genet* **4**, e1000139.
9. Dohm, J. C., Lottaz, C., Borodina, T. and Himmelbauer, H. (2008) Substantial biases in ultra-short read data sets from high-throughput DNA sequencing. *Nucleic Acids Res* **36**, e105.
10. Hillier, L. W., Marth, G. T., Quinlan, A. R., Dooling, D., Fewell, G., Barnett, D., Fox, P., Glasscock, J. I., Hickenbotham, M., Huang, W., Magrini, V. J., Richt, R. J., Sander, S. N., Stewart, D. A., Stromberg, M., Tsung, E. F., Wylie, T., Schedl, T., Wilson, R. K., and Mardis, E. R. (2008) Whole-genome sequencing and variant discovery in *C. elegans. Nat Methods* **5**, 183–188.
11. Harismendy, O., Ng, P. C., Strausberg, R., L., Wang, X., Stockwell, T. B., Beeson, K. Y., Schork, N. J., Murray, S. S., Topol, E. J., Levy, S. and Frazer, K. A. (2009) Evaluation of next generation sequencing platforms for population targeted sequencing studies. *Genome Biol* **10**, R32.

Chapter 2

Whole-Genome Sequencing of Unculturable Bacterium Using Whole-Genome Amplification

Yuichi Hongoh and Atsushi Toyoda

Abstract

More than 99% of microorganisms on the earth are unculturable with known culturing techniques. The emergence of metagenomics with high-throughput sequencing technologies has enabled researchers to capture a comprehensive view of a complex bacterial community which comprises both culturable and unculturable species. However, the function of an individual species remains difficult to elucidate in a conventional metagenomic study, which generates numerous genomic fragments of unidentifiable origins at a species or genus level. This limitation hampers any in-depth investigations of the community and its unculturable bacterial members. Recently, as an alternative or compensatory approach, genomics targeting a single unculturable bacterial species in a complex community has been proposed. In this approach, whole-genome amplification technique using Phi29 DNA polymerase is applied to obtain a sufficient quantity of DNA for genome sequence analysis from only a single to a thousand bacterial cells. It is expected that a combination of the conventional metagenomics and this single-species-targeting genomics provides a great progress in understanding of the ecology, physiology, and evolution of unculturable microbial communities.

Key words: Whole-genome amplification, Phi29 DNA polymerase, Pyrosequence, Uncultivable, Uncultured, Environmental genomics, Metagenomics, Single-cell genomics, Termite, Symbiosis

1. Introduction

Culture-independent, molecular approaches have enabled researchers to explore the world of unculturable microorganisms since the 1990s. Molecular tools such as clone analysis and fluorescent in situ hybridization analysis, targeting small subunit (SSU) rRNA, have been applied to a wide variety of environmental samples and revealed the existence of an enormous number of largely unknown, uncultured assemblages of microorganisms.

Young Min Kwon and Steven C. Ricke (eds.), *High-Throughput Next Generation Sequencing: Methods and Applications*, Methods in Molecular Biology, vol. 733, DOI 10.1007/978-1-61779-089-8_2, © Springer Science+Business Media, LLC 2011

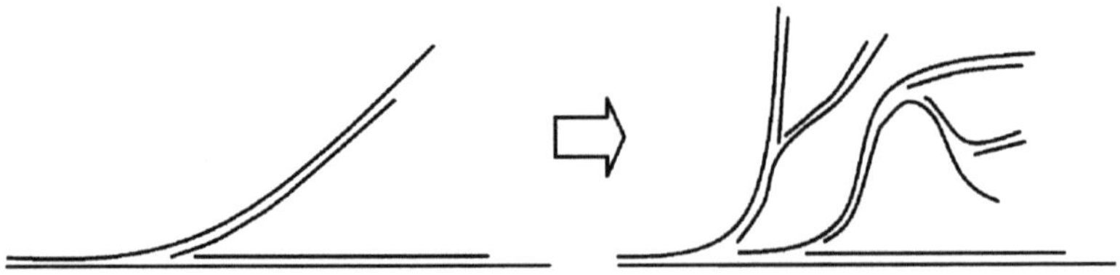

Fig. 1. Outline of multiple displacement amplification. The whole genome regions can be amplified by the action of Phi29 DNA polymerase with random hexamers.

However, their functions have been mostly inaccessible until the emergence of metagenomics.

Metagenomics with high-throughput sequencing technologies is a powerful tool to uncover the "black box" of bacterial communities, which generally consist of both culturable and unculturable species. In a conventional metagenomic study, DNA is extracted from a heterogeneous microbial community, fragmented, and sequenced to generate tens of megabases to several gigabases in total. Metagenomics reveals the functions of a bacterial community as a whole and the diversity of various genes within the community, and it is also useful to find novel genes encoding enzymes industrially or medically applicable.

However, the function of an individual bacterial species remains largely unknown because the origins of most fragments generated in such metagenomic studies are generally unidentifiable beyond a class level. Recently, to compensate this limitation in metagenomics, genomics targeting a single species of unculturable bacteria in a complex community has been proposed. This novel approach applies whole-genome amplification (WGA) using Phi29 DNA polymerase to prepare sufficient DNA quantities for the genome sequence analysis (1) (Fig. 1). To date, a couple of studies reported draft-genome sequences from single bacterial cells (2, 3), and two studies reported the complete genome sequences from 10^2 to 10^3 cells (4, 5), whereas $\geq 10^{10}$ cells are required for a conventional genome sequence analysis.

Here, we describe a protocol to acquire the complete genome sequence from 10^2 to 10^3 cells of a single bacterial species. We use an endosymbiotic bacterial species found inside the cells of protists (single-celled eukaryote) in termite gut as an example. This method is applicable to other samples with certain modifications as long as 10^2 to 10^3 cells of a single bacterial species or strain can be collected. In the case that only a single or a few bacterial cells are collectable, the sequencing analysis will result in draft status with numerous contigs remaining (2, 3, 6, 7).

2. Materials

2.1. Micromanipulation of Microbial Cells

1. An inverted phase-contrast microscope equipped with two sets of the micromanipulator TransferMan NK2 and CellTram Vario (Eppendorf, Hamburg, Germany) (see Note 1).
2. Glass capillary: 15 μm diameter (Eppendorf) (see Note 2).
3. Glass capillary: 30–100 μm diameter (custom-made, Eppendorf).
4. 1.0% Nonidet P-40 (NP-40) (see Note 3): Sterilize with a 0.22 μm filter and the ultra violet (UV). For UV-sterilization, put the solution in a plastic tube in a UV-crosslinker for 10 min. Store in single-use aliquots at –20°C.
5. Trager's solution U (sol U) (see Note 4): 37 mM NaCl, 9.2 mM $NaHCO_3$, 5.1 mM $Na_3C_6H_8O_7{\cdot}2H_2O$, 13 mM KH_2PO_4, 0.75 mM $CaCl_2$, 0.40 mM $MgSO_4$. Sterilize with a 0.22-μm filter and UV. Store in single-use aliquots at –20°C.

2.2. Cell Lysis and WGA

1. GenomiPhi HY DNA amplification kit (GE Healthcare, Hemel Hempstead, UK) (see Note 5). Store the enzyme at –70°C. The other components can be stored at –20°C.
2. Lysis buffer (LB) (2×): 400 mM KOH, 100 mM dithiothreitol, 10m Methylenediaminetetraacetic acid (EDTA)·2Na·$2H_2O$. Sterilize with UV-irradiation and store at –20°C. It can be stored for 2 weeks.
3. Neutralization buffer (NB): 600 mM Tris–HCl, pH 7.5, 400 mM HCl (final pH 6.0). Sterilize with UV-irradiation and store at –20°C.
4. Tris–EDTA (TE) buffer: 10 mM Tris–HCl, pH 8.0, and 1 mM EDTA.

2.3. Purity Check

1. *Taq* DNA polymerase: e.g., *EX-Taq* polymerase (Takara, Tokyo, Japan).
2. Proof-reading DNA polymerase: e.g., Phusion (Finnzymes, Espoo, Finland).
3. PCR primers: bacteria-specific, 27 F (5′- AGRGTTT GATYMTGGCTCAG) and 1492R (5′- GGHTACCTTGTT ACGACTT); Archaea-specific, A25F (5′- CYGKTTGATC CTGSCRG) and A1385R (5′- GGTGTGTGCAARGAGCA); Eukarya-specific, E18F (5′- GATCCMGGTTGATYCTGCC) and E1772R (5′- CWDCBGCAGGTTCACCTAC).

3. Methods

Because WGA can amplify a small amount of contaminant DNA, the handling of samples and reagents should be conducted with extreme caution. Wear disposable gloves, clean up, and bleach the benches and equipments to eliminate DNA contamination sources, use UV-sterilized filter tips and plastic tubes, and perform experiments under a laminar flow cabinet.

3.1. Collection of Bacterial Cells by Micromanipulation

1. Apply 2 μl of 1× LB into a sterile, 0.2-ml PCR tube and incubate on ice or a PCR-cooler.
2. Attach a glass capillary of 30–100 μm diameter to one of the two micromanipulators. The diameter of the capillary depends on the size of the host protist cells harboring the target bacterial symbiont.
3. Mark the position corresponding to the depth of a 0.2-ml PCR tube on a glass capillary of 15 μm diameter (Fig. 2), and attach it to the other micromanipulator.
4. Put the lid of a sterile, plastic Petri dish, upside down, on the stage of the inverted microscope. Apply 50 μl of sol U, five to ten times, onto the inner surface of the lid (Fig. 3).
5. Dissect a termite and remove the gut with forceps.
6. Puncture the dilated portion of the hindgut in one of the 50-μl drops of sol U on the lid.
7. Remove the gut from the suspension.
8. Dip the tip of the 30–100-μm diameter capillary and carefully collect several cells of the host protist species. There is no need to eliminate other species of protists and free-swimming bacteria at this step.

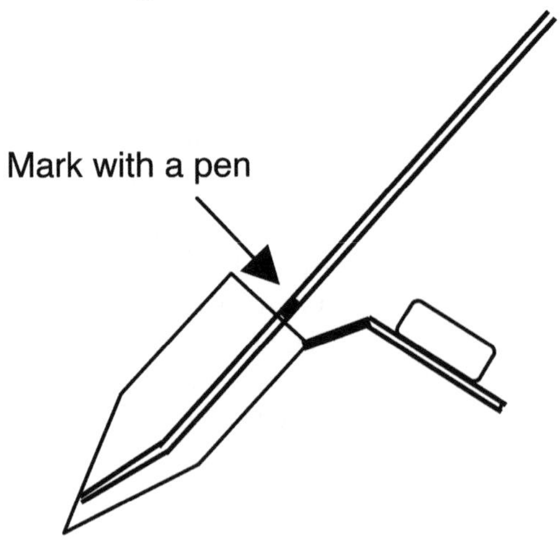

Fig. 2. Collection of bacterial cells by micromanipulation into a 0.2-ml PCR tube.

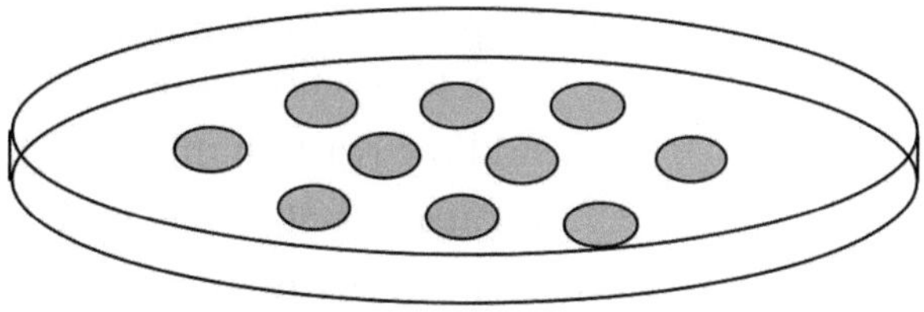

Fig. 3. Drops of buffer on the inner surface of the lid of a Petri dish.

9. Release the collected cells into another 50 μl drop of sol U.
10. Collect the host protist cells.
11. Wash the cells by repeating steps 9 and 10 more than four times.
12. Change the capillary to a new one with the same diameter.
13. Apply 50 μl of 1% NP-40 onto the inner side of the lid of a new Petri dish (see Notes 3 and 6).
14. Collect a single cell of the host protist with the capillary and keep it within its tip.
15. Replace the lid of the Petri dish on the stage by the lid with 1% NP40 prepared in step 13.
16. Dip the tip of the 15-μm diameter capillary in the 1% NP-40 drop and adjust the pressure inside the capillary by handling the CellTram Vario. Be careful not to completely expel 1% NP-40 within the capillary. Bubbles make the phase-contrast image unclear. Keep the capillary tip within the eye field of the microscope.
17. Dip the tip of the 30–100-μm diameter capillary near the tip of the 15-μm diameter capillary in the 1% NP-40 drop and carefully release the host protist cell.
18. Within a few minutes, the endosymbiotic and ectosymbiotic bacterial cells will detach or leak out from the host protist cell.
19. Collect the bacterial cells as many as possible with the 15-μm diameter capillary.
20. Release the collected cells into LB in the PCR tube. Use the mark on the capillary in order not to break the fine tip of the glass capillary.

3.2. Whole-Genome Amplification

1. Incubate the collected bacterial cells in 1× LB on ice or a PCR-cooler for 10 min (see Note 7).
2. Prepare and perform the negative control reaction in parallel with the sample reaction.
3. Add 1.0 μl NB and mix briefly.
4. Add 22 μl sample buffer and mix briefly.

5. Prepare reaction mixture: mix 2.5 μl enzyme solution and 22.5 μl reaction buffer by pipetting.
6. Add the reaction mixture to the sample mixture.
7. Incubate at 30°C for 2.5 h.
8. Stop the reaction by incubating at 65°C for 10 min.
9. Check the amplification by agarose gel electrophoresis (see Note 8).
10. Purify DNA by ethanol precipitation.
11. Dissolve the precipitate in 50 μl TE.
12. Measure the DNA concentration.

3.3. Checking Purity of Sample

Before the genome sequence analysis, the purity of the WGA sample must be checked by clone analysis of genes such as SSU rRNA. If amplification of the genes by PCR from the negative control reaction is observed, the sample should not be used for further analyses.

1. Prepare 1:200-fold dilution of the purified WGA sample dissolved in TE and of the negative control reaction without purification. Use 1/10 volume as the template for the following PCR.
2. Perform PCR using primer sets specific to eubacterial, archaeal, and eukaryotic SSU rRNA genes, respectively, and also those specific to protein-coding genes such as *hsp60* and *gyrB*. Always perform the negative control experiment for PCR. PCR conditions: 95°C 30 s, 25 cycles of (95°C 10 s, 50°C 30 s, 72°C 2 min), 72°C 4 min.
3. If there is no PCR amplification from the negative control sample either for WGA or PCR, and only PCRs using bacteria-specific primers generate products from the WGA sample, move to the next step.
4. Perform PCR amplification of eubacterial 16S rRNA genes with a proof-reading DNA polymerase.
5. Clone and sequence the PCR products with standard methods.
6. Check whether the genomes of the target bacterial species predominate in the sample and also evaluate within-species variations of the target bacterium.

3.4. Genome Sequencing Using 454 GS FLX

After the purity check of the WGA sample, prepare a library for the pyrosequencer 454 GS FLX, exactly following the protocol distributed by Roche Diagnostics (see Note 9). It includes nebulization, end-polishing using T4 DNA polymerase, and adapter ligation. Because WGA generally accompanies amplification bias among the genome regions (Fig. 4), a deeper redundancy is necessary for acquiring a complete genome sequence. First, try

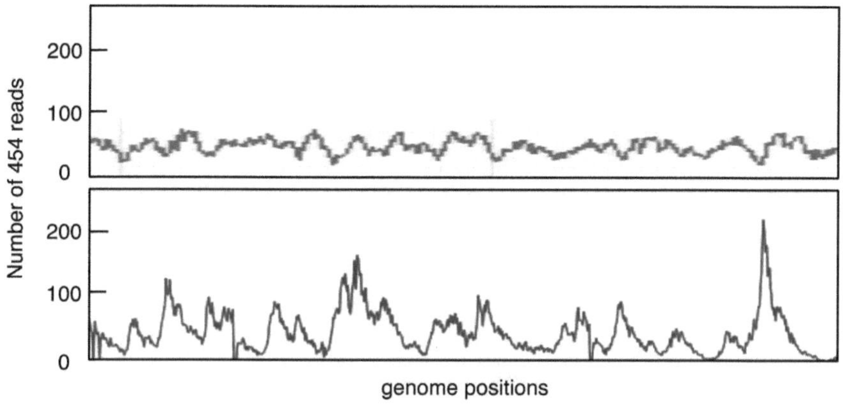

Fig. 4. Bias among genome regions caused by whole-genome amplification (WGA). Typical patterns in genome sequence analyses without (*above*) and with WGA (*below*).

pyrosequencing in a small scale and check the amplification bias and the purity. If the bias appears too large, it may be better to try another sample. Finishing can be done with a standard gap-closing procedure using a Sanger method.

4. Notes

1. It is possible to conduct the whole procedure with a single micromanipulator set, but the risk for making mistakes is much higher. When only a single set is available, carefully dip the tip of the 15-μm diameter capillary, as mentioned in step 16 of Subheading 3.1, at a position distant from the released host protist cell in the 1% NP-40 drop in order not to blow away the cell.
2. If one needs to collect only a single or a few bacterial cells, use a glass capillary of 4 μm diameter, though the handling is more difficult.
3. NP-40 is necessary to rupture the membrane structure of the host protist, and also useful to break the surface tension of the drop on the lid of the Petri dish. This makes the phase-contrast image clearer and is considerably helpful for the collection of bacterial cells by micromanipulation.
4. Sol U buffer is specifically adjusted for the protists in termite gut (8). One should use buffer adequate to each type of samples.
5. One can use other kits for WGA such as REPLI-g Midi (Qiagen, Düsseldorf, Germany). In any case, one must first check the purity of the purchased kit in a preliminary experiment; occasionally there are lots containing non-negligible amount

of DNA contaminants. Although those lots are available for a standard use recommended by the manufacturers, they should not be used in WGA from less than 10^4 bacterial cells.

6. There are bacterial species sensitive to detergent (e.g., spirochetes). In that case, use buffer or sterile water, instead of NP-40, to collect the bacterial cells, although a very small amount of NP-40 should be added (a touch by a tip is enough) to the drop in order to obtain a clearer phase-contrast image.
7. There are bacterial species of which cells are not degradable under this alkaline condition. It is recommended to confirm the degradability in a preliminary experiment.
8. Possibly, the negative control may also generate considerable WGA products even without exogenous DNA contaminants. This can be caused by the amplification from primer dimers and indigenously contaminated DNA. Unless PCR amplification of genes from the negative control for WGA is observed, one can use the WGA sample.

   It is suggested that the background amplification can be suppressed by an addition of trehalose (9). Alternatively, the use of GenomiPhi v2 or REPLI-g UltraFast Mini also diminishes the background amplification. Because these methods generate lesser amounts of WGA products, a second amplification step using GenomiPhi HY or REPLI-g Midi should be performed to obtain an enough amount of DNA (15–50 μg) for the genome sequence analysis.
9. For the genome sequence analysis, we recommend a hybrid use of 454 GS FLX and Illumina GA and do not recommend mate-pair analysis with a long insert, as suggested in a recent literature (7). WGA generates chimeric sequences (10) and the rate of their formation greatly increases in a traditional library construction for a Sanger method (11). Thus, the handling of data generated with a Sanger method needs caution. If one needs to prepare a library for a Sanger method, a S1 nuclease treatment is recommended, which reportedly decreases the frequency of chimeras (11). In any case, increase of the sequence depth by using high-throughput sequencing technologies is still considered the best way to eliminate chimeras and other artifacts in the assembling process.

Acknowledgment

The authors would like to thank Drs. M. Ohkuma, M. Hattori, and other coworkers for supporting our studies.

References

1. Dean, F. B., Hosono, S., Fang, L., Wu, X., Faruqi, A. F., Bray-Ward, P., et al. (2002) Comprehensive human genome amplification using multiple displacement amplification, *Proc. Natl. Acad. Sci. U. S. A.* **99**, 5261–5266.
2. Marcy, Y., Ouverney, C., Bik, E. M., Losekann, T., Ivanova, N., Martin, H. G., et al. (2007) Dissecting biological "dark matter" with single-cell genetic analysis of rare and uncultivated TM7 microbes from the human mouth, *Proc. Natl. Acad. Sci. U. S. A.* **104**, 11889–11894.
3. Woyke, T., Xie, G., Copeland, A., Gonzalez, J. M., Han, C., Kiss, H., et al. (2009) Assembling the marine metagenome, one cell at a time, *PLoS One* **4**, e5299.
4. Hongoh, Y., Sharma, V. K., Prakash, T., Noda, S., Taylor, T. D., Kudo, T., et al. (2008) Complete genome of the uncultured Termite Group 1 bacteria in a single host protist cell, *Proc. Natl. Acad. Sci. U. S. A.* **105**, 5555–5560.
5. Hongoh, Y., Sharma, V. K., Prakash, T., Noda, S., Toh, H., Taylor, T. D., et al. (2008) Genome of an endosymbiont coupling N2 fixation to cellulolysis within protist cells in termite gut, *Science* **322**, 1108–1109.
6. Podar, M., Abulencia, C. B., Walcher, M., Hutchison, D., Zengler, K., Garcia, J. A., et al. (2007) Targeted access to the genomes of low-abundance organisms in complex microbial communities, *Appl. Environ. Microbiol.* **73**, 3205–3214.
7. Rodrigue, S., Malmstrom, R. R., Berlin, A. M., Birren, B. W., Henn, M. R., and Chisholm, S. W. (2009) Whole genome amplification and de novo assembly of single bacterial cells, *PLoS One* **4**, e6864.
8. Trager, W. (1934) The cultivation of a cellulose-digesting flagellate, *Trichomonas termopsidis*, and of certain other termite protozoa, *Biol. Bull.* **66**, 182–190.
9. Pan, X., Urban, A. E., Palejev, D., Schulz, V., Grubert, F., Hu, Y., et al. (2008) A procedure for highly specific, sensitive, and unbiased whole-genome amplification, *Proc. Natl. Acad. Sci. U. S. A.* **105**, 15499–15504.
10. Lasken, R. S., and Stockwell, T. B. (2007) Mechanism of chimera formation during the Multiple Displacement Amplification reaction, *BMC Biotechnol.* **7**, 19.
11. Zhang, K., Martiny, A. C., Reppas, N. B., Barry, K. W., Malek, J., Chisholm, S. W., and Church, G. M. (2006) Sequencing genomes from single cells by polymerase cloning, *Nat. Biotechnol.* **24**, 680–686.

Part II

Gene Expression Analysis

Chapter 3

RNA Sequencing and Quantitation Using the Helicos Genetic Analysis System

Tal Raz, Marie Causey, Daniel R. Jones, Alix Kieu, Stan Letovsky, Doron Lipson, Edward Thayer, John F. Thompson, and Patrice M. Milos

Abstract

The recent transition in gene expression analysis technology to ultra high-throughput cDNA sequencing provides a means for higher quantitation sensitivity across a wider dynamic range than previously possible. Sensitivity of detection is mostly a function of the sheer number of sequence reads generated. Typically, RNA is converted to cDNA using random hexamers and the cDNA is subsequently sequenced (RNA-Seq). With this approach, higher read numbers are generated for long transcripts as compared to short ones. This length bias necessitates the generation of very high read numbers to achieve sensitive quantitation of short, low-expressed genes. To eliminate this length bias, we have developed an ultra high-throughput sequencing approach where only a single read is generated for each transcript molecule (single-molecule sequencing Digital Gene Expression (smsDGE)). So, for example, equivalent quantitation accuracy of the yeast transcriptome can be achieved by smsDGE using only 25% of the reads that would be required using RNA-Seq. For sample preparation, RNA is first reverse-transcribed into single-stranded cDNA using oligo-dT as a primer. A poly-A tail is then added to the 3′ ends of cDNA to facilitate the hybridization of the sample to the Helicos® single-molecule sequencing Flow-Cell to which a poly dT oligo serves as the substrate for subsequent sequencing by synthesis. No PCR, sample-size selection, or ligation steps are required, thus avoiding possible biases that may be introduced by such manipulations. Each tailed cDNA sample is injected into one of 50 flow-cell channels and sequenced on the Helicos® Genetic Analysis System. Thus, 50 samples are sequenced simultaneously generating 10–20 million sequence reads on average for each sample channel. The sequence reads can then be aligned to the reference of choice such as the transcriptome, for quantitation of known transcripts, or the genome for novel transcript discovery. This chapter provides a summary of the methods required for smsDGE.

Key words: Single-molecule sequencing, smsDGE, Expression analysis

Young Min Kwon and Steven C. Ricke (eds.), *High-Throughput Next Generation Sequencing: Methods and Applications*, Methods in Molecular Biology, vol. 733, DOI 10.1007/978-1-61779-089-8_3, © Springer Science+Business Media, LLC 2011

1. Introduction

There is a critical need for methodology providing accurate and efficient digital gene expression (DGE) signatures of cells and tissues across a large dynamic range. For that purpose, ultra high-throughput cDNA sequencing is increasingly used for transcriptome characterization. Most commonly, RNA-Seq methodology has been used with the goal of generating the sequence of the entire transcriptome with sufficient coverage to enable identification of sequence variance, discovery of novel transcripts, and expression analysis (1, 2). A caveat in using RNA-Seq for quantitation is the requirement for high-sequence coverage of the transcriptome. This requirement is the result of the heterogeneity in transcript length whereby long transcripts generate more sequence reads than short ones, thus biasing detection of long over short transcripts (3). To overcome this bias, tens of millions of reads are typically needed for sensitive quantitation.

Here, we describe smsDGE, an unbiased, ultra high-throughput sequencing method for transcript quantitation using the Helicos® Genetic Analysis System. To avoid the length bias, only a single read is generated from each transcript molecule regardless of its length. So, for example, to achieve the same level of quantitation of 95% of the yeast transcriptome, RNA-Seq would require 40 million reads, whereas smsDGE would only require ten million (4). Sequence reads may be aligned to the characterized transcriptome, or to the full genome enabling both profiling of known transcripts and de novo discovery. The methodology can be used for any organism with poly-adenylated mRNA. Here, we focus on smsDGE of yeast as an example. As a note, we add that since prokaryotic mRNA lacks a poly-A tail, preparations such as ribosomal RNA depletion and in vitro polyadenylation using a poly-A polymerase would be required for smsDGE adaptation (5–7).

In this chapter, we describe the process by which smsDGE samples are prepared involving no PCR amplification, size selection, or adaptor ligation. A poly-A tail is added to the single-stranded cDNA molecules to facilitate the sample's hybridization to the oligo-dT coated Sequencer flow-cell's surface where the sequencing reaction takes place. The minimal sample manipulation reduces the opportunity for the introduction of bias. In addition, the preparation is highly stranded (i.e. the single-stranded cDNA is strand specific). This is especially relevant in yeast expression analysis, since transcription from both DNA stands often overlaps.

2. Materials

2.1. cDNA Preparation

1. DNase 1 recombinant, RNase-free reagents (Roche, Mannheim, Germany).
2. RNase-free water.
3. RNeasy MinElute cleanup kit (Qiagen, Valencia, CA).
4. SuperScriptIII kit (Invitrogen™, Carlsbad, CA).
5. dTU25V primer (50 μM): 5′-TTTUTTTUTTTUTTTUTT TUTTTUUV-3′.
6. RNase H (2 U/μl provided with the SuperScript III kit, Invitrogen™, Carlsbad, CA).
7. USER enzyme (1 U/μl New England Biolabs®, Ipswich, MA).
8. RNase If (50 U/μl New England Biolabs®, Ipswich, MA).
9. AMPure® beads (Agencourt® Biosciences, Beverly, MA).
10. 100% Ethanol.

2.2. cDNA Poly-A Tailing and 3′ Dideoxy Blocking

1. Helicos® DGE assay reagent kit (Helicos BioSciences, Cambridge, MA).
2. Terminal Transferase kit (New England Biolabs®, Ipswich, MA).
3. Biotin-11-ddATP (1 mM, Perkin Elmer®, Waltham, MA).
4. 4–20% TBE gel (e.g. Novex TBE gels, Invitrogen™, Carlsbad, CA).
5. TBE buffer.
6. Gel-loading dye (e.g. 10× BlueJuice™, Invitrogen™, Carlsbad, CA).
7. 25 bp DNA ladder (e.g. by Invitrogen™, Carlsbad, CA).
8. SYBR gold nucleic acid gel stain (Invitrogen™, Carlsbad, CA).

3. Methods

This protocol is optimized for samples of 2–8 μg of total RNA, or 0.1–1 μg of poly-A selected mRNA. Total or poly-A RNA preps may contain residual genomic DNA fragments which can interfere with sequencing. It is, therefore, recommended to treat RNA with DNase I prior to sequencing to remove any genomic DNA contamination. A variety of methods and DNase I enzymes are commercially available and may be used. As an example, we include a DNase I treatment protocol here. Finally, we recommend an

RNeasy MinElute column purification. This treatment is optional. Advantages include improved sample purity, speed and ease of purification after DNase I treatment, and sample concentration. Disadvantages are the loss of transcripts <200 nt in length.

3.1. DNase Treatment of RNA

1. Add 5 μl of the DNaseI-10× incubation buffer (supplied with the DNase I kit) to up to 40 μg of the RNA sample.
2. Add RNase-free water up to 48 μl.
3. Add 2 μl of the DNase I-recombinant enzyme (20 U).
4. Mix the reaction gently. Do not vortex DNase I since it is highly sensitive to mechanical denaturation.
5. Incubate at 37°C for 40 min.
6. Clean the reaction up using the RNeasy MinElute cleanup kit according to manufacturer's instructions.

3.2. cDNA Synthesis

To prepare the cDNA, use the reagents provided with the SuperScript III reverse transcription kit (SSIII kit) and the dTU25V primer.

1. Prepare master-mix 1 by mixing 1 μl primer dTU25V (50 μM), with 1 μl of dNTP mix (10 mM, SSIII kit) per sample.
2. Prepare master-mix 2 by mixing 2 μl of the 10× reaction buffer, 4 μl $MgCl_2$ (25 mM), 0.4 μl DTT (0.1 M), 1 μl RNaseOut, and 1 μl SuperScript III RT enzyme per sample.
3. Add RNase-free water to the RNA up to a volume of 9.6 μl in a thermocycler tube (if the RNA volume is larger than 8 μl, the reaction may be doubled in volume by correspondingly doubling all reagents).
4. Add 2 μl of master-mix 1 to each of the RNA samples, denature them at 65°C for 5 min in a thermocycler, and place them directly on ice for 2 min. Add 8.6 μl of master-mix 2 to each sample, mix well, and place in a thermocycler for 5 min at 40°C, 50 min at 55°C, and finally 5 min at 85°C.

3.3. RNA and Primer Digestion

1. Add 1 μl RNase H (SSIII kit) to each sample and incubate at 37°C for 15 min.
2. To digest the dTU25V primer, add 1 μl USER enzyme to each sample and incubate for an additional 15 min at 37°C.
3. Finally, digest single-stranded RNA by adding 1 μl RNase If and incubate for an additional 15 min at 37°C.

3.4. Sample Cleanup

1. Warm the AMPure® beads to room temperature.
2. Prepare fresh 70% EtOH (500 μl per sample).
3. Transfer the samples to a 1.5-ml tube and add water (see Note 1) up to a volume of 40 μl.

4. Vortex the AMPure® beads well and add 52 μl of the beads to the sample.
5. Incubate the sample and bead mixture at room temperature for 15 min mixing occasionally by flicking the tube.
6. Briefly centrifuge the samples to collect the bead slurry, and place on a Dynal® magnet for 3 min.
7. Carefully aspirate the supernatant keeping the tubes on the magnet.
8. Add 200 μl of the fresh 70% EtOH to the sample (there is no need to mix the beads with the EtOH or disrupt the pellet).
9. Aspirate the supernatant and repeat the EtOH wash once more.
10. Aspirate the supernatant, briefly spin the tubes to collect the left over EtOH and remove with a fine pipette.
11. Place the tube with its lid open in a 37°C heat block until the bead pellet gets a cracked appearance (1–2 min). Avoid over-drying (see Note 2).
12. Resuspend the bead pellet in 15 μl of water.
13. Place on the magnet for 3 min and transfer the eluate into a fresh tube. The eluate contains the sample.
14. Add another 15 μl of water and resuspend the beads.
15. Place on the magnet for 3 min, elute, and add to the first eluate.
16. Measure the sample concentration by OD_{260} spectrophotometry on a NanoDrop. Sample concentration is expected to be in the range of 2–4 ng/μl.

3.5. cDNA Poly-A Tailing

The cDNA is poly-A tailed by terminal transferase to facilitate its capture on the HeliScope™ Sequencer's flow-cell surface. The length of the poly-A tail generated in the reaction is a function of both the sample's molarity and the amount of dATP added to the reaction. To prevent variability in poly-A tail length due to variable sample molarity, a fixed amount of a tailing-oligo is added to the reaction. This tailing-oligo contains dUTP, and is digested away after the tailing reaction.

1. Prepare a master mix by combining 5 μl of the 10× reaction buffer, 5 μl $CoCl_2$ (2.5 mM), 2 μl terminal transferase enzyme (terminal transferase kit), and 3.5 μl Helicos® Poly-A Tailing dATP (Helicos® DGE kit, see Note 3). *Mix very well* by pipetting the full reaction up and down a few times (see Note 4).
2. Place 40–60 ng of the sample in a thermocycler tube (see Note 5).
3. Add 7.5 μl of the poly-A tailing-control oligonucleotide (Helicos® DGE kit, see Note 6), and adjust the reaction volume to 34.5 μl with water.

4. Denature the sample and the tailing-control at 95°C for 5 min and place directly on ice for 2 min.
5. Add 15.5 μl of the master mix to each sample and *mix very well* by pipetting the full volume up and down a few times.
6. Incubate in a thermocycler for 1 h at 42°C, followed by 10 min at 70°C.

3.6. Assessing the Poly-A Tail Length

The length of the poly-A tail added to cDNA samples should fall within the specified range. Appropriately tailed cDNA is designed to be long enough to facilitate hybridization to the HeliScope™ Sequencer flow-cell, yet short enough to allow the removal of the tailing-oligo in the final sample cleanup. Gel electrophoresis is used to assess tail length. The cDNA sample itself may not be visible on the polyacrylamide gel. However, the tailing-oligo included in the sample can be visualized and is used as an approximate estimation for the cDNA sample's poly-A tail length.

1. Load 5 μl of each sample and of the tailing-control (each mixed with gel-loading dye) onto a 4–20% gradient TBE polyacrylamide gel. The gradient gel is necessary for the visualization of the tailed-oligo on the gel, which will otherwise be very faint).
2. Load a 25-bp DNA ladder as a size marker (if the Invitrogen™ reagents are used, mix 1 μl of the ladder with 4 μl BlueJuice™ and 15 μl water. Then, load 1 μl of the diluted ladder on the gel).
3. Run the gel for 45 min at 180 V.
4. Stain the gel in SYBR Gold nucleic acid gel stain: In a tray, mix 10 μl SYBR gold stain in 100 ml water. Place the gel in the tray on a shaker for 10 min (see Note 7).
5. Destain the gel in water, on a shaker, for 10–15 min changing the water every 2 min.
6. Interpreting the gel image: Fig. 1 shows the gel visualization of samples with variable tail lengths. The tailed tailing-oligo has a slower migration on the gel than the double-stranded 25 bp ladder. The ladder, therefore, cannot be used to strictly assess the length of the poly-A tail. The poly-A tail length is specified above each gel lane. Optimal tail lengths are between 70 and 150 nt (corresponding to between the 225 and 500 bp ladder bands).
7. The tailing-oligo control sample has lower molarity than the samples and, therefore, is expected to have longer poly-A tails. The control reaction often has tails longer than 150 nt and will typically be visualized on the gel above the 500-bp ladder band.

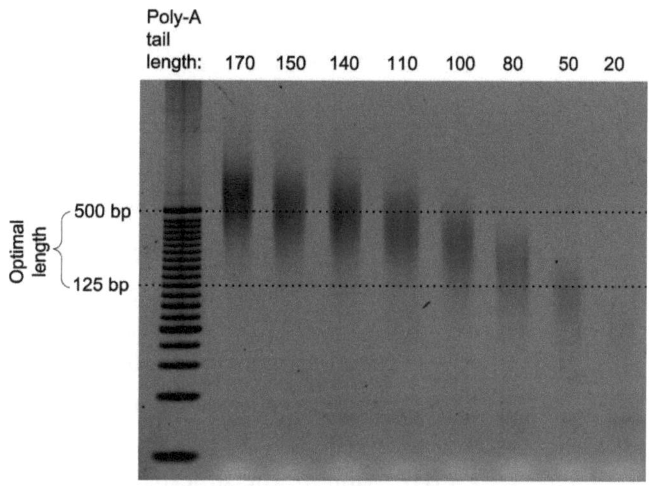

Fig. 1. Poly-A tail length assessment on a polyacrylamide gel. The optimal poly-A tail length is between 70 and 150 nt long. The tailed tailing-oligo spiked into the cDNA sample is visible on the gel, and is used as a proxy for the poly-A tail length of the full sample. Tail lengths are indicated over each gel lane (the size of the single-stranded tailing-oligo is different than the double-stranded 25-bp ladder predicts).

8. Infrequently, the tail length will fall outside the recommended 70 nt length. Short tails may be adjusted to the correct length by additional tailing conditions (see Note 8), however, long tails are indicative of samples having low molarity and, therefore, are expected to have poorer performance.

3.7. 3′ Dideoxy Blocking

After the poly-A tailing, a dideoxy nucleotide is added to the 3′ end of the tailed molecules to prevent cDNA 3′ incorporation of nucleotides during the sequencing reaction. The 3′ blocking reaction is followed by a USER digestion of the tailing-oligo to help oligo removal during the final cleanup.

1. Denature the samples at 95°C for 5 min and place directly on ice for 2 min.
2. Add 0.4 μl biotin-11-ddATP.
3. Add 2 μl (40 U) terminal transferase.
4. Incubate at 37°C for 1 h, and at 70°C for 10 min.
5. To digest the tailing-oligo, add 1 μl USER enzyme.
6. Incubate at 37°C for 30 min.

3.8. Final CleanUp

1. Warm the AMPure® beads to room temperature.
2. Prepare fresh 70% EtOH (500 μl per sample).
3. Transfer the samples to a 1.5-ml tube and add water up to a volume of 55 μl.

4. Vortex the AMPure® beads well and add 72 μl of the beads to the sample.
5. Incubate the sample and bead mixture at room temperature for 15 min mixing occasionally by flicking the tube.
6. Briefly centrifuge the samples to collect the bead slurry, and place on a Dynal® magnet for 3 min.
7. Carefully aspirate the supernatant keeping the tubes on the magnet.
8. Add 200 μl of the fresh 70% EtOH to the sample (there is no need to mix the beads with the EtOH or disrupt the pellet).
9. Aspirate the supernatant and repeat the EtOH wash once more.
10. Aspirate the supernatant, briefly spin the tubes to collect left over EtOH and remove with a fine pipette.
11. Place the tube with its lid open in a 37°C heat block until the bead pellet gets a cracked appearance (1–2 min). Avoid overdrying.
12. Resuspend the bead pellet with 20 μl of TE (or water).
13. Place on the magnet for 3 min and elute into a fresh tube. The eluate contains the sample.
14. Add another 20 μl of TE and resuspend the beads.
15. Place on the magnet for 3 min, elute, and add to the first eluate.
16. The sample is ready for quantification and sequencing.

3.9. Assessing Sample Concentration and Sequencing

Sample concentration is assessed using the HeliScope™ OptiHyb assay. The OptiHyb is typically performed prior to sample loading on the HeliScope™ Sequencer by the instrument operator and is not described here in detail. The assay should be performed using 4 μl of the DGE samples mixed into 96 μl hybridization buffer in the OptiHyb well (a 1:25 dilution). The general assay principal is:

1. First, an oligo titration is made using a poly-A tailed standard oligo of known molar concentration (Helicos®). This titration is used to establish a standard curve.
2. Next, the poly-A tailed sample is hybridized to unused wells in the assay plate.
3. Sample concentration is assessed by a horseradish peroxidase (HRP) chemiluminescence reaction in which the sample's 3′ dideoxy-biotin-dATP is bound to streptavidin and an HRP-conjugate generates an absorbance signal. A plate reader is used for absorbance measurement.
4. Sample molarity is assessed by comparing the absorption levels recorded to the oligo standard curve.

The HeliScope™ Sequencer has two flow-cells, each with 25 channels. Thus, 50 different samples may be sequenced simultaneously. Samples are injected into the flow-cell using the HeliScope™ Sample Loader. They are then allowed to hybridize to the oligonucleotides coating the channel's surface which, in turn, serve as primers in the sequencing reaction (8). A standard sequencing run is approximately 8 days long but can be adjusted based on user needs. Image data is processed by the accompanying HeliScope™ Analysis Engine in near real time. Once the run completes the .SRF data files are converted to .SMS files which are used for subsequent data analysis.

3.10. Data Analysis

HeliScope™ Sequencer data analysis can be done using a UNIX system with at least 5 GB per CPU core. The HeliSphere data analysis pipeline is an open-source software written in the Python programming language (available for download at: http://open.helicosbio.com/mwiki/index.php/Releases). The software is updated frequently and is accompanied by specific instructions for running the DGE pipeline. Download both the HeliSphere package and the examples, and follow the HeliSphere User's Guide documentation available at http://open.helicosbio.com/mwiki/index.php/Docs.

The pipeline uses the raw-read input file to generate transcriptome (or genome) aligned sequence reads and to report transcript counts. Note that for DGE data, read alignment should be done only to the "forward" strand. This is especially important for yeast DGE where transcription from both DNA strands often overlaps.

3.10.1. Data Summary Reports

A number of data summary reports are generated by the DGE pipeline. The main report of interest is the *DGE summary* report (example in Table 1):

1. Group: Flow-cell number (1 or 2), and channel number (1–25).
2. Raw reads: These are the unfiltered sequence reads. This number mostly provides a general measure of sample density on the flow-cell surface. It is also used to assess general sequencing quality by comparison to filtered read counts.
3. Filtered reads: These are mostly filtered for read length, but also for some sequence context (e.g. base addition order). For yeast DGE, it is recommended to filter out reads shorter than 24 nucleotides.
4. Aligned reads: These are the number of aligned reads scoring 4 or higher. The alignment score is calculated as follows:

$$\text{score} = (5m - 4e)/l$$

where m is the number of matched nucleotides, e is the number of errors, and l is the read length. For example if a read is 36

Table 1
Example of a smsDGE summary output table

Group	Raw	Filtered	Aligned	%filtered	%aligned	MeanLen	Counted	mRNA	%mRNA	%rRNA	%chrM	%oligo	Del	Ins	Sub	Error
fc1.ch05	41,077,266	22,567,621	12,538,630	55%	56%	33	12,419,599	8,398,130	67.6	14.5	17.6	0.1	2.7%	1.3%	0.4%	4.4%
fc1.ch06	42,486,998	25,117,811	11,082,170	59%	44%	34	10,981,795	8,163,796	74.3	11.7	13.4	0.5	3.0%	1.4%	0.4%	4.8%

nucleotides long and matches the reference in 33 of the 36 positions the score will be: $((33\times5)+(36-33)\times4)/36=4.9$.

5. %filtered: filtered reads/raw reads.
6. %aligned: aligned reads/filtered reads.
7. MeanLen: This means the read mean length. HeliScope™ read length varies from very short to as long as 60 nt. The average read length is usually around 33–36 nt.
8. Counted: This is the number of aligned reads scoring 4.3 or higher.
9. mRNA: Counted – the number of ribosomal RNA reads – the number of mitochondrial reads – the number of reads aligning to the poly-A tailing-oligo.
10. %mRNA, %rRNA, %chrM, and %oligo: mRNA/Counted, ribosomal RNA reads/Counted, mitochondrial reads/Counted, and poly-A tailing-oligo reads/Counted, respectively.
11. Del, Ins, Sub, and Error: per base deletion, insertion, substitution, and total error rates, respectively.

3.10.2. Transcript Count Files

Once sequence reads are aligned to the transcriptome, transcript counts are reported in a count.txt file. Since reads may be assigned to more than one transcript, four types of transcript counts are performed: Unique, All, Frac, and RMC. They are described as follows:

Count files are named: <flow-cell number>.<channel number>. count.txt:

Count files have the following fields:

1. Gene name: The transcript name as it appears in the reference library selected (for yeast this will typically be the SGD transcript annotations).
2. Med length: Transcript length.
3. Unique: Uniquely mapping reads [reads mapping to only a single transcript with no other alignments within the maxDelta score below (as specified in the config file). e.g. with maxDelta of 0.3, a read alignment of a 4.7 score will be considered unique if no other alignments scored 4.4 or higher].
4. All: Both unique and nonunique mapping reads.
5. Frac: The read count/the number of alignments (e.g. if a read maps to two transcripts each transcript will receive half a count).
6. RMC: A read counting, probability based, method designed to assess transcript abundance using ambiguously aligned reads. Ambiguous reads are assigned to transcripts based both on the transcript abundance and alignment score (4).

7. RPM: Reads per million. (RMC × total reads)/10^6. In DGE one read is produced per transcript molecule, so the RPM measure is equivalent to transcripts per million.
8. AvgScore: The average score for all reads assigned to the transcript using RMC.

4. Notes

1. Distilled water should be used for all sample preparation steps. It is referred to here as "water."
2. The AMPure® beads may also be dried by leaving the tubes open at room temperature (drying time will vary).
3. It is recommended to use the Helicos® reagents for tailing, since the tailing-oligo and dATP lot concentrations are optimized to generate the expected tail length.
4. The terminal transferase enzyme is very viscous, and poor mixing of both the master mix and the samples can result in poor tailing.
5. This protocol is optimized for poly-A tailing of 40–60 ng of cDNA. More or less cDNA may be used, but requires optimization of the amount of dATP used in the reaction.
6. The DGE kit has three different poly-A tailing-oligos. Any of them may be used for the tailing reaction.
7. It is best to avoid touching the center of the gel. Handle the gel with clean gloves to avoid stains.
8. Short tail length may be an indication of high sample molarity, or it may indicate poor terminal transferase activity. If it is the latter, the control reaction is expected to have short poly-A tails as well. If tail length is too short, it can be corrected as follows. Return the tailing reaction to the thermocycler and denature at 95°C for 5 min. Place directly on ice for 2 min and then add 2 μl terminal transferase. Incubate the reaction at 42°C for 1 h followed by 10 min at 70°C (there is no need to add dATP, since a sufficient amount remains in the reaction from the first tailing to generate a moderate increase in tail length upon retailing).

References

1. Nagalakshmi, U., Wang, Z., Waern, K., Shou, C., Raha, D., et al. (2008) The transcriptional landscape of the yeast genome defined by RNA sequencing. *Science* **320**, 1344–1349.
2. Mortazavi, A., Williams, B. A., McCue, K., Schaeffer, L., and Wold, B. (2008) Mapping and quantifying mammalian transcriptomes by RNA-Seq. *Nat Methods* **5**, 621–528.
3. Oshlack, A., and Wakefield, M. J. (2009) Transcript length bias in RNA-seq data confounds systems biology. *Biol. Direct* **9**, 4–14.

4. Lipson, D., Raz, T., Kieu, A., Jones, D. R., Giladi, E. et al. (2009) Quantification of the yeast transcriptome by single-molecule sequencing. *Nat. Biotechnol* **27**, 652–658.
5. Frias-Lopez, J., Shi, Y., Tyson, G. W., Coleman, M. L., Schuster, S. C. et al. (2008) Microbial community gene expression in ocean surface waters. *Proc. Natl. Acad. Sci. U.S.A.* **105**, 3805–3810.
6. Perkins, T. T., Kingsley, R. A., Fookes, M. C., Gardner, P. P., James, K. D., et al. (2009) A strand-specific RNA-Seq analysis of the transcriptome of the typhoid bacillus *Salmonella typhi*. *PLoS Genet.* **5**, e1000569.
7. Yoder-Himes, D. R., Chain, P. S., Zhu, Y., Wurtzel, O., Rubin, E. M., et al. (2009) Mapping the *Burkholderia cenocepacia* niche response via high-throughput sequencing. *Proc. Natl. Acad. Sci. U.S. A.* **106**, 3976–3981.
8. Harris, T. D., Buzby, P. R., Babcock, H., Beer, E., Bowers, J., et al. (2008) Single-molecule DNA sequencing of a viral genome. *Science* **320**, 106–109.

Chapter 4

Transcriptome Profiling Using Single-Molecule Direct RNA Sequencing

Fatih Ozsolak and Patrice M. Milos

Abstract

Methods for in-depth characterization of transcriptomes and quantification of transcript levels have emerged as valuable tools for understanding cellular physiology and human disease biology, and have begun to be utilized in various clinical diagnostic applications. Today, current methods utilized by the scientific community typically require RNA to be converted to cDNA prior to comprehensive measurements. However, this cDNA conversion process has been shown to introduce many biases and artifacts that interfere with the proper characterization and quantitation of transcripts. We have developed a direct RNA sequencing (DRS) approach, in which, unlike other technologies, RNA is sequenced directly without prior conversion to cDNA. The benefits of DRS include the ability to use minute quantities (e.g. on the order of several femtomoles) of RNA with minimal sample preparation, the ability to analyze short RNAs which pose unique challenges for analysis using cDNA-based approaches, and the ability to perform these analyses in a low-cost and high-throughput manner. Here, we describe the strategies and procedures we employ to prepare various RNA species for analysis with DRS.

Key words: RNA sequencing, Single-molecule sequencing, Transcriptome profiling, Polyadenylation site mapping

1. Introduction

The emergence of microarray (1–4) and high-throughput DNA/cDNA sequencing technologies (5–10) and their application to understanding biological processes and human disease initially provided a relatively simplistic view of transcriptomes which has since been replaced with a larger, more complicated view of genome-wide transcription. We now have a much more comprehensive view of the genome in which a large fraction of transcripts emanate from unannotated parts of the genome (reviewed in (11)), and has highlighted our limited, yet rapidly emerging,

Young Min Kwon and Steven C. Ricke (eds.), *High-Throughput Next Generation Sequencing: Methods and Applications*, Methods in Molecular Biology, vol. 733, DOI 10.1007/978-1-61779-089-8_4, © Springer Science+Business Media, LLC 2011

knowledge of the transcriptome and the intimate role RNA plays in health and disease (12–17). New technologies and methods, which offer unique approaches to transcriptome characterization and quantitation, with particular emphasis on minimizing the inherent biases seen with existing methods and the ability to work with minute quantities of cellular RNA are critical to fully explore transcriptome biology.

DNA/cDNA sequencing has eliminated some of the technical challenges posed by earlier hybridization-based microarray strategies, including limited dynamic range of detection and the relatively high background due to cross-hybridization, but several fundamental shortcomings still remain which prevent us from understanding the "true" nature of transcriptomes. One limitation of cDNA-based approaches is the tendency of various reverse transcriptases (RT) to generate spurious second-strand cDNA due to their DNA-dependent DNA polymerase activities (18–20). This is thought to occur through either a hairpin loop at the 3′ end of the first-strand cDNA or by specific or nonspecific re-priming, involving either RNA fragments or primers used for the first-strand synthesis. This effect confounds analyses aimed at identifying the strand of the genomic DNA which gives rise to the RNA (e.g. detecting sense vs. antisense transcripts) (21). Another cDNA limitation, known as template switching (22–25), occurs during the process of reverse transcription in which the nascent cDNA being synthesized can sometimes dissociate from the template RNA and re-anneal to a different stretch of RNA with a sequence similar to the initial template. This event creates an artifactual cDNA that comprises the 5′ region of the initial template attached to the 3′ region of the second template. In addition to causing difficulties in RNA quantification, template switching causes problems in the identification of exon–intron boundaries and true chimeric transcripts. RTs are also known to synthesize cDNAs in a primer independent manner, thought to be caused by self-priming due to RNA secondary structure, resulting in the generation of random cDNAs (26, 27). While this self-priming may only occur at a frequency between 2 and 10% of cDNA molecules, these self-primed products are thought to be a major source of error in achieving accurate detection and quantification of RNA expression. Furthermore, RTs are error-prone due to their lack of proofreading mechanisms (28, 29) and yield low quantities of cDNA, necessitating the use of large quantities of input RNA and relatively high levels of amplification. Finally, most commercial technologies for gene expression/whole transcriptome analyses add further artifacts to their measurements by requiring double-stranded cDNA for subsequent amplification by PCR, which can eliminate information regarding which DNA strand is encoding the RNA. While strand-specific libraries can be prepared, they are laborious with many steps (30) or involve RNA–RNA ligation, which is highly inefficient (31).

Since almost all RNA analysis technologies in use today suffer from the limitations briefly summarized above, there is a great need for a technology that can eliminate the difficulties associated with reverse transcription, amplification, ligation, and other cDNA synthesis-based artifacts. To address these difficulties, we have developed the first direct RNA sequencing (DRS) technology (32) using single-molecule sequencing enabled by the Helicos® Genetic Analysis System. A short read technology, DRS currently produces alignable reads up to 55 nts in length with a mean read length of 33–34 nts. Each DRS run contains two flow cells with 50 independent channels, and produce between 800,000 and 8,000,000 aligned reads (≥25 nts) per channel on an average, depending on the requested run time and fields of view (FOV) (e.g. imaging quantity) per channel. The DRS sample preparation step involves only polyadenylation of 3′ ends of RNA molecules without the need for complicated and potentially biased steps such as ligation or PCR amplification of cDNAs (Fig. 1). For several applications, such as gene expression profiling of polyA+ RNA species or polyadenylation site mapping, the natural poly-A tails that are already present on the RNAs provide the hybridization template sufficient for sequencing. This ensures that any biases that may be introduced by sample preparation steps are reduced or eliminated. The simplicity of DRS sample preparation, requiring femtomole-level RNA quantities, combined with its strand-specific and quantitative nature free from reverse transcription-associated artifacts will make DRS the method of choice for transcriptome analyses. As such, we here provide the readers with a detailed description of the methods utilized for our early studies of DRS.

2. Materials

2.1. Reagents

1. Poly(A) Tailing Kit (Ambion). This kit contains *Escherichia coli* PolyA polymerase (2 U/μl), 5× *E. coli* PolyA polymerase buffer, 10 mM ATP solution, and 25 mM $MnCl_2$.
2. 10 mM 3′dATP (Axxora/Jena Biosciences).
3. Nuclease-free water (Ambion).
4. 5 mg/ml Linear Acrylamide (Ambion).
5. 5 M Ammonium acetate (Ambion).
6. Phenol/Chloroform/Isoamyl alcohol.
7. 100% Ethanol.
8. 70% Ethanol.
9. Helicos RNA sequencing kit (Helicos BioSciences Corporation).

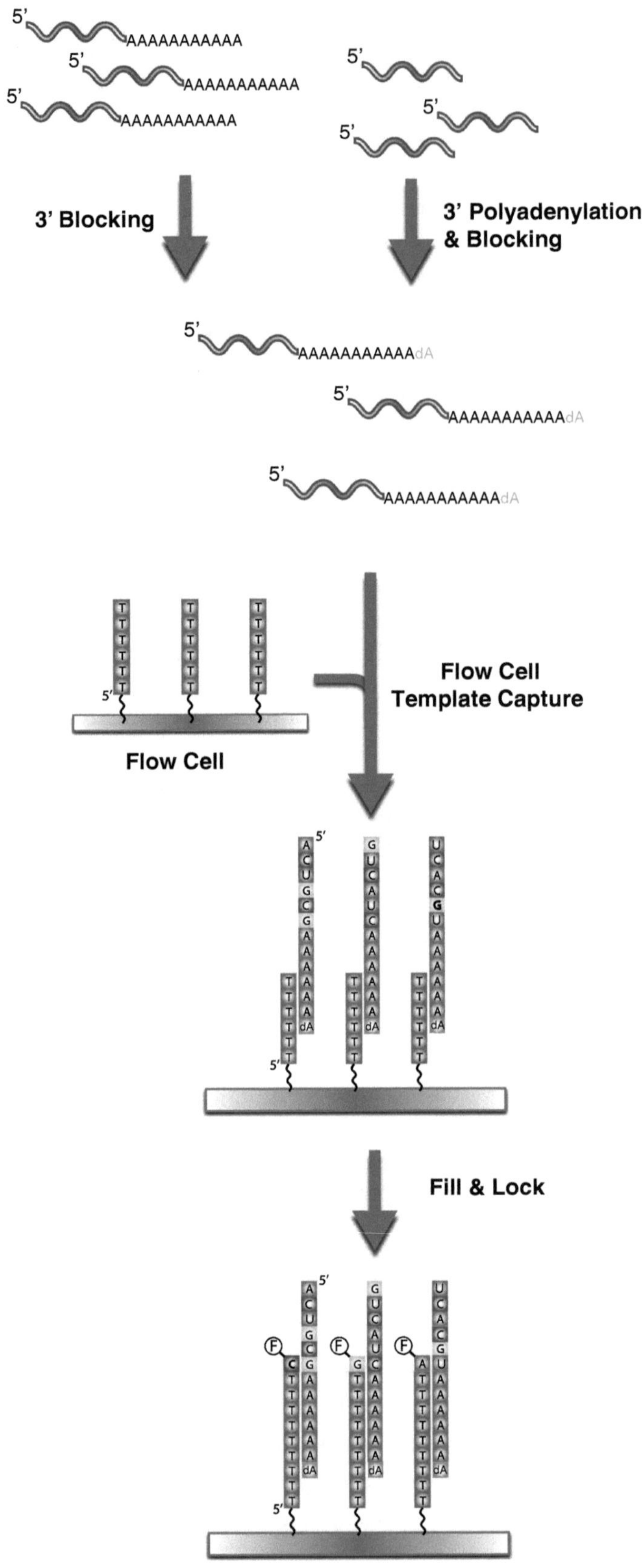
3' Blocking
3' Polyadenylation & Blocking
Flow Cell
Template Capture
Flow Cell
Fill & Lock

2.2. Equipment

1. Thermal cycler (of choice).
2. Refrigerated microcentrifuge.
3. Aluminum blocks (VWR).
4. HeliScope™ Single Molecule Sequencer (Helicos BioSciences Corporation).

3. Methods

We here outline the protocols routinely employed to prepare RNA samples for DRS using the Helicos® Genetic Analysis System. In the first section, we describe the methodology utilized for gene expression profiling and polyadenylation site mapping applications. For these studies, we benefit from the naturally occurring polyadenylation of RNA and require only 3′ blocking followed directly by flow cell hybridization and subsequent sequencing by synthesis ((32); Subheading 3.1). In many cases, a broader view, or qualitatively different view, of the transcriptome requires the RNA molecules to be synthetically polyadenylated prior to hybridization and sequencing. These applications are exemplified by studies involving small RNA sequencing or whole transcriptome analyses, where in each case an additional 3′ poly-adenylation step is performed (Subheading 3.2).

3.1. Gene Expression Analyses and Polyadenylation Site Mapping with DRS

Proceed with this method if the RNA of interest is polyadenylated or part of the total RNA in which the RNA species of interest are polyadenylated.

1. Prepare RNA to be blocked in nuclease-free water in a 22 µl volume (see Notes 1 and 2).

Fig. 1. DRS sample preparation. RNA species which contain a 3′ poly-A tail require 3′ end blocking as described. Other RNA species are enzymatically 3′ polyadenylated and 3′ blocked. The blocking step is performed to prevent "downward" nucleotide additions to the 3′ end of the template during the sequencing process (details of the sequencing strategy and chemistry have been described previously (32)). Polyadenylated RNA is captured on the sequencing flow cell surfaces coated with poly(dT) oligonucleotides through hybridization. A "fill" step is performed with dTTP and polymerase, and then the templates are "locked" in position with fluorescently labeled proprietary Virtual Terminator™ (VT)-A, -C and -G sequencing nucleotide analogs. VT analogs are nucleotides used for sequencing, containing a fluorescent dye and chemically cleavable groups that prevent the addition of another nucleotide. These "fill and lock" steps correct for any misalignments that may be present in poly-A/T duplexes, and ensure that the sequencing starts in the template rather than the poly-A tail.

2. Heat the RNA at 85°C for 1 min in a thermocycler, followed by rapid cooling in a prechilled aluminum block kept in an ice and water slurry (~0°C, see Note 3).
3. Add the following reagents in the indicated order while keeping the denatured sample in the cooled aluminum block: 8 μl of 5× *E. coli* PolyA polymerase buffer; 4 μl of 25 mM $MnCl_2$; 2 μl of 10 mM 3′dATP, and 4 μl of 2 U/μl *E. coli* PolyA polymerase. Mix well by pipetting gently up and down at least five times without vortexing. Incubate for 1 h at 37°C (see Notes 4 and 5).
4. Transfer the samples to a 1.5-ml tube (see Note 6). Add 260 μl nuclease-free water and 300 μl phenol/chloroform/isoamyl alcohol. Vortex vigorously for 30 s.
5. Centrifuge at room temperature for 5 min at maximum speed (~16,000 × *g*). Transfer ~280 μl from the upper aqueous phase to a fresh 1.5-ml tube.
6. Add 30 μl 5 M ammonium acetate, 4 μl 5 mg/ml linear acrylamide, and 900 μl 100% ethanol. Incubate for a minimum of 30 min at −80°C.
7. Centrifuge down at 4°C for 30 min at top speed (~16,000 × *g*). Remove the supernatant with a 1,000-μl pipette tip. Avoid touching the pellet with the pipette tip.
8. Wash the pellet once by adding 500 μl of 70% ethanol. Centrifuge the tube at 4°C for 5 min at top speed (~16,000 × *g*) followed by removal of the supernatant. Keep the pellet at room temperature for 10 min to allow the evaporation of remaining ethanol and water.
9. Resuspend the pellet in 20 μl water.

3.2. Small RNA and Transcriptome Profiling with DRS

The following method is utilized when the RNA of interest is not polyadenylated and thus requires the addition of a poly-A tail for subsequent hybridization and sequencing.

1. Prepare 5 pmol of RNA to be tailed and blocked in nuclease-free water in a 24 μl volume (see Notes 7–9).
2. Heat the RNA at 85°C for 1 min in a thermocycler, followed by rapid cooling in a prechilled aluminum block kept in an ice and water slurry (~0°C, see Note 3).
3. Add the following reagents in the indicated order while keeping the sample on the cooled aluminum block: 8 μl of 5× *E. coli* PolyA polymerase buffer; 4 μl of 25 mM $MnCl_2$; 1 μl of 1 mM ATP, and 3 μl of 2 U/μl *E. coli* PolyA polymerase. Mix well by gently pipetting up and down at least five times without vortexing. Incubate for 10 min at 37°C.
4. While keeping the sample on the 37°C thermocycler block, add 2 μl of 10 mM 3′-dATP to the sample (total volume is

now 42 μl) and mix thoroughly. Incubate the sample for an additional 50 min at 37°C.

5. Sample cleanup can be performed as described in steps 4–9 of Subheading 3.1.

3.3. Flow Cell Hybridization and Single-Molecule Sequencing

3.3.1. Flow Cell Hybridization

Following the methods described in Subheading 3.1 or 3.2 above, the RNA molecules are ready for flow cell hybridization and subsequent sequencing. Hybridization of samples to Helicos flow cell channels is performed in 15–75 μl volume and chosen by the user. The RNA samples are mixed 50:50 with 2× hybridization buffer provided in the Helicos RNA Sequencing Kit (Helicos BioSciences Corporation). The volume of nuclease-free water to be used to resuspend the RNA sample should be determined considering the input RNA quantity and the volume of hybridization cocktail preferred to be used. In general, 0.5–2 fmol of RNA material is required to optimally load each sequencing channel. Following hybridization, the RNA molecules are "filled and locked" (Fig. 1) and are ready for sequencing. The detailed protocols and reagents for flow cell preparation and sequencing are provided in the Helicos RNA Sequencing Kit (Helicos BioSciences Corporation).

3.3.2. Single-Molecule Direct RNA Sequencing

Following hybridization, the Helicos flow cells are moved to the HeliScope™ Single Molecule Sequencer and sequencing by synthesis initiated using prespecified scripts which offer the user the opportunity to run one or two flow cells and define the numbers of FOV from 110 FOV to the standard condition of 1,100 FOV for maximal sequencing yield.

3.4. Data Analysis

The various Unix-based programs and pipelines for the filtering, alignment and downstream analyses of the HeliScope DNA sequencing and DRS data can be downloaded freely at http://open.helicosbio.com/mwiki/index.php/Releases. The detailed descriptions of the programs and functions available are described at http://open.helicosbio.com/helisphere_user_guide/2009-R1/index.html.

Briefly, an initial filtering step is performed on the raw DRS reads before initiating their alignment to reference sequences. This filtering step involves the following read selection steps:

1. DRS generates reads between 6 and 60 nts in length. Depending on the experimental goals, the reference sequence complexity, and size, a user-defined minimum read length cutoff is employed to remove short reads that cannot be aligned reliably. While we routinely use reads ≥25 nts for alignment to human and mouse genomes, for smaller genomes such as *Saccharomyces cerevisiae* and *E. coli*, DRS reads as short as 18 nts can be used.

2. Any 5′ polyT stretches in DRS reads are trimmed. The likely cause of such homopolymeric stretches is the infrequent incomplete "fill" which may occur during dTTP addition step (Fig. 1). Therefore, polyT trimming is preferred to minimize potential misalignment events.
3. Because of flow cell surface imperfections and/or imaging errors, artifactual reads that have a repetition of the base-addition order sequence (CTAG) may appear. Such reads are eliminated during the filtering step.

Total raw base error rate for DRS is currently in the range of 4–5%, dominated by missing base errors typical of single molecule sequencing (2–3%), while the insertion rate is 1–2% and the substitution error rate is 0.1–0.3%. Given that the majority of sequencing errors are due to indels, an aligner that is tolerant to these types of errors should be employed. We highly recommend the use of the indexDPgenomic aligner freely available from the Helicos BioSciences HeliSphere (http://open.helicosbio.com/mwiki/index.php/Releases). While multiple aligners are available and can successfully align Helicos sequence reads, including BWA (33) and SHRiMP (34), use of these aligners will result in a significant reduction in actual aligned reads due to their reduced ability to deal effectively with indels.

4. Notes

1. RNAs isolated with various techniques and commercial kits can be used. We have successfully used total RNA isolated with Trizol® (Invitrogen) and RNeasy Plus Mini Kit (Qiagen). The removal of any potential contaminating genomic DNA from the RNA samples is strongly recommended. RNAs enriched for the polyA+ species can also be used instead of total RNA. RNA quantification can be performed using UV absorbance. For the quantification of low-quantity RNA samples, in our experience, the Quant-iT™ RiboGreen® Reagent (Invitrogen, R11490) provides the most accurate results.
2. The HeliScope™ Single Molecule Sequencer generally requires 2–20 ng total RNA to load each of its 50 channels at optimal levels for DRS. The required total RNA quantity is determined by the number of polyA+ RNA molecules in each sample, which varies depending on the cell type and organism being studied. The Optihyb assay system from Helicos BioSciences allows determination of polyA+ RNA molarity in a sample. We recommend HeliScope Sequencer users to employ this assay to achieve the optimal results.

3. It is essential to chill the block to 0°C in an ice and water slurry, and to cool the sample as quickly as possible to 0°C to minimize re-annealing of the denatured RNA strands.
4. The blocking reaction can be used for RNA sample quantities as high as 10 pmol.
5. A cleanup step is not required if the sample will be used for sequencing within several days of tailing/blocking. After the completion of the tailing/blocking step, the sample should be supplemented with 5 μl 100 mM EDTA prior to storage at −80°C until HeliScope DRS flow cell preparation. For long-term storage of the blocked sample, we recommend that the sample be cleaned prior to storage.
6. The cleanup step can be performed using various approaches. We have successfully used the Qiagen RNeasy MinElute Cleanup Kit (Qiagen, 74204) following manufacturer's instructions and the standard phenol/chloroform method described here.
7. The RNA molarity should be estimated using the concentration information and RNA size distribution. RNA quantities less than the recommended 5 pmol level can be used (down to 0.05 pmol) with this protocol, but ATP and 3′ dATP levels should be adjusted accordingly to maintain the 200:1 ATP:template RNA and 4,000:1 3′dATP:template RNA ratios.
8. For small RNA profiling applications, enrichment for small RNA species can be performed using various methods. We have successfully used the miRNeasy (Qiagen, 217004) and the mirVana (Ambion, AM1560) kits following the manufacturer's guidelines to obtain and profile <200 nts small RNA species. Other gel-based RNA size selection methods can also be used to enrich for RNA species with a more limited size distribution, but has not been tested for DRS.
9. For whole transcriptome applications, RNA species generally need to be fragmented prior to polyadenylation and sequencing. Fragmentation can be done in multiple ways, such as divalent cation-catalyzed hydrolysis (14). Depending on the fragmentation method used, phosphatase treatment of the RNA prior to 3′ polyadenylation and blocking may be necessary to make 3′ RNA ends available for tailing.

Acknowledgments

We thank our colleagues at the Helicos BioSciences Corporation for technical assistance and discussions.

References

1. Fodor, S. P., Read, J. L., Pirrung, M. C., Stryer, L., Lu, A. T., and Solas, D. (1991) Light-directed, spatially addressable parallel chemical synthesis, *Science* **251**, 767–773.
2. Lennon, G. G., and Lehrach, H. (1991) Hybridization analyses of arrayed cDNA libraries, *Trends Genet* 7, 314–317.
3. Shalon, D., Smith, S. J., and Brown, P. O. (1996) A DNA microarray system for analyzing complex DNA samples using two-color fluorescent probe hybridization, *Genome Res* **6**, 639–645.
4. Southern, E. M., Maskos, U., and Elder, J. K. (1992) Analyzing and comparing nucleic acid sequences by hybridization to arrays of oligonucleotides: evaluation using experimental models, *Genomics* **13**, 1008–1017.
5. Bennett, S. T., Barnes, C., Cox, A., Davies, L., and Brown, C. (2005) Toward the 1,000 dollars human genome, *Pharmacogenomics* **6**, 373–382.
6. Deamer, D. W., and Branton, D. (2002) Characterization of nucleic acids by nanopore analysis, *Acc Chem Res* **35**, 817–825.
7. Levene, M. J., Korlach, J., Turner, S. W., Foquet, M., Craighead, H. G., and Webb, W. W. (2003) Zero-mode waveguides for single-molecule analysis at high concentrations, *Science* **299**, 682–686.
8. Margulies, M., Egholm, M., Altman, W. E., Attiya, S., Bader, J. S., Bemben, L. A., Berka, J., Braverman, M. S., Chen, Y. J., Chen, Z., Dewell, S. B., Du, L., Fierro, J. M., Gomes, X. V., Godwin, B. C., He, W., Helgesen, S., Ho, C. H., Irzyk, G. P., Jando, S. C., Alenquer, M. L., Jarvie, T. P., Jirage, K. B., Kim, J. B., Knight, J. R., Lanza, J. R., Leamon, J. H., Lefkowitz, S. M., Lei, M., Li, J., Lohman, K. L., Lu, H., Makhijani, V. B., McDade, K. E., McKenna, M. P., Myers, E. W., Nickerson, E., Nobile, J. R., Plant, R., Puc, B. P., Ronan, M. T., Roth, G. T., Sarkis, G. J., Simons, J. F., Simpson, J. W., Srinivasan, M., Tartaro, K. R., Tomasz, A., Vogt, K. A., Volkmer, G. A., Wang, S. H., Wang, Y., Weiner, M. P., Yu, P., Begley, R. F., and Rothberg, J. M. (2005) Genome sequencing in microfabricated high-density picolitre reactors, *Nature* **437**, 376–380.
9. Shendure, J., Porreca, G. J., Reppas, N. B., Lin, X., McCutcheon, J. P., Rosenbaum, A. M., Wang, M. D., Zhang, K., Mitra, R. D., and Church, G. M. (2005) Accurate multiplex polony sequencing of an evolved bacterial genome, *Science* **309**, 1728–1732.
10. Valouev, A., Ichikawa, J., Tonthat, T., Stuart, J., Ranade, S., Peckham, H., Zeng, K., Malek, J. A., Costa, G., McKernan, K., Sidow, A., Fire, A., and Johnson, S. M. (2008) A high-resolution, nucleosome position map of C. elegans reveals a lack of universal sequence-dictated positioning, *Genome Res* **18**, 1051–1063.
11. Kapranov, P., Willingham, A. T., and Gingeras, T. R. (2007) Genome-wide transcription and the implications for genomic organization, *Nat Rev Genet* **8**, 413–423.
12. Denoeud, F., Aury, J. M., Da Silva, C., Noel, B., Rogier, O., Delledonne, M., Morgante, M., Valle, G., Wincker, P., Scarpelli, C., Jaillon, O., and Artiguenave, F. (2008) Annotating genomes with massive-scale RNA sequencing, *Genome Biol* **9**, R175.
13. Marioni, J. C., Mason, C. E., Mane, S. M., Stephens, M., and Gilad, Y. (2008) RNA-seq: an assessment of technical reproducibility and comparison with gene expression arrays, *Genome Res* **18**, 1509–1517.
14. Mortazavi, A., Williams, B. A., McCue, K., Schaeffer, L., and Wold, B. (2008) Mapping and quantifying mammalian transcriptomes by RNA-Seq, *Nat Methods* **5**, 621–628.
15. Nagalakshmi, U., Wang, Z., Waern, K., Shou, C., Raha, D., Gerstein, M., and Snyder, M. (2008) The transcriptional landscape of the yeast genome defined by RNA sequencing, *Science* **320**, 1344–1349.
16. Sultan, M., Schulz, M. H., Richard, H., Magen, A., Klingenhoff, A., Scherf, M., Seifert, M., Borodina, T., Soldatov, A., Parkhomchuk, D., Schmidt, D., O'Keeffe, S., Haas, S., Vingron, M., Lehrach, H., and Yaspo, M. L. (2008) A global view of gene activity and alternative splicing by deep sequencing of the human transcriptome, *Science* **321**, 956–960.
17. Wilhelm, B. T., Marguerat, S., Watt, S., Schubert, F., Wood, V., Goodhead, I., Penkett, C. J., Rogers, J., and Bahler, J. (2008) Dynamic repertoire of a eukaryotic transcriptome surveyed at single-nucleotide resolution, *Nature* **453**, 1239–1243.
18. Gubler, U. (1987) Second-strand cDNA synthesis: classical method, *Methods Enzymol* **152**, 325–329.
19. Gubler, U. (1987) Second-strand cDNA synthesis: mRNA fragments as primers, *Methods Enzymol* **152**, 330–335.
20. Spiegelman, S., Burny, A., Das, M. R., Keydar, J., Schlom, J., Travnicek, M., and Watson, K.

(1970) DNA-directed DNA polymerase activity in oncogenic RNA viruses, *Nature* **227**, 1029–1031.

21. Wu, J. Q., Du, J., Rozowsky, J., Zhang, Z., Urban, A. E., Euskirchen, G., Weissman, S., Gerstein, M., and Snyder, M. (2008) Systematic analysis of transcribed loci in ENCODE regions using RACE sequencing reveals extensive transcription in the human genome, *Genome Biol* **9**, R3.
22. Cocquet, J., Chong, A., Zhang, G., and Veitia, R. A. (2006) Reverse transcriptase template switching and false alternative transcripts, *Genomics* **88**, 127–131.
23. Mader, R. M., Schmidt, W. M., Sedivy, R., Rizovski, B., Braun, J., Kalipciyan, M., Exner, M., Steger, G. G., and Mueller, M. W. (2001) Reverse transcriptase template switching during reverse transcriptase-polymerase chain reaction: artificial generation of deletions in ribonucleotide reductase mRNA, *J Lab Clin Med* **137**, 422–428.
24. Roy, S. W., and Irimia, M. (2008) When good transcripts go bad: artifactual RT-PCR 'splicing' and genome analysis, *Bioessays* **30**, 601–605.
25. Roy, S. W., and Irimia, M. (2008) Intron missplicing: no alternative?, *Genome Biol* **9**, 208.
26. Haddad, F., Qin, A. X., Bodell, P. W., Zhang, L. Y., Guo, H., Giger, J. M., and Baldwin, K. M. (2006) Regulation of antisense RNA expression during cardiac MHC gene switching in response to pressure overload, *Am J Physiol Heart Circ Physiol* **290**, H2351–2361.
27. Haddad, F., Qin, A. X., Giger, J. M., Guo, H., and Baldwin, K. M. (2007) Potential pitfalls in the accuracy of analysis of natural sense-antisense RNA pairs by reverse transcription-PCR, *BMC Biotechnol* 7, 21.
28. Roberts, J. D., Preston, B. D., Johnston, L. A., Soni, A., Loeb, L. A., and Kunkel, T. A. (1989) Fidelity of two retroviral reverse transcriptases during DNA-dependent DNA synthesis in vitro, *Mol Cell Biol* **9**, 469–476.
29. Varadaraj, K., and Skinner, D. M. (1994) Denaturants or cosolvents improve the specificity of PCR amplification of a G+C-rich DNA using genetically engineered DNA polymerases, *Gene* **140**, 1–5.
30. Cloonan, N., Forrest, A. R., Kolle, G., Gardiner, B. B., Faulkner, G. J., Brown, M. K., Taylor, D. F., Steptoe, A. L., Wani, S., Bethel, G., Robertson, A. J., Perkins, A. C., Bruce, S. J., Lee, C. C., Ranade, S. S., Peckham, H. E., Manning, J. M., McKernan, K. J., and Grimmond, S. M. (2008) Stem cell transcriptome profiling via massive-scale mRNA sequencing, *Nat Methods* **5**, 613–619.
31. Lister, R., O'Malley, R. C., Tonti-Filippini, J., Gregory, B. D., Berry, C. C., Millar, A. H., and Ecker, J. R. (2008) Highly integrated single-base resolution maps of the epigenome in Arabidopsis, *Cell* **133**, 523–536.
32. Ozsolak, F., Platt, A. R., Jones, D. R., Reifenberger, J. G., Sass, L. E., McInerney, P., Thompson, J. F., Bowers, J., Jarosz, M., and Milos, P. M. (2009) Direct RNA sequencing, *Nature* **461**, 814–818.
33. Li, H., and Durbin, R. (2009) Fast and accurate short read alignment with Burrows-Wheeler transform, *Bioinformatics* **25**, 1754–1760.
34. Rumble, S. M., Lacroute, P., Dalca, A. V., Fiume, M., Sidow, A., and Brudno, M. (2009) SHRiMP: accurate mapping of short color-space reads, *PLoS Comput Biol* **5**, e1000386.

Chapter 5

Discovery of Bacterial sRNAs by High-Throughput Sequencing

Jane M. Liu and Andrew Camilli

Abstract

sRNA-Seq is an unbiased method that allows for the discovery of small noncoding RNAs in bacterial transcriptomes through direct cloning and massively parallel sequencing by synthesis. Small bacterial transcripts are enriched from a total RNA preparation and modified with 5′ and 3′ linkers that allow for downstream amplification and sequencing. This protocol includes a treatment that depletes small RNA fractions of tRNAs and 5S rRNA, thereby enriching the starting pool for non-tRNA/rRNA sequences. This protocol can be readily modified to target different RNA species for depletion or to change the size range of RNAs to be sequenced. Thus, sRNA-Seq represents a comprehensive, versatile cloning protocol that may be applicable to the cloning of small RNAs of any size range from any organisms.

Key words: sRNAs, Transcriptome, Massively parallel sequencing, tRNA/rRNA depletion, Direct cloning

1. Introduction

Regulatory RNAs are ubiquitous in nature – they have now been found in all branches of life and in all organisms investigated. In bacteria, small (~30–500 nt) noncoding RNAs (sRNAs) are members of regulatory circuits involved in diverse processes including quorum sensing, carbon metabolism, stress responses, and virulence (1–3). Many of the best-characterized sRNAs base pair with mRNAs transcribed from distinct loci and affect the translation and/or stability of the targeted mRNA (4). Alternatively, several sRNAs affect downstream gene expression by binding protein transcription factors directly and inhibiting their activity (5, 6). Although most sRNAs characterized to date fall into one of these two categories, ones that are transcribed antisense to, or even within, annotated protein-encoding genes have also been described (7–10).

Young Min Kwon and Steven C. Ricke (eds.), *High-Throughput Next Generation Sequencing: Methods and Applications*, Methods in Molecular Biology, vol. 733, DOI 10.1007/978-1-61779-089-8_5, © Springer Science+Business Media, LLC 2011

Both experimental and computational approaches have proven useful in identifying sRNAs in diverse species. These methods often focus on transcripts within intergenic regions (IGRs), where many sRNAs have been identified, or they focus on sRNAs of a specific size range. A more general method, sRNA-Seq, involves direct cloning and massively parallel sequencing by synthesis (MPSS), providing an unbiased approach that allows for interrogation of the entire sRNA repertoire in any bacterium. Using sRNA-Seq, we have recently reported the discovery of over 2,000 sRNAs in the *Vibrio cholerae* transcriptome (9). These results strongly suggest that there are many sRNAs remaining to be discovered in bacteria, even in *Escherichia coli*, where the majority of the sRNA discoveries have been made. This method includes a treatment that depletes total RNA fractions of highly abundant tRNAs and small subunit rRNA, thereby enriching the sample for sRNA transcripts with novel functionality. Similar methods of direct cloning and MPSS have been extremely powerful in the microRNA field (11), and we anticipate that sRNA-Seq will likewise allow for comprehensive sRNA identification in many bacteria.

2. Materials

2.1. General Materials

1. Disposable, RNase-free pipette tips, polypropylene 1.5-mL nonstick microcentrifuge tubes and thin-wall microfuge tubes for PCR.
2. RNase-Zap for preparing materials to be RNase-free (Ambion, Austin, TX).
3. Mini-PROTEAN 3 Electrophoresis System for vertical gel electrophoresis (Bio-Rad, Hercules, CA). The glass plates, combs, and chambers should be treated with RNase-Zap prior to use.
4. 10× TBE: 108 g Tris base, 55 g boric acid, 40 mL 0.5 M EDTA pH 8.0, 960 mL ultrapure H_2O. Filter-sterilize and store at room temperature. From this stock, prepare 1 L, filter-sterilized 1× TBE, as needed.
5. Polyacrylamide gels: 30% w/v acrylamide, 0.8% w/v bis-acrylamide solution (37.5:1) (this is a neurotoxin when unpolymerized), urea, *N,N,N,N′*-tetramethyl-ethylenediamine (TEMED).
6. Ammonium persulfate: prepare 1.6 or 10% (w/v) solutions in ultrapure water and store at 4°C. Use within 1 month of solution preparation.

7. Century and Decade Markers (with 10× PNK Buffer and Cleavage Reagent) (Ambion, Austin, TX). Store Century Markers in 5-μL aliquots and all RNA markers at −80°C.
8. 100-bp quick-load DNA ladder (NEB, Ipswich, MA). Store at 4°C.
9. Loading Buffer II (Ambion, Austin, TX). Store at −20°C and then thaw on ice prior to use.
10. SYBR Gold (Invitrogen, Carlsbad, CA). Store at −20°C. Prior to use, allow the dye to thaw completely at room temperature. This dye is light sensitive.
11. Sodium chloride (NaCl): prepare a 0.4 M solution from a room-temperature 5 M solution and ultrapure water. Filter-sterilize the 0.4 M solution immediately before use.
12. Nanosep tubes: either 100K or 0.2 μm size is appropriate (VWR, West Chester, PA).
13. Glycogen (5 mg/mL) (Ambion, Austin, TX). Store at −20°C in 100-μL aliquots.
14. 100 and 70% ethanol are prepared with highest-grade ethanol and ultrapure water. Store at −20°C.

2.2. Total RNA Preparation

1. GS-3 tubes and Nalgene Oak Ridge phenol-resistant SS34 tubes (VWR, West Chester, PA). The SS34 tubes should be treated with RNase-Zap prior to use.
2. AE Buffer: 50 mM NaOAc pH 5.2, 10 mM EDTA pH 8.0, prepared with ultrapure water and filter-sterilized.
3. SDS: prepare a 20% w/v solution with ultrapure water and filter-sterilize; store at room temperature.
4. Acid phenol:chloroform (5:1) pH 4.5 (Ambion, Austin, TX). Store at 4°C.
5. Phase Lock Gel tubes (50 mL size) (Eppendorf, Hamburg, Germany).
6. Chloroform. Store at room temperature.
7. Sodium acetate (NaOAc): a 3 M solution is made with ultrapure water and adjusted to pH 5.2 with acetic acid. Filter-sterilize the final solution and store at room temperature.
8. Isopropanol. Store at room temperature.
9. DEPC-H_2O (Ambion, Austin, TX). Store at room temperature.

2.3. Addition of 3′ Linkers

1. miRNA cloning Linker 1 (5′-rAppCTGTAGGCACCATCAAT/3ddC/-3′) and 2 (5′-rAppCACTCGGGCACCAAGGA/3ddC/-3′) (IDT, Coralville, IA). They are resuspended at 100 μM concentration in IDTE buffer (10 mM Tris–HCl pH 8.0, 0.1 mM EDTA pH 8.0) and stored at −20°C.

2. T4 RNA Ligase (10 U/μL) (Promega, Madison, WI) and diluted tenfold in 1× Ligase Buffer immediately prior to use.
3. 5× Ligase Buffer: 250 mM HEPES pH 8.3, 50 mM $MgCl_2$, 16.5 mM DTT, 50 μg/mL BSA, 41.5% glycerol. Store buffer at −20°C in 100-μL aliquots.

2.4. Depletion of tRNAs and rRNAs

1. Single-stranded DNA oligonucleotides to deplete tRNAs are ~30 nt long and complementary to the 3′-ends of the tRNAs. The tRNA sequences for many organisms can be found using the Genomic tRNA Database (http://gtrnadb.ucsc.edu/). Our lab has written a script that designs a set of oligonucleotides complementary to a given sequence set, allowing mismatches at the user's discretion (https://sciviz.tufts.edu/confluence/display/CamillLaboratory/Camilli+Lab+Supplementary+Data+Site;jsessionid=9785E20CFBEA6DDCF7EB3DC51A7D73BE). Two to three mismatches allow for one oligonucleotide sequence to potentially base-pair with several different tRNAs, ultimately reducing the number of oligonucleotide sequences that need to be designed and synthesized. Alternatively, by eye, one can align the 3′ ends of tRNAs and design by hand oligonucleotides that are complementary to the tRNAs. Because the 3′ ends of many tRNAs are identical, or highly similar, fewer oligonucleotides than tRNAs need to be designed. We designed 25 oligonucleotides that are sufficient for depleting all 98 tRNAs in the *V. cholerae* transcriptome. Depletion oligonucleotides are kept as 100 μM solutions in 1 mM Tris–HCl pH 8.0 at −20°C.
2. Oligonucleotides to deplete the 5S rRNA are ~30 nt long and complementary to the 3′ end of the 5S rRNA and additional regions of the transcript. For *V. cholerae*, we designed four oligonucleotides to deplete the 5S rRNA. These oligonucleotides were kept as 100 μM solutions in 1 mM Tris–HCl pH 8.0 at −20°C.
3. Oligo Mix: as a starting point, mix all depletion oligonucleotides 1:1, to create a final 100 μM solution. Store the Oligo Mix at −20°C in 100-μL aliquots.
4. 2× Depletion Buffer: 100 mM Tris–HCl pH 7.8, 600 mM KCl, 20 mM $MgCl_2$, 20 mM DTT.
5. RNase H (NEB, Ipswich, MA).

2.5. Reverse Transcription

1. dNTPs (100 mM) (NEB, Ipswich, MA). Mix 1:1:1:1 and dilute tenfold with ultrapure water. The solution is stored at −20°C.
2. RT/REV Primer (5′-GATTGATGGTGCCTACAG) is resuspended in 1 mM Tris–HCl pH 8.0 to make a 100 μM solution and stored at −20°C. A working solution of 10 μM is made as needed.

3. SuperScript III Reverse Transcriptase (with 10× buffer and 0.1 M DTT) (Invitrogen, Carlsbad, CA).
4. RNase Inhibitor (Ambion, Austin, TX).
5. ExoSAP-IT (USB, Cleveland, OH).
6. Phenol:chloroform:IAA (25:24:1). Store at 4°C.

2.6. PCR

1. *Taq* DNA Polymerase (with 10× Buffer) (NEB, Ipswich, MA).
2. SIII forward primer (5′-GCCTCCCTCGCGCCATCAGGAT TGATGGTGCCTACAG-3′) and SIII reverse primer (5′GCC TTGCCAGCCCGCTCAGGTCCTTGGTGCCCGAG TG-3′) are resuspended in 1 mM Tris–HCl pH 8.0 to make a 100 μM solution and stored at −20°C. A working solution of 10 μM is made, as needed. Underlined sequences are regions used by 454 to prime sequencing reactions.
3. 6× DNA Loading Dye: 0.03% bromophenol blue, 0.03% xylene cyanol FF, 15% Ficoll 400, 10 mM Tris–HCl pH 7.5, and 50 mM EDTA pH 8.0.
4. *n*-Butanol. Store at room temperature.
5. Micro Bio-Spin 30 (Bio-Rad, Hercules, CA) columns may be used for buffer exchange. These should be stored at 4°C.
6. TOPO TA Cloning kit for Sequencing (Invitrogen, Carlsbad, CA).
7. QIAprep 96 Turbo Miniprep Kit (Qiagen, Valencia, CA).

3. Methods

The sRNA-Seq method is illustrated in Fig. 1. Representative gels from the various steps in the method are shown in Fig. 2. All the typical precautions against RNase contamination should be observed throughout the sRNA-Seq cloning protocol. Gloves should be worn at all times and changed frequently. Materials and solutions should be kept protected from dust. Any solid materials not supplied as RNase-free should be treated so that they are free from nucleases.

RNA transcripts in a size range predetermined by the investigator are gel-purified, enriched, and cloned through the addition of oligonucleotide linkers to both the 5′ and 3′ ends of the transcript. The depletion step takes place after addition of a 3′ linker and effectively removes tRNAs and 5S rRNA from downstream analysis by separating these transcripts from the linker that is necessary for reverse transcription, amplification, and sequencing. Readout of each individual clone is accomplished through high-throughput 454 pyrosequencing (a robust means of MPSS).

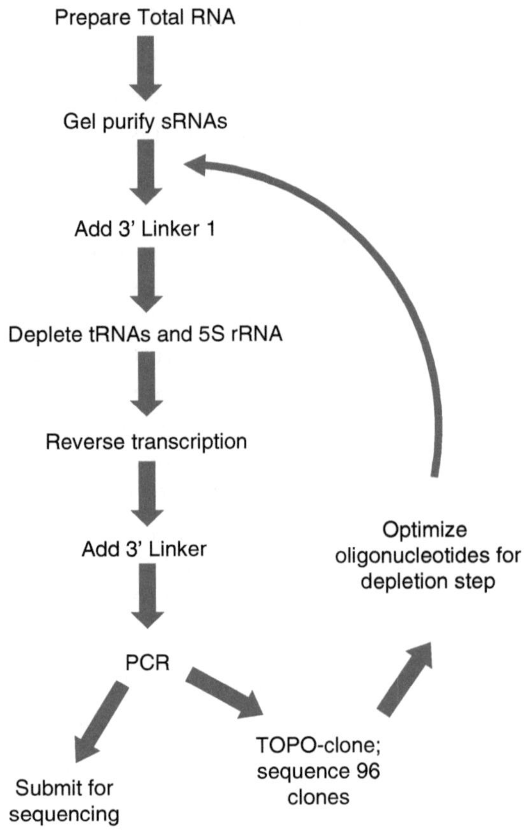

Fig. 1. Flowchart depicting the sRNA-Seq method. First-time users are strongly encouraged to run through the procedure several times to become familiar with the methodology. PCR products from these pilot experiments should be TOPO-cloned, and clones may be sequenced in 96-well format, providing important information that can be used to optimize the procedure, particularly the depletion step.

The direct cloning portion of the sRNA-Seq protocol, once optimized by the individual, takes less than 1 week to complete. Initially, we suggest running through the method 1–2 times to become familiar with the steps and using small-scale sequencing (96-well format) to optimize the protocol for the user's needs.

In the following sections, we provide an example of cloning 80–200 nt sRNAs and the use of 454 (Roche) sequencing. The long read lengths (~300 nt) of 454 sequencing render it highly applicable for sequencing sRNAs without the need for paired-end MPSS. Nevertheless, this method should be easily adaptable to other size ranges of sRNAs, as well as other methods of MPSS, such as Solexa (Illumina) or SOLiD (ABI) paired-end sequencing.

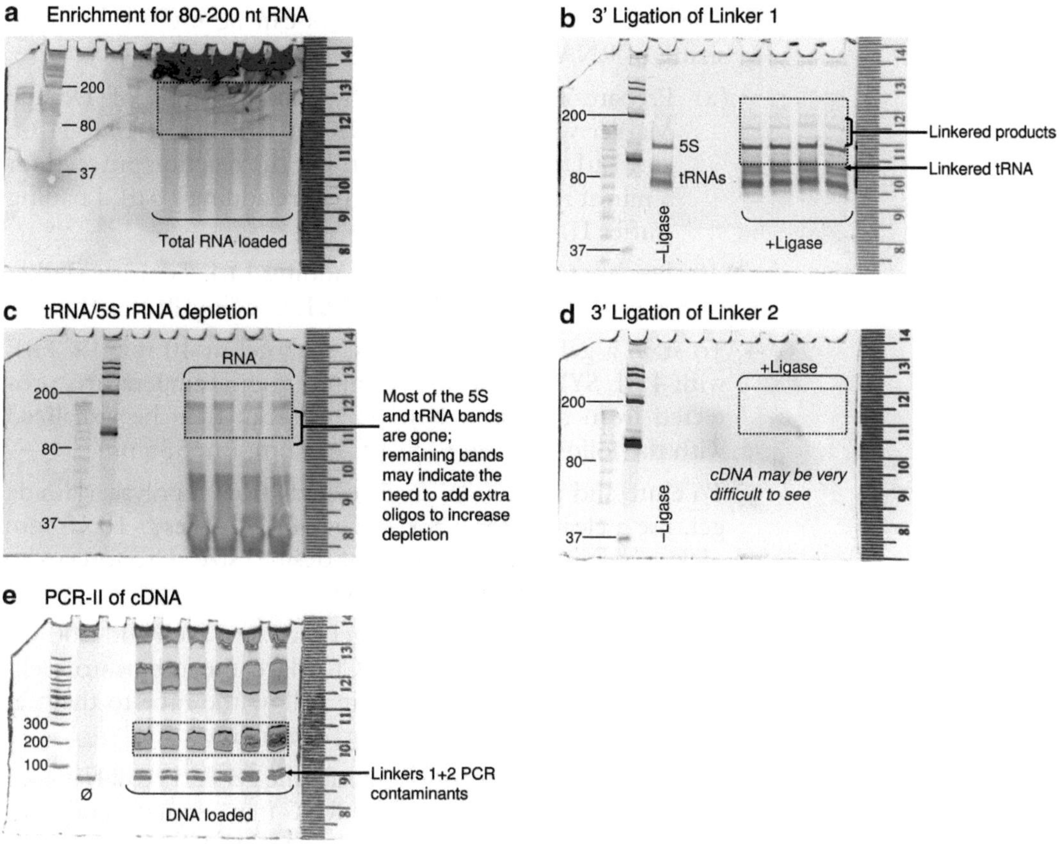

Fig. 2. Examples of gels run out for direct cloning of sRNAs in *V. cholerae*. All gels were stained with SYBR Gold for 30 min. Regions in *dotted boxes* were gel-purified. In gels (**a–d**), Decade Marker is loaded in the first lane on the *left* and Century Marker, plus 1 pmol of a 37-mer, is loaded in the second lane. In gel (**e**), 100 bp DNA ladder is loaded; a control (Ø) consisting of PCR-amplified (Linker 2–TA–Linker 1) is loaded next to the ladder.

3.1. General Protocols

1. To prepare a 1-mm thick, 10% polyacrylamide TBE-Urea (denaturing) minigel, wear personal protective equipment, then mix 1 mL 10× TBE with 3.33 mL 30% w/v acrylamide:0.8% w/v bis-acrylamide (37.5:1), 2.14 mL H_2O, 4.2 g urea, 330 μL 1.6% ammonium persulfate, and 5 μL TEMED. Pour the gel and add a ten-well comb. The gel should polymerize in about 30 min (see Note 1). Before loading gels with samples, prerun the gel for 30 min at 300 V in 1× TBE. Run all samples at 200 V.
2. To prepare a 1-mm thick, 12% TBE (native) minigel, mix 1 mL 10× TBE with 4 mL 30% w/v acrylamide:0.8% w/v bis-acrylamide (37.5:1), 4.93 mL H_2O, 70 μL 10% ammonium persulfate, and 5 μL TEMED. Pour the gel and add a ten-well comb. The gel should polymerize in about 30 min (see Note 1). Before loading gels with samples, prerun the gel for 30 min at 300 V in 1× TBE. Run all samples at 200 V.

3. The Decade and Century Markers can be used to estimate sizing of RNA (see Note 2).
   (a) Prepare Decade Marker by mixing 1 μL of Decade Marker with 1 μL 10× PNK buffer, 7 μL DEPC-H_2O, and 1 μL Cleavage Reagent. Let the reaction proceed for 5 min at room temperature before adding 10 μL Loading Buffer II.
   (b) Prepare Century Marker by adding 1 μL Century Marker to 4 μL DEPC-H_2O and 10 μL Loading Buffer II.
4. To stain a gel for nucleic acids, soak gel in 40 mL 1× TBE with 4 μL SYBR Gold for 30 min at room temperature, protected from the light. Nucleic acids can then be visualized with the following filter set: Ex 465 nm, Em 535 nm.
5. To elute and precipitate nucleic acids from a polyacrylamide gel, use a clean razor blade to cut out gel slices (~10×5 mm each) containing RNA/DNA of desired size. Use forceps to place gel slices into 1.5-mL tubes.
   (a) For each tube, crush the gel slice and then add 400 μL filter-sterilized 0.4 M NaCl. Place the tube into well-crushed dry ice for ~30 min. Allow the tube to thaw, at room temperature, rotating, overnight.
   (b) The following morning, transfer the gel slurry to a Nanosep tube and spin at 16,000×*g* for 3 min. Remove and keep the flow-through. If there is any residual liquid remaining with the gel pieces, repeat the spin. Ethanol-precipitate the RNA by adding 8 μL glycogen and 1 mL ice-cold 100% ethanol, and placing the tube on well-crushed dry ice for ~30 min. Spin the tube at 16,000×*g*, for 30 min at 4°C. (For PCR products, wash pellet with 500 μL ice-cold 70% ethanol and spin for 5 min at 16,000×*g*.) Decant the liquid carefully and use a "Kimwipe" to wick off any excess liquid around the rim of the tube. Air-dry the tube, inverted, for 5–10 min at room temperature.

3.2. Prepare Total RNA

1. Grow up 300 mL of bacteria in desired medium to desired OD_{600}. Spin down the bacteria in a GS-3 tube at 6,000×*g* for 10 min at 4°C. Decant the supernatant. Add 12 mL AE Buffer to the tube and resuspend the pellet by vortexing and/or pipetting up and down. Transfer the mixture to an SS34 tube.
2. Add 1 mL 20% SDS, 12 mL acid phenol:chloroform (5:1). Vortex to mix. Incubate the tube for 10 min in 65°C water bath, vortexing for 10 s every minute. Incubate the tube for 5 min on ice. Spin the tube for 15 min at 12,000×*g* in an SS34 rotor. In the meantime, prespin a 50-mL Phase Lock Gel tube, for 5 min at 1500×*g*.

3. After the 15-min spin, the aqueous and organic layer should have separated, with a white precipitate in the middle layer. Carefully remove the top (aqueous) layer with a glass pipette and transfer to the prespun Phase Lock Gel tube (see Note 3).
4. Add 13 mL chloroform to the Phase Lock Gel tube. Cap the tube and invert the tube several times to mix. Spin the tube for 10 min at 1500 × *g* in a tabletop centrifuge at room temperature. Pour the supernatant (on top of the white gel) into a new SS34 tube.
5. Add 1 mL 3 M NaOAc (pH 5.2), 50 μL glycogen, and 10 mL room-temperature isopropanol. Spin the tube for 40 min at ≥15,000 × *g* in an SS34 rotor.
6. Wash the pellet with 4 mL ice-cold 70% ethanol and spin again for 20 min at ≥15,000 × *g*. Pour off the ethanol and shake to remove residual liquid. There should be a clear or milky-white pellet that is the RNA. Place the tube, without the cap, in a speedvac with the rotor off on low heat for 15 min to remove residual ethanol.
7. Resuspend the pellet in DEPC-H_2O (see Note 4).
8. Quantitate the RNA in a spectrophotometer at OD_{260}. Pure RNA has an OD_{260}/OD_{280} ratio around 2.0 (see Note 5).
9. Run 1 μg of the RNA on a 1% agarose gel (a native gel is fine here). The gel should show three distinct bands for the 23S rRNA, 16S rRNA, and the 5S rRNA and tRNAs (which will run together). The 23S rRNA band should be approximately twice as bright as the 16S rRNA band.
10. Store the RNA in 500-μg aliquots at −80°C until ready to use for the sRNA cloning steps (see Note 6).

3.3. Enrich for Transcripts 80–200 nt Long

1. Prepare a 10% TBE-Urea minigel.
2. Mix 500 μg total RNA with 1–2 volumes of Loading Buffer II. Heat the RNA and the markers at 65°C for 5 min just prior to loading onto the gel.
3. Load the markers into the leftmost lanes of the prerun 10% TBE-Urea gel. Split the RNA equally into the remaining lanes, leaving the lane next to the markers empty, if possible. Run the gel for 35 min at 200 V (see Note 7).
4. Stain the gel with SYBR Gold.
5. Visualize the RNA and cut out the RNA between 80 and 200 nt (Fig. 2a) (see Note 8).
6. Elute the RNA overnight and precipitate the RNA with glycogen and ethanol. Resuspend the RNA with DEPC-H_2O (combining all the tubes into one) into a final volume of 20 μL.

3.4. Add on 3′ Linker 1

1. Prepare a 10% TBE-Urea minigel.
2. Using the RNA from Subheading 3.3, set up three ligation reactions, one of which serves as a negative control. Save the remaining unused RNA at −80°C for future cloning, if needed.
   (a) Set up two of the same reactions: to 5 μL of RNA, add 1 μL Linker 1, 2 μL 5× Ligase Buffer, 1 μL DEPC-H_2O, and 1 μL T4 RNA Ligase (1 U/μL; see Note 9).
   (b) For the negative control, to 1 μL RNA, add 2 μL 5× Ligase Buffer and 7 μL DEPC-H_2O.
   (c) Let reactions proceed for 2 h at room temperature.
3. Add 20 μL Loading Buffer II to each +Ligase reaction; add 10 μL Loading Buffer II to the −Ligase reaction. Heat the sample and the markers at 65°C for 5 min. Load the +Ligase reactions across the four rightmost lanes of the pre-run 10% TBE-Urea gel. Load the markers and the −Ligase sample in lanes on the left. Run the samples at 200 V for 45 min.
4. Stain the gel with SYBR Gold, visualize the RNA, and cut out the RNA between 98 and 218 nt (Fig. 2b).
5. Elute the linkered-RNA overnight.

3.5. Deplete Linkered-RNA of tRNAs and 5S rRNA

1. Prepare a 10% TBE-Urea minigel.
2. Take the overnight elution and ethanol-coprecipitate the RNA and depletion oligonucleotides by adding 15 μL Oligo Mix, 8 μL glycogen, 1 mL ice-cold 100% ethanol, and placing the tube on well-crushed dry ice for ~30 min (see Note 10). Finish the ethanol precipitation as usual.
3. Resuspend the dry pellets of linkered-RNA with 1× Depletion Buffer and combine the samples into one tube with a final volume of 10 μL. Transfer the sample to a PCR tube.
4. Using a thermocycler, heat up the sample at 65°C for 5 min, then cool to 37°C, 0.1°C/s. When the sample reaches 37°C, add 0.5 μL RNase H and incubate the reaction at 37°C for 30 min. Repeat this step once.
5. Add 20 μL Loading Buffer II to the sample. Heat the sample and markers at 65°C for 5 min. Load the sample across the four rightmost lanes of the prerun 10% TBE-Urea gel. Load the markers in lanes on the left. Run the gel at 200 V for 45 min.
6. Stain the gel with SYBR Gold, visualize the RNA, and cut out the RNA between 98 and 218 nt (Fig. 2c) (see Note 11).
7. Elute the RNA overnight and precipitate the RNA with glycogen and ethanol. Resuspend the RNA with DEPC-H_2O, combining all tubes, to a final volume of 10 μL.

3.6. Reverse Transcription of sRNA-Enriched Pool

1. In a PCR tube, to the 10 μL of RNA, add 1 μL 10 mM dNTPs, 10 μL RT/REV primer (10 μM), and 1 μL DEPC-H_2O. Heat the mixture at 65°C for 5 min; transfer the tube to ice.
2. To the sample, add 4 μL 5× RT Buffer, 1 μL 0.1 M DTT, 1 μL RNase Inhibitor, and 1 μL Reverse Transcriptase. Incubate the reaction at 50°C for 1 h; inactivate the enzyme at 70°C for 20 min.
3. Add 8 μL ExoSAP-IT to the sample and incubate the reaction at 37°C for 30 min.
4. Transfer the reaction to a new 1.5-mL tube. Clean up the reaction by adding 25 μL phenol:chloroform:IAA (25:24:1) to the sample. Vortex the sample for 30 s and spin the tube for 3 min at 16,000 × *g* at room temperature.
5. Remove the aqueous layer (20 μL) and ethanol precipitate the cDNA by adding 2 μL 3 M NaOAc (pH 5.2), 8 μL glycogen, and 90 μL ice-cold ethanol. Place the tube into well-crushed dry ice for ~30 min and then spin down the sample for 30 min, at 16,000 × *g*, at 4°C. Decant the liquid from the tube and air-dry the sample for 10 min before resuspending the pellet in 11 μL DEPC-H_2O.

3.7. Add on 3′ Linker 2

1. The addition of the second linker proceeds just as with the addition of the first linker. Prepare a 10% TBE-Urea minigel.
2. Using the cDNA from Subheading 3.6, set up three ligation reactions, one of which serves as a negative control.
   (a) Set up two of the same reactions: to 5 μL of cDNA, add 1 μL Linker 2, 2 μL 5× Ligase Buffer, 1 μL DEPC-H_2O, and 1 μL T4 RNA Ligase (1 U/μL) (see Note 9).
   (b) For the negative control, to 1 μL cDNA, add 2 μL 5× Ligase Buffer and 7 μL DEPC-H_2O.
   (c) Let reactions proceed for 2 h at room temperature.
3. Add 20 μL Loading Buffer II to the +Ligase reactions; add 10 μL Loading Buffer II to the –Ligase reaction. Heat the sample and the markers at 65°C for 5 min. Load the +Ligase reactions across the four rightmost lanes of the prerun 10% TBE-Urea gel. Load the markers and the –Ligase sample in lanes on the left. Run the samples at 200 V for 50 min.
4. Stain the gel with SYBR Gold, visualize the linkered-cDNA, and cut out the cDNA between 116 and 236 nt (Fig. 2d) (see Note 12).
5. Elute the cDNA overnight and precipitate the cDNA with glycogen and ethanol. Resuspend the cDNA with DEPC-H_2O, combining all tubes, to a final volume of 10 μL.

3.8. Prepare the Samples for Sequencing

1. Prepare a 12% TBE gel.
2. For the first PCR (PCR-I), mix 3 μL of the linkered-cDNA with 38.5 μL dH_2O, 5 μL 10× Standard Taq Buffer, 1 μL 10 mM dNTPs, 1 μL 10 μL SIII forward primer (10 μM), 1 μL SIII reverse primer (10 μM), and 0.5 μL Taq. Incubate the reaction at 95°C for 10 min, followed by 25 cycles of 95°C for 30 s, 52°C for 30 s, 72°C for 30 s. After a final extension at 72°C for 5 min, add 30 μL 6× DNA Loading Dye and run the sample across four lanes in a prerun 12% TBE gel at 200 V for 47 min. Use a 100-bp ladder (5 μL) for a marker.
3. Stain the DNA with SYBR Gold, visualize the DNA, and excise the region between 154 and 274 bp (see Note 13).
4. Elute the DNA overnight and precipitate the DNA with glycogen and ethanol. Resuspend the pellet in 30 μL IDTE pH 7.5.
5. For the second PCR (PCR-II), set up 8 × 50 μL reactions. For each reaction, mix 1 μL of the PCR-I product with 40.5 μL dH_2O, 5 μL 10× Standard Taq Buffer, 1 μL 10 mM dNTPs, 1 μL 10 μL SIII forward primer (10 μM), 1 μL SIII reverse primer (10 μM), and 0.5 μL Taq. Incubate the reaction at 95°C for 10 min, followed by 15 cycles of 95°C for 30 s, 52°C for 30 s, 72°C for 30 s (see Note 14). After a final extension at 72°C for 5 min, combine all eight samples and concentrate the sample with *n*-butanol, followed by buffer exchange (see Note 15); the final volume should be 80–100 μL. Add an equal volume of 6× DNA Loading Dye and run the sample across six lanes in a prerun 12% TBE gel at 200 V for 47 min. Use a 100-bp ladder (5 μL) for a marker.
6. Stain the DNA with SYBR Gold, visualize the DNA, and excise the region between 154 and 274 bp (Fig. 2e) (see Note 13).
7. Elute the DNA and precipitate the DNA with glycogen and ethanol. Resuspend the pellet in 20 μL IDTE pH 7.5 (see Note 16).
8. The sample is now ready to be quantified and then sequenced.
   (a) If available, run some of the sample (1 μL) on a bioanalyzer (see Note 17).
   (b) TOPO clone 2 μL of the sample and sequence 48–96 clones to check that transcripts are of appropriate size, and tRNA/rRNA depletion is adequate (see Note 18).
   (c) If depletion of tRNA/rRNA is adequate, sequence sample by 454 with Primer B (see example of reads in Fig. 3).

GTCCTTGGTGCCCGAGTGGGGGGCCCTGGTCCTCCCGCAACACTAGTTCGTGAACCTGGTC
AGATCCGGAAGGAAGCAGCCACAGCGGATGATGTGTGTGCCGGGATGTGGCTGGGGTCTCC
GCAAAACTGTAGGCACCATCAATC*CTGATGGCGCGAGGGAGGC*
110 nt; BLASTN: IGR 1067/8, 4.5S RNA

GTCCTTGGTGCCCGAGTGTAAGATGTTCTATCTTTCAGACCTTTTGTTTCACGTTATTGG
ATTAGGCTGATTCAGCCGCCCCAGTCACCATTTGACTGGGGCGTTTTTTAACTGTAGGCA
CCATCAATC*CTGATGGCGCGAGGGAGGC*
93 nt; BLASTN: ORF VC0110

GTCCTTGGTGCCCGAGTGGCAGATGTTTTGTGGAGCCTCAACTCCAATACAGAACATTCA
GGGGGAGTAGTGCCGAGGTGAATCAAAGTTGTGGCTTTGGTTTATCGGTTGAACGGGCTGAA
TCCCTTCAACTGTCATCAGCTCGAATCTGATGAAGAGCTTCTGAGGGAAATCTTTCAC
ACTGTAGGCACCATCAATC*CTGATGGCGCGAGGGAGGC*
163 nt; BLASTN: IGR VC0391/2

Fig. 3. Example sequences (read 5′ to 3′) from sRNA-Seq, sequenced by 454 using Primer B. *Underlined sequences* correspond to Linkers 2 and 1. Sequence in *italics* is part of SIII forward primer and allows for 454 sequencing using Primer A. Sequences between *underlined* linkers were used for BLASTN analysis, the results of which are listed below each sequence.

4. Notes

1. To ensure complete polymerization of the gels, pour the gels the day before and after 30 min of polymerization, wrap the gel (with its comb intact) in a paper towel (wet with 1× TBE) and plastic wrap. Store the gel at 4°C overnight.
2. The Decade Marker can be used to set the lower boundary for RNA sizing, while the Century Marker can be used to set the upper boundary. We also found it useful to add 1 pmol of an oligonucleotide of known size (we used a 37-mer) to the Century Marker to help calibrate the sizing of the Decade Marker bands on the gel.
3. When separating the aqueous layer (with the RNA) from the organic (phenol:chloroform) layer, you do not need to remove the entire aqueous layer. Leaving behind some of the aqueous layer will ensure that you do not accidentally carry over the white precipitate into the Phase Lock Gel tube.
4. To resuspend the pellet of RNA, begin by adding approximately 500 μL and gently pipetting up and down. You may need to heat the pellet at 65°C, briefly, to help dissolve the pellet. The RNA may be very viscous, which will make quantification difficult. Add more DEPC-H_2O, if necessary, to reduce viscosity and to dissolve the pellet completely. At this point, keeping a high concentration of RNA will reduce the volume needed to load onto the gel for sRNA enrichment.

However, if the RNA is too viscous, then it is difficult to accurately quantitate the sample.

5. For the RNA, an OD_{260}/OD_{280} ratio between 1.8 and 2.0 is fine. Using this protocol, we generally have RNA preps that are between 5 and 10 μg/μL; you will likely need to make a 1:10 dilution to get an accurate OD_{260} reading.
6. We have successfully used this total RNA preparation protocol to isolate RNA (including sRNAs) from *E. coli*, *V. cholerae*, and *Streptococcus pneumoniae*. This protocol was originally designed for use with *Saccharomyces cerevisiae* (http://derisilab.ucsf.edu).
7. The exact timing of how long to run a gel will need to be determined empirically. When samples are run on a gel for longer periods of time, a higher degree of resolution is achieved. However, this increases the size of the gel that must be cut out, which makes the downstream gel elution and RNA precipitation steps more laborious. Also, when we have set out to clone sRNAs 20–200 nt long, we split the samples into two sets of gels. We used a 10% gel to clone 80–200 nt sRNAs and a 15% gel to clone 20–80 nt sRNAs.
8. We visualize the gel with a fluorescent ruler and then, at the bench, we use the ruler to guide our cutting of the gel pieces. To account for both minor errors in how the samples may run and experimental error during cutting, we tend to cut a slightly broader range (50–250 nt) during the sRNA enrichment step to ensure that the desired window is obtained.
9. T4 RNA Ligase generally has some trace ATP contamination that is carried over during purification of the protein. This ATP can charge the 5′ end of phosphorylated RNA/DNA, which may result in undesired RNA–RNA ligations or circularization of RNA. To avoid these contaminating products, a 1:10 dilution of the T4 RNA Ligase (in 1× Ligase Buffer) is used to dilute out the ATP. Linker 1 and 2 are both 5′-preadenylated and 3′-modified such that the only ligations that should occur within the +Ligase reactions should be between the 5′ end of the linker and a 3′-OH. Any sRNAs that have a 3′-modification will, therefore, not be cloned.
10. We find that it is important that the Oligo Mix is added to the linkered-RNA prior to ethanol precipitation. The amount of Oligo Mix to use is something that should be tested empirically by the researcher; we have had success cloning *V. cholerae* sRNAs using 1.5 nmol of a mixture containing 29 different oligonucleotides. Most of the oligonucleotides are mixed in a 1:1 ratio. With *V. cholerae*, we observed through pilot sequencing (see Note 18) that several tRNAs were overrepresented in the final cDNA library and that tRNA Ser(GCT)

was highly abundant. Oligonucleotides complementary to the overrepresented tRNAs were twice as abundant in the final Oligo Mix used (2:1 ratio). For the highly abundant Ser(GCT), we added 0.5 nmol of the anti-Ser(GCT) oligonucleotide directly to the linkered-RNA, prior to ethanol precipitation, in addition to the 1.5 nmol Oligo Mix, and successfully depleted the Ser(GCT) and all other tRNAs to ~25% of the final pool of cDNAs.

11. After the first ligation reaction, when running the linkered-RNA on a polyacrylamide gel, several sharp bands should be visible after staining (Fig. 2b): bands representing the 5S rRNA and linkered-5S rRNA, as well as bands for the tRNAs and linkered-tRNA. After depletion, most of these sharp bands should be replaced instead by diffuse smears (Fig. 2c). The presence of sharp bands may indicate that the depletion step did not work completely, although this should be confirmed with pilot sequencing (see Note 18).
12. After the second 3′ linker ligation reaction, the sample will appear very faint when run on a gel (Fig. 2d). You may see hardly any cDNA on the gel after staining. This makes the use of markers particularly important as they will guide you to cut out the correct region of the gel.
13. The PCR products, when run out on a gel, may contain some sharp bands (Fig. 2e). These bands should be avoided whenever possible when cutting the gel to recover DNA. They represent primer-dimer contaminants, linkers with only 1–2 bases in between them, or may indicate that the depletion step did not work well.
14. Two rounds of PCR are used to prevent any one (or set of) sequence to take over the entire reaction. The number of cycles for each round of PCR may need to be determined empirically; in our sequencing reactions with *V. cholerae*, we had success with 25 cycles followed by a 1/30 dilution of the DNA and another 15 cycles.
15. To concentrate a sample using *n*-butanol, add 1–3 volumes of *n*-butanol to an aqueous sample. Vortex the sample and then spin it at 16,000 × *g* for 3–5 min. Remove and discard the top organic layer. Add another 1–3 volumes of *n*-butanol and repeat the spin. Through repeated addition and removal of *n*-butanol, the aqueous layer, with the cDNA, should decrease in volume. After concentration, the aqueous layer should be passed through a buffer-exchange column (e.g., Micro Bio-Spin 30) to remove salts and any residual *n*-butanol.
16. IDTE buffer is used when submitting samples for 454 sequencing.

17. Running the samples out on a bioanalyzer will provide more accurate information about the concentration of your sample than using either a Nanodrop or other UV spectrophotometer. You will expect to see a broad curve over the range of desired transcript size and not sharp peaks.
18. The best way to optimize your Oligo Mix (see Note 10) is to TOPO-clone the final cDNA library and sequence at least 48 (96, if possible) clones and confirm that tRNAs and rRNA are being depleted. A QIAprep 96 Turbo Miniprep Kit is very useful for preparing the clones for sequencing. If depletion is not adequate, then the oligonucleotide ratios can be altered or additional depletion oligonucleotides can be added, and the process repeated iteratively. The pilot sequencing experiments may even point out abundant sRNAs, providing some experimental data prior to the large-scale sequencing.

Acknowledgments

We thank Kip Bodi for writing the script for designing tRNA-depletion oligonucleotides. This work was supported by Award Number K12GM074869 (J.M.L.) and A145746 (A.C.) from the National Institute of General Medical Sciences and National Institute of Health, respectively. A.C. is a Howard Hughes Medical Institute investigator. The content is solely the responsibility of the authors and does not necessarily represent the official views of the National Institute Of General Medical Sciences or the National Institutes of Health.

References

1. Gorke, B., and Vogel, J. (2008) Noncoding RNA control of the making and breaking of sugars. *Genes & Development* **22**, 2914–2925.
2. Gottesman, S. (2005) Micros for microbes: non-coding regulatory RNAs in bacteria. *Trends in Genetics* **21**, 399–404.
3. Romby, P., Vandenesch, F., and Wagner, E. G. H. (2006) The role of RNAs in the regulation of virulence-gene expression. *Current Opinion in Microbiology* **9**, 229–236.
4. Aiba, H. (2007) Mechanism of RNA silencing by Hfq-binding small RNAs. *Current Opinion in Microbiology* **10**, 134–139.
5. Liu, M. Y., Gui, G. J., Wei, B. D., Preston, J. F., Oakford, L., Yuksel, U., Giedroc, D. P., and Romeo, T. (1997) The RNA molecule CsrB binds to the global regulatory protein CsrA and antagonizes its activity in Escherichia coli. *Journal of Biological Chemistry* **272**, 17502–17510.
6. Wassarman, K. M., and Storz, G. (2000) 6S RNA regulates E-coli RNA polymerase activity. *Cell* **101**, 613–623.
7. Fozo, E. M., Hemm, M. R., and Storz, G. (2008) Small Toxic Proteins and the Antisense RNAs That Repress Them. *Microbiology and Molecular Biology Reviews* **72**, 579–589.
8. Kawano, M., Reynolds, A. A., Miranda-Rios, J., and Storz, G. (2005) Detection of 5′- and 3′-UTR-derived small RNAs and cis-encoded antisense RNAs in Escherichia coli. *Nucleic Acids Research* **33**, 1040–1050.
9. Liu, J. M., Livny, J., Lawrence, M. S., Kimball, M. D., Waldor, M. K., and Camilli, A. (2009) Experimental discovery of sRNAs in Vibrio

cholerae by direct cloning, 5S/tRNA depletion and parallel sequencing. *Nucleic Acids Research* **37**, e46.

10. Padalon-Brauch, G., Hershberg, R., Elgrably-Weiss, M., Baruch, K., Rosenshine, I., Margalit, H., and Altuvia, S. (2008) Small RNAs encoded within genetic islands of Salmonella typhimurium show host-induced expression and role in virulence. *Nucleic Acids Research* **36**, 1913–1927.

11. Ruby, J. G., Jan, C., Player, C., Axtell, M. J., Lee, W., Nusbaum, C., Ge, H., and Bartel, D. P. (2006) Large-scale sequencing reveals 21U-RNAs and additional microRNAs and endogenous siRNAs in C-elegans. *Cell* **127**, 1193–1207.

Chapter 6

Identification of Virus Encoding MicroRNAs Using 454 FLX Sequencing Platform

Byung-Whi Kong

Abstract

MicroRNAs are a class of small noncoding RNA molecules that play a pivotal role in the regulation of gene expression at the posttranscriptional level. Most large double-stranded DNA viruses, mainly the herpesvirus family, are known to express miRNAs. Viral miRNAs can regulate both viral- and cellular transcripts. By eliminating cloning steps for large number of Sanger sequencing reactions, recent development of massively parallel next-generation sequencing methods has accelerated identification of small RNA species expressed from viruses, prokaryotes, and eukaryotes. The miRNAs expressed from infectious laryngotracheitis virus (ILTV), which is an alphaherpesvirus belonging to the herpesviridae family and which causes an acute respiratory disorder in chicken, were identified by small RNA enrichment and the 454 FLX sequencing method.

Key words: Viral microRNA, Herpesvirus, FLX sequencing

1. Introduction

MicroRNAs are a class of small (approximately 22 nucleotides; nt) noncoding RNAs transcribed from the genomes of all multicellular organisms and some viruses. MiRNAs are predicted to play a critical role in the regulation of many genes, especially for signaling pathways involved in development, cellular differentiation, proliferation, apoptosis, and oncogenesis (1). To date, most viral miRNAs have been derived exclusively from dsDNA viruses, mainly of the herpesvirus family, including Epstein–Barr virus (EBV), Kaposi's Sarcoma-associated herpesvirus (KSHV), human cytomegalovirus (hCMV), and HSV-1 but also from simian polyomaviruses and human adenovirus (2). In addition, two different serotypes of avian oncogenic herpesvirus, Marek's disease virus (MDV; *gallid herpesvirus 2*), were reported to express MDV-specific miRNAs (3).

Young Min Kwon and Steven C. Ricke (eds.), *High-Throughput Next Generation Sequencing: Methods and Applications*, Methods in Molecular Biology, vol. 733, DOI 10.1007/978-1-61779-089-8_6,

Previously, identification of small RNAs was dependent on concatermerizaton, plasmid vector cloning, and huge number of Sanger sequencing processes. However, recent development of massively parallel next-generation sequencing methods can produce massive number of short sequence reads (35–200 bp) by a single sequencing reaction. It is suitable for the study of the identification of small RNA species. Though the method described here is focused on the herpesvirus [infectious laryngotracheitis virus (ILTV)] encoding miRNAs isolated from virus-infected cultured cells, most small RNA species identified are host cellular miRNAs. Since the size of virus genome is much smaller compared to host eukaryotic cellular genome, the identification of genomic location of microRNA expression is easier than that in eukaryotic genome.

2. Materials

2.1. RNA Extraction

1. TRIzol Reagent (Invitrogen, Carlsbad, CA) (see Note 1).
2. Chloroform.
3. DNase I (New England BioLabs, Ipswich, MA).
4. Isopropyl alcohol and 75% ethyl alcohol.
5. Agarose.

2.2. MicroRNA Cloning

1. MicroRNA cloning kit (miRCat™; Integrated DNA Technologies, Coralville, IA) includes the following:
   - 3′ cloning linker: 5′-rAppCTGTAGGCACCATCAAT/3d dC/-3′.
   - 5′ cloning linker: 5′-TGGAATrUrCrUrCrGrGrGrCrAr CrCrArArGrGrU-3′.
   - miSPIKE™ internal size control RNA oligonucleotides: 5′-rCrUrCrArGrGrArTrGrGrCrGrGrArGrCrGrGrUr-CrU-3′.
   - Forward PCR primer: 5′-TGGAATTCTCGGGCACC-3.
   - Reverse transcription/PCR primer: 5′-GATTGATGGTG CCTACAG-3′.
   - RNase/DNase/pyrogen-free water (nuclease-free water).
   - TE (pH 7.5) buffer.
   - 10× ligation buffer.
   - Ligation enhancer.
   - 10 mM ATP.
   - 10 mg/ml glycogen.

- 3 M NaOAc (pH 5.2).
- T4 RNA ligase (5 U/μl).
- T4 DNA ligase (30 U/μl).

2. SuperScript™ III reverse transcriptase kit (Invitrogen, Carlsbad, CA) includes the following:
   - Reverse transcriptase.
   - 5× First-strand buffer.
   - 0.1 M Dithiothreitol (DTT).
   - RNase-OUT™ (40 U/ml).
3. 10 mM dNTP mix for reverse transcription reaction.
4. PCR reagents includes the following:
   - Taq DNA polymerase (5 U/ml).
   - 10× reaction buffer.
   - 2.5 mM dNTP mix.
5. PCR primers containing 454 FLX sequencing primer sequences (underlined) for the reamplification of enriched miRNAs: forward 5′-GCCTCCCTCGCGCCATCAGTGGAATTCTCGGGCACC-3′ and reverse 5′-GCCTTGCCAGCCCGCTCAGGATTGATGGTGCCTACAG-3′.
6. DTR desalting columns (Edge Biosystems, Gaithersburg, MD).
7. Disposable pestles for 1.5-ml tubes (Kimble–Kontes Glass Co., Vineland, NJ).
8. Phenol:chloroform:isoamyl alcohol (25:24:1) (EMD chemicals, Gibbstown, NJ).
9. 100% Ethyl alcohol.

2.3. TBE–Urea PAGE

1. Gel electrophoresis system.
2. Precasting 15% TBE–Urea gel.
3. 2× loading buffer (Bio-Rad, Hercules, CA).
4. Low Range ssRNA ladder and microRNA marker (New England BioLabs, Ipswich, MA).
5. GelStar® Nucleic Acid Stain (Lonza BioScience, Rockland, ME).
6. 1× TBE buffer is prepared from 10× stock buffer.
7. Electric power supply.
8. UV light box.
9. 1.5-ml microcentrifuge tubes.

2.4. Sequence Data Analysis

1. Microsoft Excel.
2. Microsoft Access.

3. Web-based program for sequence format conversion: http://hcv.lanl.gov/content/sequence/FORMAT_CONVERSION/form.html.
4. Blast search: http://blast.ncbi.nlm.nih.gov/Blast.cgi.

3. Methods

3.1. RNA Extraction

1. Total RNA was extracted from virus-infected cultured cells at 2 days postinfection (dpi) using TRIzol Reagent following the manufacturer's instructions. Cell culture media was discarded and remaining cells were washed three times with phosphate-buffered saline (PBS) and 1 ml TRIzol Reagent is applied directly to the cell plate. Using cell scraper, cells are collected and transferred into 1.5-ml microcentrifuge tube. Most cells are lysed in TRIzol Reagent by 30 s vortexing. Then, steps from the addition of 200 μl chloroform are followed as per manufacturer's instruction.
2. Isolated total RNA (50 μl total volume) was treated with DNase I (5 U total) and incubated in 37°C for 1 h.
3. DNase I-treated RNA was repurified using 1 ml TRIzol following the manufacturer's instructions.
4. Concentration of RNA was determined by spectrophotometry. Calculate the RNA concentration by applying the conversion that 1 OD at 260 equals 40 μg/ml RNA. And, determine the RNA quality by checking the value ratio of 260/280. If the ratio is over 1.6, RNA quality is considered to be good to move to the next step.
5. RNA purity was determined by the fractionation onto 1% agarose gel to visualize decent bands for 28S, 18S, and 5S ribosomal RNA.

3.2. Small RNA Enrichment

Basically, miRCat™ kit from Integrated DNA Technologies was utilized for microRNA cloning methods with modifications.

1. 100 μg total RNA (e.g., 33.3 μl for 3 μg/ml conc. RNA) is mixed with the 1 μl (10 pmol) miSPIKE™ size control RNA, which is a 21-mer RNA oligo.
2. Add same volume (e.g., 33.3 μl) of 2× Urea sample buffer to the mixture of total RNA and miSPIKE™ size control RNA and incubate at 70°C for 5 min.
3. Cool down samples on ice and prepare TBE–Urea PAGE.
4. Prepare the tank running buffer (1× TBE) by diluting 100 ml of the 10× TBE with 1 L of water in a measuring cylinder. Mix well with magnetic stirrer bar.

5. Place the assembly of precasting 15% TBE–Urea gel units into running tank, fill in the tank with running buffer and use a 3-ml syringe fitted with a 22-gauge needle to wash the sample loading wells with running buffer.
6. Load samples into wells. When sample volumes are over 30 μl, load samples into multiple wells. Connect a power supply. The gel is run at 200 V for 30 min when the bromophenol blue bands reach to the bottom of the gel.
7. Prepare 2× GelStar™ staining buffer by diluting 10 μl of 10,000× stock in 50 ml of 1× TBE.
8. Disassemble gel units, slice left top corner of gel to mark the gel orientation, put gels into 2× GelStar™ staining buffer, and incubate on the rocking stand for 30 min.
9. Place the gel on a medium wavelength (312 nm) UV light box, select RNA fragments on the ~21 nt bands specified by miSPIKE™ size control RNA, cut the gel 2 mm above and below the control band (Fig. 1a), and put gel slice in a sterile 1.5-ml microcentrifuge tube.
10. Crush gel slice with a Kimble–Kontes glass rod and add 200 μl sterile, nuclease-free water and continue to crush the gel into a fine slurry. Incubate the tube at 70°C for 10 min.
11. Prepare DTR column by centrifugation at 800 ×*g* for 3 min to discard preserving buffer and replace a sterile 1.5-ml microcentrifuge collection tube.

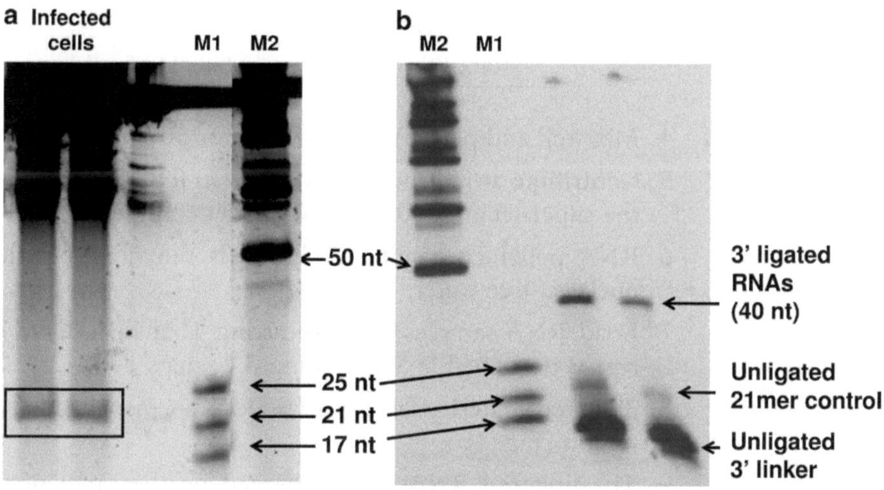

Fig. 1. PAGE RNA purification gels. (**a**) The total RNA extracted from ILTV-infected chicken embryo kidney (CEK) cells was mixed with 21-mer size control RNA, miSPIKE™, and fractionated onto 15% 1× TBE–Urea (7 M) polyacryamide gel. Gels were stained with 2× GelStar® Nucleic Acid Stain in 1× TBE buffer. Gels were visualized by GelDoc system (Bio-Rad, Hercules, CA). A *black lined box* indicates the ~21 nt small RNAs enriched with size control nucleotides. M1 represents microRNA size marker and M2 indicates small range single-stranded RNA size marker. (**b**) The 3′ linker-ligated RNAs were fractionated onto PEGE at the same condition described above.

12. Transfer entire volume of gel slurry onto the DTR column and centrifuge at 800 × *g* for 3 min. Discard DTR column.
13. Add 3 μl of 10 mg/ml glycogen, 25 μl of 3 M NaOAc (pH 5.2), and 900 μl ice cold 100% ethyl alcohol. Mix by inversion and incubate at −80°C for 30 min.
14. Centrifuge tubes at full speed (16,000 × *g*) for 10 min, discard the supernatant, and dry RNA pellet.
15. RNA pellet is rehydrated with 10 μl nuclease-free water.

3.3. MicroRNA Cloning

1. Rehydrated RNA is ligated with 3′ RNA linker by the following reaction:

Recovered small RNA fraction	10 μl
3′ RNA linker (50 μM)	1 μl
10× ligation buffer	2 μl
Ligation enhancer	6 μl
T4 RNA ligase (1 U/μl)	1 μl
Total	20 μl

2. Incubate these reactions at 22°C for 2 h.
3. Add following reagents into the reaction:

TE (pH 7.5)	80 μl
Glycogen	3 μl
3 M NaOAc (pH 5.2)	10 μl (1/10 vol.)
100% Ethyl alcohol	250 μl (2.5 vol.)

4. Mix well and place tube at −80°C for 30 min.
5. Centrifuge at full speed (~16,000 × *g*) for 10 min and discard the supernatant, and dry RNA pellet completely.
6. RNA pellet containing 3′ linker is rehydrated with 10 μl nuclease-free water.
7. Load RNA samples into precasting 15% TBE–Urea gel and run as described in Subheading 3.2, steps 1–6.
8. 3′ linker-ligated small RNA fractions are visualized at ~40 bp region (Fig. 1b).
9. The linkered RNAs are recovered by the same way as the purification of the entire small RNA fraction by repeating steps 7–15 in Subheading 3.2.
10. Rehydrated 3′ linkered RNA is ligated with 5′ RNA linker by the following reaction:

Recovered 3′ linkered RNA fraction	8 μl
5′ RNA linker (50 μM)	1 μl
10× ligation buffer	2 μl
Ligation enhancer	6 μl
10 mM ATP	2 μl
T4 RNA ligase (1 U/μl)	1 μl
Total	20 μl

11. Incubate these reactions at 22°C for 2 h.
12. Repeat steps 3–6.

3.4. Reverse Transcription and PCR

1. Recovered linkered RNA fractions are reverse transcribed by the following reactions:

Recovered linkered RNA fraction	9 μl
dNTPs (10 mM)	1 μl
RT primer (10 μM)	1 μl
Total	11 μl

Incubate at 65°C for 5 min for RNA denaturation

Place on ice and add following reagents:

5× First-strand buffer	4 μl
0.1 M DTT	1 μl
RNase-OUT™ (40 U/μl)	1 μl
SuperScript™III RT enzyme (200 U/μl)	1 μl
Total	20 μl

2. Incubate at 50°C for 1 h followed by 15 min at 70°C.
3. Six parallel PCR amplifications are set up by the following reaction:

Reverse transcription reaction	3 μl
Distilled water	34.5 μl
10× PCR buffer	5 μl
dNTPs (2.5 mM)	5 μl
Forward primer (10 μM)	1 μl
Reverse primer (10 μM)	1 μl
Taq polymerase (5 U/μl)	0.5 μl
Total	50 μl

PCR conditions:

- 95°C for 10 min
- 95°C for 30 s (1)
- 52°C for 30 s (2)
- 72°C for 30 s (3)
- 72°C for 5 min
- 35 cycles of (1) through (3).

4. Run 5 μl of each reaction on a 2% agarose gel to check the PCR quality. Expected amplicon size is ~64 bp (Fig. 2). Remaining 45 μl of each reactions is pooled in a 1.5-ml tube (see Note 2).
5. Add an equal volume (270 μl) of phenol:chloroform:isoamyl alcohol (25:24:1), vortex this reaction, and centrifuge at full speed (~1,600 × *g*) for 5 min.
6. Transfer the upper (aqueous) phase to a new 1.5-ml microcentrifuge tube.
7. Add 1/10 vol. (27 μl) of 3 M NaOAc (pH 5.2) and 3 vol. (900 μl) of cold 100% ethyl alcohol (see Note 3).
8. Place the tube at –80°C for 20 min.
9. Centrifuge at full speed (16,000 × *g*) for 10 min.
10. Pour off the supernatant and wash the pellet in 900 μl of ice cold 70% ethyl alcohol.
11. Pour off the supernatant and dry pellet.
12. Add 10 μl nuclease-free water to resolve the pellet.

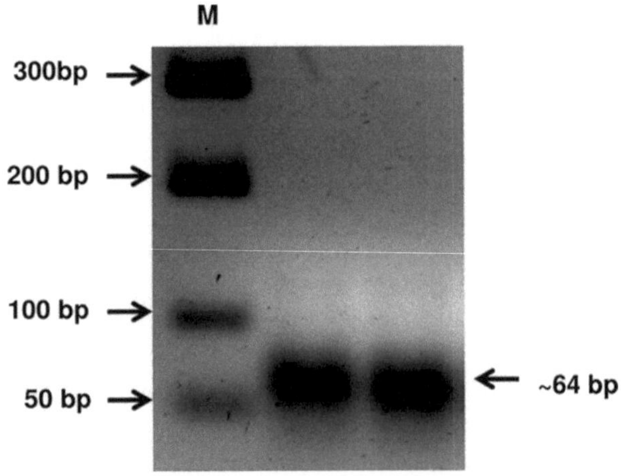

Fig. 2. Initial amplification to produce ~64-bp amplicon linkered in both 3′- and 5′- ends. PCR products with both forward PCR primer and reverse transcription/PCR primers were analyzed by 2% agarose gel. M represents DNA size marker.

13. Take 1 μl amplicon and set up multiple PCR reamplification reactions with primers containing FLX sequencing primer sequences as described in Subheading 3.4, step 3.
14. Pool and purify PCR amplicons using phenol:chloroform: isoamyl alcohol (25:24:1) method as described above.

3.5. 454 FLX Sequencing and Data Analysis

1. Determine the concentration of purified small RNA amplicons and send 50 ng of samples to sequencing center (see Notes 4 and 5).
2. After obtaining sequence reads, primer sequences are eliminated from raw data using replace function (Ctrl F) by Microsoft Excel.
3. Pool all sequence reads into one column and save it as .xls file (e.g., microRNA.xls).
4. Open the Excel file with Microsoft Access program and check names of both data sheet (file name) and column.
5. Choose "Create" function, click "Query Design" and close "Show Table" window.
6. Click right mouse button, choose "SQL View", and write the following commands in the window (in case of that Data sheet name is "sheet1" and column name is "Column1"):

```
SELECT Count(sheet1.column) AS [count], sheet1.Column1
FROM sheet1
GROUP BY sheet1.Column1;
```

7. Save the query as a new name and close the window. Then, open newly generated query file from the left option panel.
8. Copy read counts and sequence reads, and paste them to Microsoft Excel. And, sort read sequences alphabetically.
9. Sorted sequences are indexed by the format converter from the following Web page: http://hcv.lanl.gov/content/sequence/FORMAT_CONVERSION/form.html.
10. A FASTA format of .txt file will be generated containing unique read sequences.
11. Sequences are analyzed by TWO BLAST function against known virus genome sequences (i.e., ILTV; GenBank accession number: NC_006623). Up to 3,000 sequence reads can be analyzed per a single TWO BLAST function.
12. Sequences showing homologies are chosen and microRNAs fulfilling the following criteria are considered as mature microRNAs encoded from viral genome:
    (a) Homologies shown in at least 18 nucleotides
    (b) At least two read counts
    (c) Determine orientations between sense and antisense (see Note 6)

13. Secondary structures for viral microRNA precursor molecules are predicted by mFold software from the following Web page: http://mobyle.pasteur.fr/cgi-bin/portal.py?form=mfold.About 40 nucleotides at each 5′- and 3′ flanking sequences including a candidate miRNA sequences (total ~100 nucleotides) are collected from viral genome information, paste sequences in the given window of the mFold Web page. Then, click "run" button. Of the given results, secondary structures matching the other arm of the hairpin with the lowest free energy and allowing for G-T base paring are chosen (see Note 7).

4. Notes

1. Other RNA isolation kits, such as RNeasy Mini kit from Qiagen or mirVANA™ miRNA isolation kit from Applied Biosystems, can be utilized.
2. Initial RT-PCR reaction to produce ~64-bp amplicon (Fig. 3, lane 2) with forward primer and reverse transcription/PCR primer (Subheading 3.4, steps 1–4) can also generate false-positive results of 43-bp amplicon (Fig. 3, lane 1) indicating that linkers are ligated to each other. Though 3′ linker-ligated band was excised from PEGE gels (Fig. 1b), unligated 3′ linkers could be contaminated from the processes in handling gel slices. Therefore, the strict precaution is needed to prevent samples from possible contaminations which can lead to production of large amount of false positives during PCR

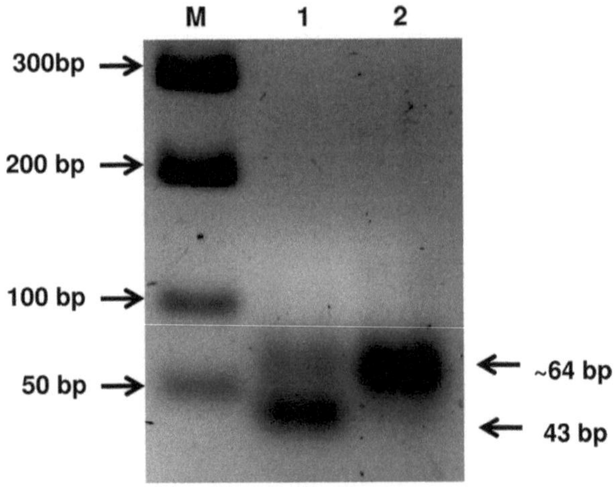

Fig. 3. Negative amplification for the ligation between linkers only. PCR products with both forward and reverse transcription/PCR primers were analyzed by 2% agarose gel. M represents DNA size marker, *lane 1* showed the negative amplification for ligated products between linkers only without small RNAs, and *lane 2* showed the positive ~64-bp amplicon containing small RNAs.

reactions. After the first PCR, amplicons should be analyzed by gel electrophoresis, and if needed, amplicons showing ~64-bp bands need to be purified from gels.

3. QIAquick® Nucleotide cleanup columns from Qiagen (Valencia, CA) can be utilized to purify amplicons for both ~64-bp initial PCR products and ~100-bp FLX PCR products.
4. Before sending the next-generation sequencing analysis, amplified libraries can be prescreened for the quality control by simple TA cloning method (pGEM-T easy system from Promega or pCR-Topo system from Invitrogen) and ~10 positive bacterial colonies can be preanalyzed by Sanger sequencing method.
5. There are several different options for the next-generation sequencing, such as Illumina or Helicos. Since the nucleotide length of microRNA is relatively short (~22 nucleotides), Illumina sequencing method can be utilized to get more sequence read counts, reaching millions of sequence reads. A specialized sequence analytic tool is needed to analyze much higher number of sequence reads. The results of identification of ILTV encoding microRNA using Illumina sequencing platform are available in ref. 4.
6. When microRNA sequences were analyzed by the forward primer (5′ linker side), microRNA sequence reads obtained are always sense oriented (5′ to 3′). Therefore, the direction of microRNA expression in the viral genome can be determined depending on the microRNA orientations.
7. The results generated from this assay can be found in ref. 5.

Acknowledgments

We thank Dr. Joon Jin Song, University of Arkansas, for the support in sorting sequence reads. This work was supported by Arkansas Agricultural Experimental Station.

References

1. Bushati, N.and Cohen, S.M. (2007) microRNA functions. *Annu Rev Cell Dev Biol* **23**, 175–205
2. Cullen, B.R. (2006) Viruses and microRNAs. *Nat Genet* **38 Suppl**, S25–30
3. Burnside, J., Ouyang, M., Anderson, A., Bernberg, E., Lu, C., Meyers, B.C., Green, P.J., Markis, M., Isaacs, G., Huang, E., and Morgan, R.W. (2008) Deep sequencing of chicken microRNAs. *BMC Genomics* **9**, 185
4. Waidner, L.A., Morgan, R.W., Anderson, A.S., Bernberg, E.L., Kamboj, S., Garcia, M., Riblet, S.M., Ouyang, M., Isaacs, G.K., Markis, M., Meyers, B.C., Green, P.J., and Burnside, J. (2009) MicroRNAs of Gallid and Meleagrid herpesviruses show generally conserved genomic locations and are virus-specific. *Virology* **388**, 128–36
5. Rachamadugu, R., Lee, J.Y., Wooming, A., and Kong, B.W. (2009) Identification and expression analysis of infectious laryngotracheitis virus encoding microRNAs. *Virus Genes*

Chapter 7

Ribosomal RNA Depletion for Massively Parallel Bacterial RNA-Sequencing Applications

Zhoutao Chen and Xiaoping Duan

Abstract

RNA-sequencing (RNA-Seq) is a digital display of a transcriptome using next-generation sequencing technologies and provides detailed, high-throughput view of the transcriptome. The first step in RNA-Seq is to isolate whole transcriptome from total RNA. Since large ribosomal RNA (rRNA) constitutes approximately 90% RNA species in total RNA, whole transcriptome analysis without any contamination from rRNA is very difficult using existing RNA isolation methods. RiboMinus™ purification method provides a novel and efficient method to isolate RNA molecules of the transcriptome devoid of large rRNA from total RNA for transcriptome analysis. It allows for whole transcriptome isolation through selective depletion of abundant rRNA molecules from total RNA. The rRNA depleted RNA fraction is termed as RiboMinus™ RNA fraction, which is enriched in polyadenylated RNA, nonpolyadenylated RNA, preprocessed RNA, tRNA, numerous regulatory RNA molecules, and other RNA transcripts of yet unknown function. Using RiboMinus™ method to isolate RiboMinus RNA results in up to 99.0% removal of 16S and 23S rRNA molecules from 0.5 to 10 μg total bacterial RNA based on Bioanalyzer analysis. It enables efficient whole transcriptome sequencing analysis without major contamination from highly abundant rRNA. Residual rRNA accounts for less than 10% of entire transcriptome based on both SOLiD and Genome Analyzer RNA-Seq data.

Key words: Transcriptome, RNA-Seq, Ribosomal RNA, Bacteria, 16S, 23S, Next-generation sequencing, Ribominus, Polyadenylation

1. Introduction

A transcriptome is defined as a complete collection of transcribed elements of the genome (1) and contains both mRNA transcripts and non-mRNA transcripts. Microarray-based gene expression analysis was considered to be the dominant high-throughput technology for any transcriptome studies. With the maturation of next-generation sequencing technologies, RNA-sequencing (RNA-Seq) has

Young Min Kwon and Steven C. Ricke (eds.), *High-Throughput Next Generation Sequencing: Methods and Applications*, Methods in Molecular Biology, vol. 733, DOI 10.1007/978-1-61779-089-8_7, © Springer Science+Business Media, LLC 2011

become a more recent solution for high-throughput transcriptome study (2–4). RNA-Seq is the digital display of a transcriptome using next-generation sequencing technology and can provide a detailed, high-throughput view of the transcriptome.

The first step in RNA-Seq is to isolate whole transcriptome from total RNA. Since large ribosomal RNA (rRNA) constitutes approximately 90% RNA species in total RNA, whole transcriptome analysis without any contamination from rRNA is very difficult when employing existing RNA isolation methods (5–7). This suggests the need for developing procedures to remove unwanted, abundant rRNA transcripts. For eukaryotes, mRNA can be enriched based on their polyadenylation status, although it is only a partial isolation of the transcriptome. The 3′-ends of prokaryotic mRNA can be polyadenylated as in eukaryotes, but the poly(A) tracts of prokaryotic mRNA are generally shorter, ranging from 15 to 60 adenylate residues and are associated with only 2–60% of the molecules of a given mRNA species (8). This limits effectiveness of poly(A) enrichment of prokaryotic mRNA species compared to eukaryotic mRNA.

RiboMinus™ purification method provides a novel and efficient method to isolate RNA molecules of the transcriptome devoid of large rRNA from total RNA for transcriptome analysis. It is not dependent on the polyadenylation status or presence of a 5′-cap structure on the RNA. It allows for whole transcriptome isolation through selective depletion of abundant rRNA molecules from total RNA. The ribosomal RNA depleted RNA fraction is termed as RiboMinus RNA fraction, which is enriched in polyadenylated mRNA, nonpolyadenylated RNA, preprocessed RNA, tRNA, and may also contain regulatory RNA molecules such as microRNA (miRNA) and short interfering RNA (siRNA), snRNA, and other RNA transcripts of yet unknown function. Using RiboMinus™ method to isolate RiboMinus RNA results in up to 99.0% removal of 16S and 23S rRNA molecules from 0.5 to 10 μg total bacterial RNA based on Agilent Bioanalyzer analysis. It enables whole transcriptome sequencing analysis without interference from highly abundant rRNA.

2. Materials

2.1. Preparing RiboMinus™ RNA

1. Total RNA: 0.5–10 μg total RNA without notable degradation in no more than 10 μL for each reaction (see Notes 1–3). *Escherichia coli* Total RNA (Ambion, Austin, TX) was used here.
2. PureLink™ RNA Mini Kit (Invitrogen, Carlsbad, CA) or TRIzol® Reagent (Invitrogen, Carlsbad, CA).
3. 2100 Bioanalyzer (Agilent Technologies, Santa Clara, CA).

4. Agilent RNA 6000 Nano Kit (Agilent Technologies, Santa Clara, CA).
5. RiboMinus™ Eukaryote Kit for RNA-Seq (Invitrogen, Carlsbad, CA), which contains RiboMinus™ Magnetic Beads and RiboMinus™ Eukaryote Probe, Hybridization Buffer, and DEPC-treated (RNase-Free) water. Do not use the Eukaryote probe for any bacterial RNA samples. Use the bacterial 16S and 23S probe sets instead (see Note 4).
6. RiboMinus™ bacterial 16S and 23S probe sets (available from the authors free of charge).
7. Water baths or heat blocks set to 70–75°C and 37°C.
8. DynaMag™-2 Magnet (Invitrogen, Carlsbad, CA) or MagnaSep™ Magnetic Particle Separator (Invitrogen, Carlsbad, CA) or equivalent.

2.2. Concentrating RiboMinus™ RNA

1. RiboMinus™ Concentration Module (Invitrogen, Carlsbad, CA), which contains Binding Buffer (L3), Wash Buffer (W5), RNase-Free Water, Spin Column with Collection Tubes, Wash Tubes, and Recovery Tubes.
2. Water baths or heat blocks set to 70–75°C and 37°C.
3. Sterile, RNase-free microcentrifuge tubes (Major Laboratory Supplier).
4. Glycogen: 20 μg/μL (Invitrogen, Carlsbad, CA).
5. 3 M Sodium acetate, pH 5.2, prepared in RNase-free water.
6. 96–100% Ethanol.
7. 70% Ethanol.
8. Microcentrifuge capable of centrifuging >12,000 ×*g*.
9. Quant-iT™ RNA Assay Kit (Invitrogen, Carlsbad, CA).
10. 2100 Bioanalyzer (Agilent Technologies, Santa Clara, CA).
11. Agilent RNA 6000 Nano Kit (Agilent Technologies, Santa Clara, CA).

3. Methods

Total RNA is hybridized with bacterial rRNA sequence-specific 5′-biotin labeled oligonucleotide probes to selectively deplete abundant large ribosomal RNA molecules from total RNA in bacteria. The bacterial probe is an oligonucleotide probe mixture containing two probes targeting for 16S rRNA and three probes targeting for 23S rRNA of many prokaryotic species. Some example organisms that are compatible with these bacterial probes are listed in Table 1.

Table 1
Examples of organisms compatible with RiboMinus™ bacterial probes[a]

Bacillus anthracis	*Lactococcus lactis*	*Streptococcus mutans*
Bacillus subtilis	*Listeria innocua*	*Streptococcus pneumoniae*
Bacillus thuringiensis	*Listeria monocytogenes*	*Streptococcus pyogenes*
Burkholderia mallei	*Neisseria gonorrhoeae*	*Thermotoga maritima*
Burkholderia pseudomallei	*Neisseria meningitidis*	*Vibrio cholerae*
Caulobacter crescentus	*Pseudomonas aeruginosa*	*Yersinia enterocolitica*
Enterococcus faecalis	*Pseudomonas syringae*	*Yersinia pestis*
Escherichia coli	*Salmonella enterica*	
Klebsiella pneumoniae	*Shewanella oneidensis*	

[a] Contact author for probe compatibility with any organisms of interest that are not in the table

Each probe is single-stranded and contains three Locked Nucleic Acid (LNA™) monomers incorporated at specific locations. The incorporation of LNA into the oligonucleotide probe does not affect the ability of the oligonucleotide to bind RNA but increases the stability of the rRNA/probe complex (9). The 5′-end of each probe is conjugated to biotin to allow binding to streptavidin RiboMinus™ Magnetic Beads. The RiboMinus Magnetic Beads consist of 1 μm streptavidin-coated polystyrene beads with a magnetic core and can be used for the removal of probe/rRNA complexes from the sample. The beads bind to the biotin-labeled probe hybridized with rRNA or the probe alone. The size and the biotin-binding capacity of the RiboMinus Magnetic Beads are optimized for use with RiboMinus method and result in efficient depletion of rRNA using 0.5–10 μg total RNA as the starting material. The RiboMinus RNA sample is subsequently concentrated using ethanol precipitation or RiboMinus Concentration Module.

3.1. Preparing Total RNA

High-quality intact total RNA from bacteria samples is required for RiboMinus™ based ribosomal RNA depletion method (see Note 1–3). Perform DNase-treatment of the total RNA to remove DNA contamination (see Note 5). Total RNA is isolated using PureLink™ RNA Mini Kit or TRIzol® Reagent. 0.5–10 μg total RNA in no more than 10 μL is used for each reaction (see Note 6). Resuspend isolated total RNA in DEPC-treated water accordingly. Check the quality of your total RNA on an Agilent Bioanalyzer with RNA 6000 Nano kit.

3.2. Hybridization Step

The reaction listed below is for 0.5–10 μg of total RNA sample with the RiboMinus™ reagents. To process greater than 10 μg total RNA sample, divide the sample into multiple fractions, each containing less than 10 μg total RNA (see Note 6).

1. Set a water bath or heat block to 70–75°C.
2. To a sterile, RNase-free 1.5-mL microcentrifuge tube, add the following:

Total RNA (0.5–10 μg): less than 10 μL

RiboMinus™ bacterial 16S and 23S probe sets (12.5 pmol/μL): 8 μL

Hybridization Buffer: 100 μL

3. Incubate the tube at 70–75°C for 5 min to denature RNA.
4. Allow the sample to cool to 37°C slowly over a period of 30 min by placing the tube in a 37°C water bath. To promote sequence-specific hybridization, it is important to allow slow cooling. Do not cool samples quickly by placing tubes in cold water.
5. While the sample is cooling down, proceed to Subheading 3.3.

3.3. Preparing Beads

1. Resuspend RiboMinus™ Magnetic Beads in its original bottle by thorough vortexing (see Note 7).
2. Pipette 750 μL of the bead suspension into a sterile, RNase-free, 1.5-mL microcentrifuge tube.
3. Place the tube with the bead suspension on a magnetic separator for 1 min. The beads will settle to the tube side that faces the magnet. Gently aspirate and discard the supernatant (see Note 8).
4. Add 750 μL sterile, DEPC water to the beads and resuspend the beads by slow vortexing.
5. Place tube on a magnetic separator for 1 min. Aspirate and discard the supernatant.
6. Repeat steps 4 and 5 once.
7. Resuspend the beads in 750 μL Hybridization Buffer and transfer 250 μL beads to a new tube and maintain the tube at 37°C for use at a later step.
8. Place the tube with 500 μL beads on a magnetic separator for 1 min. Aspirate and discard the supernatant.
9. Resuspend beads in 200 μL Hybridization Buffer and keep the beads at 37°C until use.

3.4. Removing rRNA

1. After the incubation at 37°C for 30 min of the hybridized sample (above), briefly centrifuge the tube to collect the sample to the bottom of the tube.

2. Transfer the sample (approximately 118 μL) to the prepared RiboMinus™ Magnetic beads from step 9 (Subheading 3.3). Mix well by pipetting up and down or by low-speed vortexing.
3. Incubate the tube at 37°C for 15 min. During incubation, gently mix the contents occasionally. Briefly centrifuge the tube to collect the sample to the bottom of the tube.
4. Place the tube on a magnetic separator for 1 min to pellet the rRNA–probe complex. *Do not discard the supernatant. The supernatant contains RiboMinus™ RNA.*
5. Place the tube with 250 μL beads from step 7 (Subheading 3.3) on a magnetic separator for 1 min. Aspirate and discard the supernatant.
6. To this tube of beads, add approximately 318 μL supernatant containing RiboMinus™ RNA from step 4, above. Mix well by pipetting up and down or low-speed vortexing.
7. Incubate the tube at 37°C for 15 min. During incubation, gently mix the contents occasionally. Briefly centrifuge the tube to collect the sample to the bottom of the tube.
8. Place the tube on a magnetic separator for 1 min to pellet the rRNA–probe complex. *Do not discard the supernatant as the supernatant contains RiboMinus™ RNA.*
9. Transfer the supernatant (approximately 318 μL) containing RiboMinus™ RNA to a new tube.

3.5. Concentrating RiboMinus™ RNA

RiboMinus™ RNA must be concentrated for further use in downstream RNA-Seq applications after purifying RiboMinus™ RNA using the RiboMinus™ method. To retain all species of RNA including smaller RNA between 50 and 200 nucleotides (nt), there are two options to concentrate RiboMinus™ RNA:

1. Ethanol precipitation (Subheading 3.5.1)
2. RiboMinus™ Concentration Module using silica spin columns (Invitrogen, Carlsbad, CA) with a modified protocol (Subheading 3.5.2).

3.5.1. Concentrating RiboMinus™ RNA Using Ethanol Precipitation

1. Transfer the RiboMinus™ RNA sample into a clean, RNAse-free 1.5-mL or 2-mL microcentrifuge tube.
2. Add the following components to RiboMinus™ RNA:

   1 μL Glycogen (20 μg/μL)

   1/10th Sample (eluted RNA) volume (approximately 30 μL for this protocol) of 3 M sodium acetate

   2.5× Sample volumes (approximately 750 μL for this protocol) of 100% ethanol
3. Mix well and incubate at −80°C for a minimum of 30 min.

4. Centrifuge the tube for 15 min at ≥12,000 × *g* at 4°C. Carefully discard the supernatant without disturbing the pellet.
5. Add 500 μL 70% cold ethanol.
6. Centrifuge the tube for 5 min at ≥12,000 × *g* at 4°C. Discard the supernatant without disturbing the pellet.
7. Repeat steps 5 and 6 once.
8. Air-dry the pellet for approximately 5 min. Resuspend the RiboMinus™ RNA pellet in 10–30 μL DEPC-treated water (see Note 9).
9. Place RiboMinus™ RNA on ice and proceed to Subheading 3.6 or store RiboMinus™ RNA at −80°C until use.

3.5.2. Concentrating RiboMinus™ RNA Using Modified RiboMinus™ Concentration Module Protocol

Generally, RNA species less than 200 nt are excluded from the standard binding conditions for RNA isolation using silica spin column. Recent studies have shown the importance of these small RNA species, which include regulatory RNA molecules such as microRNA (miRNA), short interfering RNA (siRNA), snRNA, and other RNA transcripts of yet unknown function. To retain all species of RNA greater than 50 nt during the concentration step of RiboMinus™ RNA isolation, use modified silica spin column protocol with the RiboMinus Concentration Module wherein the binding of RNA is performed with higher ethanol concentration (see Note 10).

1. Before using Wash Buffer (W5) for the first time, add 6 mL of 96–100% ethanol to 1.5 mL Wash Buffer (W5) included with the kit. Check the box on the Wash Buffer label to indicate that ethanol was added. Store Wash Buffer (W5) with ethanol at room temperature.
2. Transfer the RiboMinus™ RNA (approximately 300 μL) sample to a new tube capable of holding greater than 2 mL.
3. Add 1× sample volume of Binding Buffer L3 (300 μL for this protocol) and 3× sample volumes of 96–100% ethanol (900 μL for this protocol) to a final concentration of 60% (see Note 10). Mix thoroughly by vortexing.
4. Transfer up to 600 μL of the sample (from step 2) to the spin column (with the collection tube).
5. Centrifuge at 12,000 × *g* for 1 min at room temperature. Discard the flow-through, and reinsert the Spin Column into the same Collection Tube.
6. Repeat steps 4 and 5 until the entire sample is processed.
7. Wash the column with 600 μL Wash Buffer (W5) prepared with ethanol. Centrifuge the column at 12,000 × *g* for 1 min at room temperature. Discard the flow-through.
8. Discard the collection tube and place the column into a clean collection tube, supplied with the kit.

9. Centrifuge the column at maximum speed for 2–3 min at room temperature to remove any residual Wash Buffer (W5). Place the column in a clean, 1.5-mL Recovery Tube.
10. Add 10–30 μL of RNase-free water to the center of the column. Incubate the column at room temperature for 1 min.
11. Centrifuge the column at maximum speed for 1 min at room temperature. The Recovery Tube contains purified RiboMinus™ RNA.
12. Place RiboMinus™ RNA on ice and proceed to Subheading 3.6 or store RiboMinus™ RNA at -80°C until use.

3.6. Analyzing RiboMinus™ RNA

The purified RiboMinus™ RNA is easily quantitated using UV absorbance at 260 nm or Quant-iT™ RNA Assay Kit. The RNA isolated using RiboMinus™ method should be of high quality and efficiently depleted in rRNA species (see Notes 11 and 12). The efficiency of rRNA depletion in RiboMinus™ RNA, RNA degradation, and RNA concentration can be effectively analyzed using Agilent 2100 bioanalyzer with Agilent RNA 6000 Nano Kit (Fig. 1).

3.7. Constructing RNA-Seq Libraries

Qualified RiboMinus™ RNA can be used directly to construct RNA-Seq libraries or whole transcriptome libraries according to procedure provided by next-generation sequencing platform vendors for SOLiD (Applied Biosystems), Genome Analyzer (Illumina), and Genome Sequencer (Roche/454), etc. The RNA-Seq libraries for *Bacillus anthracis*, *Caulobacter crescentus*, and *Pseudomonas syringae* purified with RiboMinus™ method have been successfully sequenced on SOLiD or Genome Analyzer. The residual rRNA in the whole transcriptome is approximately 10% of total mapped sequencing reads (Bergman N and Chang J, personal communication).

4. Notes

1. Always use proper microbiological aseptic techniques when working with RNA. Use disposable, individually wrapped, sterile plasticware and use sterile, new pipette tips and microcentrifuge tubes.
2. Wear latex gloves while handling reagents and RNA samples to prevent RNase contamination from the skin surface. Use RNase *AWAY*® Reagent (Invitrogen, Carlsbad, CA) to remove RNase contamination from surfaces.
3. Total RNA must be of high quality without notable degradation. Owing to limited probes per target, RiboMinus™ method

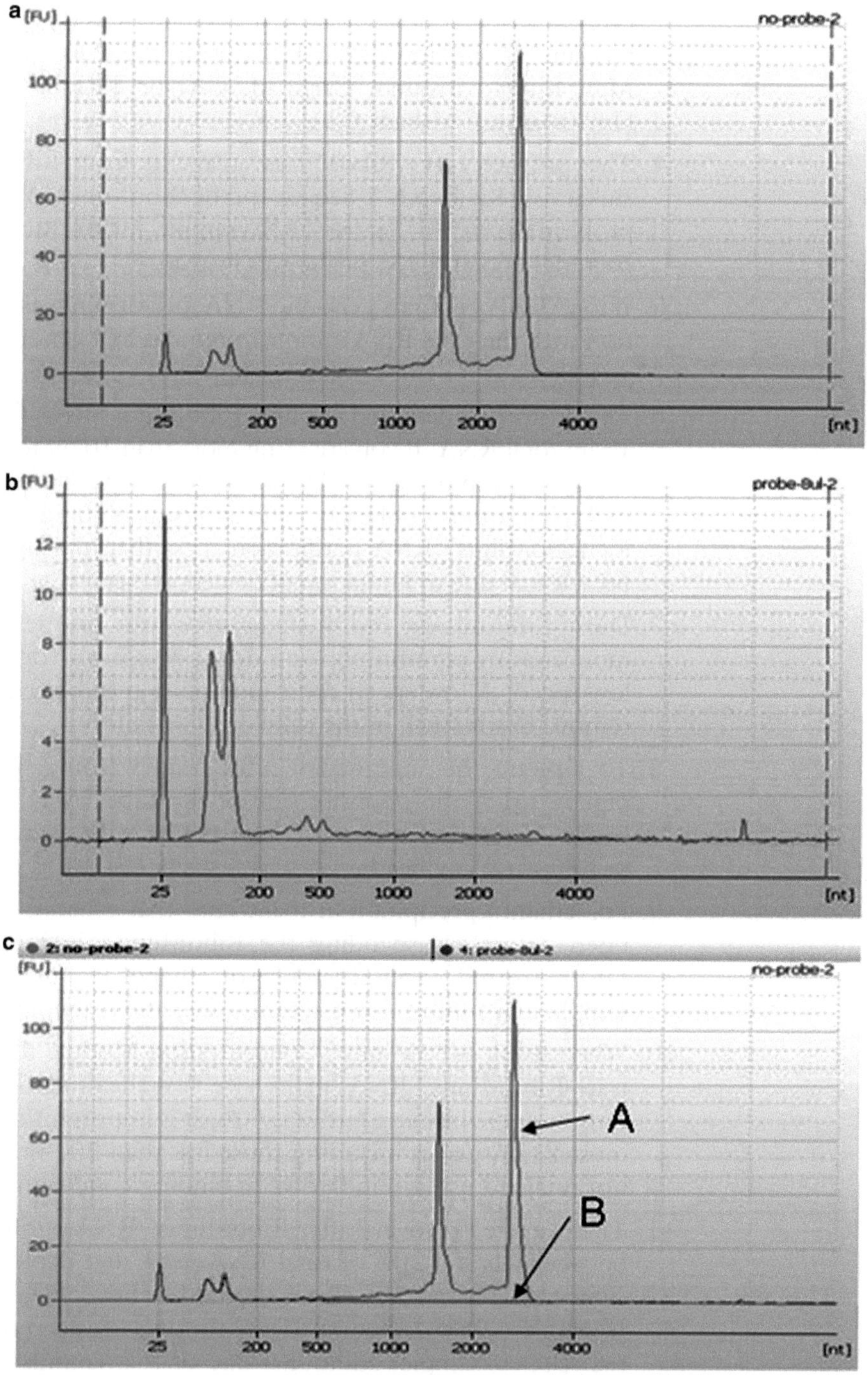

Fig. 1. 2100 Bioanalyzer electropherogram of RiboMinus™ RNA using Agilent RNA 6000 Nano kit. (**a**) Total RNA control, which is total RNA from *Escherichia coli* gone through RiboMinus™ purification procedure but without any probes. (**b**) RiboMinus RNA from 10 μg *E. coli* total RNA after RiboMinus purification with bacterial probes. (**c**) Overlay of both electropherograms. The depletion efficiency of 16S and 23S rRNA is greater than 99% based on the Bioanalyzer data.

cannot work effectively with degraded RNA sample. We recommend isolating total RNA using the PureLink™ RNA Mini Kit (Invitrogen, Carlsbad, CA) or TRIzol® Reagent (Invitrogen, Carlsbad, CA).

4. This protocol uses RiboMinus™ reagents from RiboMinus™ Eukaryote Kit for RNA-Seq except the RiboMinus™ Eukaryote probe in the kit. For bacterial RNA sample, use the RiboMinus™ bacterial 16S and 23S probe sets instead.
5. If total RNA contains genomic DNA, perform DNase I digestion with the total RNA sample to remove any genomic DNA contamination before isolating RiboMinus™ RNA.
6. The protocol is designed to purify RiboMinus™ RNA from 0.5–10 μg total RNA. If you are using more than 10 μg total RNA, incomplete removal of rRNA may occur. Divide the sample into multiple aliquots until each contains less than 10 μg total RNA for RiboMinus™ RNA purification. Less than 0.5 μg has not been tested due to limitation of detection method.
7. During the mixing and washing steps with magnetic beads, mix beads by pipetting up and down or using a vortex set to low speed. A low-speed centrifuge pulse may be required to remove beads stuck in the tube cap.
8. To aspirate the supernatant after bead washing, place the pipette tip at the opposite side of the tube, away from the beads. Carefully remove the supernatant without disturbing or removing any beads.
9. For ethanol precipitation, make sure that ethanol is evaporated before resuspending the RiboMinus™ RNA pellet in DEPC-treated water. Presence of ethanol in purified RNA sample may inhibit downstream enzymatic reactions.
10. When using RiboMinus™ Concentration Module to concentrate RiboMinus™ RNA, make sure that the ethanol concentration in the binding buffer is 60%. It is important to recover most RNA species including those less than 200 nt, typically not retained using standard silica binding conditions.
11. The expected recovery after RiboMinus™ RNA purification is between 10 and 20% of total RNA used, and depletion efficiency of 16S and 23S rRNA should be above 98% normally. If less than 98% depletion occurs routinely for certain bacterial species/sample, additional one or two rounds of probe hybridization and magnetic bead capture on RiboMinus™ RNA can further improve the removal efficiency.
12. A quantitative PCR (qPCR) method can be used to evaluate ribosomal RNA depletion efficiency after RiboMinus™ RNA purification. It will offer higher sensitivity than Bioanalyzer method. However, multiple qPCR targets per rRNA must be used to avoid false representation of the residual rRNA quantitation.

Acknowledgments

The authors thank Gary Bee and Byung-In Lee for their previous work on the RiboMinus™ method, and Dr. Jeff Chang and Dr. Nicholas Bergman for testing bacterial probe set and for providing feedback on the sequencing performance of RiboMinus RNA.

References

1. Ruan, Y., Le Ber, P., Ng, H., and Liu, E. (2004) Interrogating the transcriptome. *Trends Biotechnol.* **22**, 23–30.
2. Cloonan, N., Forrest, A. R. R., Kolle, G., Gardiner, B. B. A., Faulkner, G. J., Brown, M. K. et al. (2008) Stem cell transcriptome profiling via massive-scale mRNA sequencing. *Nature Methods* **5**, 613–619.
3. Lister, R., O'Malley, R. C., Tonti-Filippini, J., Gregory, B. D., Berry, C. C., Millar, H., et al. (2008) Highly integrated dingle-base resolution maps of the epigenome in Arabidopsis. *Cell* **133**, 523–536.
4. Tang, F., Barbacioru, C., Wang, Y., Nordman, E., Lee, C., Xu, N., et al. (2009) mRNA-seq whole-transcriptome analysis of a single cell. *Nature Methods* **6**, 377–382.
5. Cheung, A.L., Eberhardt, K.J., Fischetti, V.A. (1994) A method to isolate RNA from gram-positive bacteria and mycobacteria. *Anal Biochem.* **222**, 511–514.
6. Chirgwin, J. M., Przybyla, A. E., MacDonald, R. J., and Rutter, W. Z. (1979) Isolation of biologically active ribonucleic acid from sources enriched in ribonucleases. *Biochem.* **18**, 5294–5299.
7. Di Cello, F., Xie, Y., Paul-Satyaseela, M., and Kim, K. S. (2005) Approaches to bacterial RNA isolation and purification for microarray analysis of *Escherichia coli* K1 interaction with human brain microvascular endothelial cells. *Journal of Clinical Microbiology* **43**,4197–4199.
8. Sarkar, N. (1997) Polyadenylation of mRNA in prokaryotes. *Annu Rev Biochem* **66**, 173–197.
9. McTigue, P. M., Peterson, R. J., and Kahn, J. D. (2004) Sequence-dependent thermodynamic parameters for locked nucleic acid (LNA)-DNA duplex formation. *Biochemistry* **43**, 5388–5405.

Part III

Microbial Diversity

Chapter 8

Integrating High-Throughput Pyrosequencing and Quantitative Real-Time PCR to Analyze Complex Microbial Communities

Husen Zhang, Prathap Parameswaran, Jonathan Badalamenti, Bruce E. Rittmann, and Rosa Krajmalnik-Brown

Abstract

New high-throughput technologies continue to emerge for studying complex microbial communities. In particular, massively parallel pyrosequencing enables very high numbers of sequences, providing a more complete view of community structures and a more accurate inference of the functions than has been possible just a few years ago. In parallel, quantitative real-time PCR (QPCR) allows quantitative monitoring of specific community members over time, space, or different environmental conditions. In this review, we discuss the principles of these two methods and their complementary applications in studying microbial ecology in bioenvironmental systems. We explain parallel sequencing of amplicon libraries and using bar codes to differentiate multiple samples in a pyrosequencing run. We also describe best procedures and chemistries for QPCR amplifications and address advantages of applying automation to increase accuracy. We provide three examples in which we used pyrosequencing and QPCR together to define and quantify members of microbial communities: in the human large intestine, in a methanogenic digester whose sludge was made more bioavailable by a high-voltage pretreatment, and on the biofilm anode of a microbial electrolytic cell. We highlight our key findings in these systems and how both methods were used in concert to achieve those findings. Finally, we supply detailed methods for generating PCR amplicon libraries for pyrosequencing, pyrosequencing data analysis, QPCR methodology, instrumentation, and automation.

Key words: Microbial ecology, Pyrosequencing, Quantitative PCR, Methanogenesis, Microbial fuel cell, Human intestine, Obesity, Acetogenesis

1. Introduction

Microbial ecology is the science underlying many beneficial bioenvironmental processes that involve mixtures of microorganisms (1, 2). A prerequisite for successful management of microbial

Young Min Kwon and Steven C. Ricke (eds.), *High-Throughput Next Generation Sequencing: Methods and Applications*, Methods in Molecular Biology, vol. 733, DOI 10.1007/978-1-61779-089-8_8,

communities is the ability to define their structures, including accurately describing the relative abundance and, if possible, the function of their members.

Traditionally, culture-based methods, such as most probable number (MPN) (3, 4) and low-throughput culture-independent genomic-based methods, such as clone libraries (5), have been used to assess microbial diversity. Culture-dependent methods can skew the view of the microbial world, since most microbes are not cultured by standard techniques (6). For example, the human gut has more than 2,000 bacterial phylotypes (7), and most remain uncultured (8). Conventional Sanger sequencing-based methods, such as 16S rRNA-gene clone libraries, overcome culture-dependent bias, but still suffer from undersampling the community due to the limited number of clones that can be sequenced economically and rapidly (9).

Recent advances in sequencing technology have enabled high-throughput sequencing of microbial communities. One such massively parallel sequencing technology, pyrosequencing, is faster and more cost-effective than traditional Sanger sequencing (9). When primer bar coding is used, pyrosequencing allows simultaneous high-throughput sequencing of multiple samples in a single run (7, 10, 11). In contrast, Sanger sequencing can only be used to sequence one sample at a time. Being able to sequence multiple samples simultaneously makes pyrosequencing more cost-effective and suitable for collaborative efforts involving samples from multiple projects.

While pyrosequencing is the ideal tool for exploring the vast majority of the rare phylotypes in complex microbial communities, it is not well suited for quantitative tracking of critical phylotypes of interest within the community. For tracking key phylotypes over time or space, quantitative real-time PCR (QPCR) provides quantitative, specific, highly sensitive, and rapid values of targeted gene-copy numbers (12–14). Thus, we often combine high-throughput pyrosequencing and QPCR to study the microbial communities of a range of systems in which we need to accurately define the community structure and quantitatively track specific members of interest under different environmental conditions.

In this chapter, we first describe the principles of pyrosequencing, QPCR, and how they can be combined. Then, we present three examples in which we successfully combined high-throughput pyrosequencing to assess overall bacterial communities and QPCR to assess and quantify the presence of *Archaea* and specific methanogenic groups. The combination of both techniques helped us gain important insights into *Bacteria–Archaea* interactions in anaerobic ecosystems of importance to bioenergy and human health. Finally, we provide helpful details of the materials and methods used to carry out these two methods.

1.1. Principles of 454 Pyrosequencing and QPCR

1.1.1. 454 Pyrosequencing

Applying pyrosequencing in microbial ecology involves presequencing, pyrosequencing, and postsequencing steps. Presequencing refers to generating PCR amplicons from bacterial DNA extracted from environmental samples or human stools. Postsequencing refers to sequence analysis, such as assigning sequences to bacterial taxa. We provide a step-by-step guide for conducting pre- and postsequencing later in Subheadings 3.2–3.4. In this section, we focus on the principles of pyrosequencing.

Massively parallel pyrosequencing relies on two unique features: (1) conducting PCR on DNA-capturing beads in an emulsion of PCR reagents in oil and (2) using a PicoTiterPlate device for sequencing amplicons attached to beads (15). Pyrosequencing on a 454 GS-FLX model involves four steps: sample attachment, sample amplification by emulsion PCR, bead embedment in the wells of a PicoTiterPlate, and sequencing by synthesis. Figure 1 summarizes the work flow involved in these steps.

During the sample-attachment step, the DNA fragments in the amplicon libraries (generated from the presequencing step) bind to the beads under conditions that favor one fragment per bead.

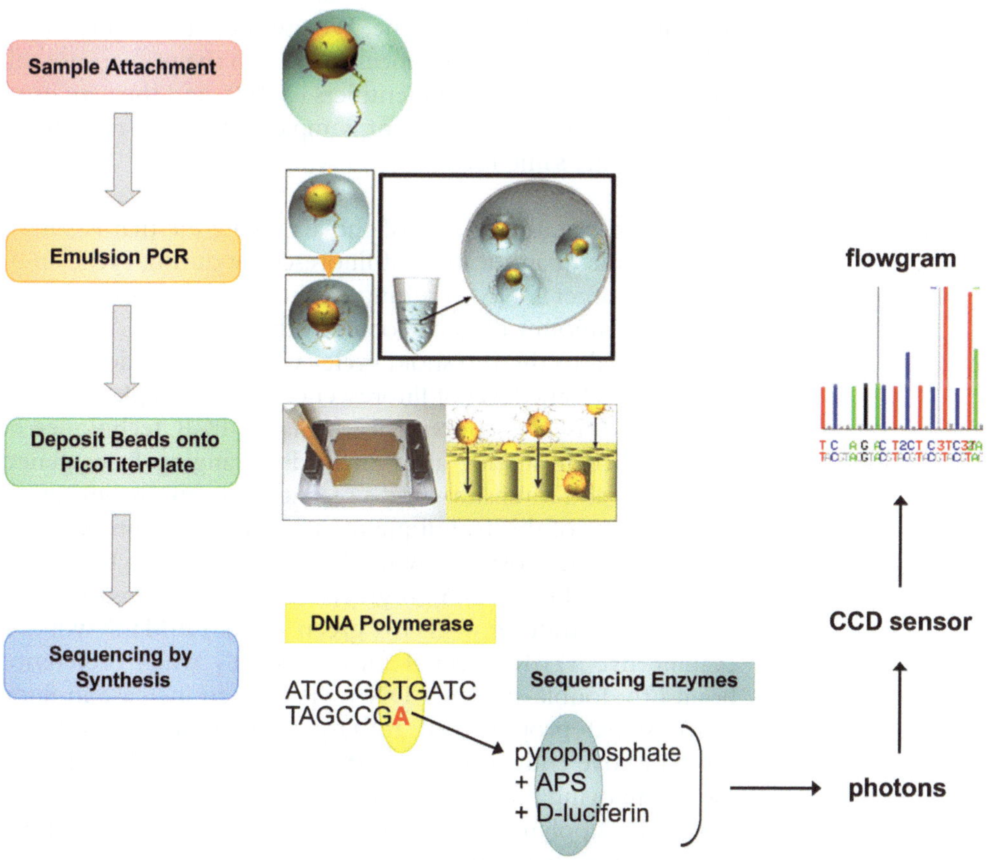

Fig. 1. Schematic drawing of the 454 pyrosequencing work flow, adapted from Gharizadeh et al. (16) and Margulies et al. (15).

The beads are then surrounded by an emulsion of PCR reagent and oil, resulting in isolated microreactors. PCR amplification in each microreactor generates ~10^7 copies from a single fragment, which is enough DNA for sequencing. The emulsion is then broken, the double-stranded amplicons are denatured, and beads carrying single-stranded DNA template are deposited to PicoTiterPlate wells, whose diameter allows only one bead per well. To do sequencing by synthesis, smaller beads carrying pyrosequencing enzymes are deposited into the wells, and nucleotide bases are fed in a fixed order (T, A, C, G) across the wells. Bases complementary to the DNA template bond, and new nucleotides are incorporated, which leads to a pyrophosphate release and generates a chemiluminescent signal that is recorded by a charge-coupled device camera (15). The collection of signals is called a "flowgram," and it gives the sequence of the target DNA molecule based on sequencing by synthesis.

A standard 8-h pyrosequencing run on a 454 GS-FLX instrument generates ~400,000 sequences of around 250 bases in length. We sometimes refer to sequences as "reads." Because the reads generated by pyrosequencing are part of the 16S rRNA gene, they are also referred to as "tags," which is short for tags of the full-length 16S rRNA gene. We use these terms interchangeably. Short nucleotide sequences fused into the primers are called bar codes or multiplexing identifiers (MID), and they are used to differentiate between different samples when we run multiple samples at the same time.

1.1.2. QPCR

Quantitative PCR is a state-of-the-art technique that is based on the real-time quantification of fluorescence during cyclic amplification of target DNA, resulting in the quantification of a target gene. The amount of gene copies in the target DNA is inversely proportional to the threshold cycle (CT), which is defined as the cycle at which the level of fluorescence from a sample in the assay is above a baseline fluorescence signal. QPCR techniques have been applied in a wide array of fields – ranging from genetics, pathology, and forensics – and a recent exhaustive review explains the relevance of this technique to environmental samples (17).

Two of the most widely used fluorescence detection chemistries during QPCR are SYBR green I and TaqMan® assays (2, 18). Figure 2 illustrates how SYBR green I and TaqMan® work. The cyanine dye SYBR Green I fluoresces only when preferentially bound to double-stranded DNA, yielding an increase in fluorescence signal proportional to target DNA amplification. The unbound dye emits very little or no fluorescence. In a TaqMan® assay employing fluorescent probes, the 5′–3′ exonuclease activity of the DNA polymerase is exploited along with fluorescent oligonucleotide probes that consist of a 3′ quencher and a 5′ reporter molecule (a fluorophore). Exonuclease cleavage of the probe

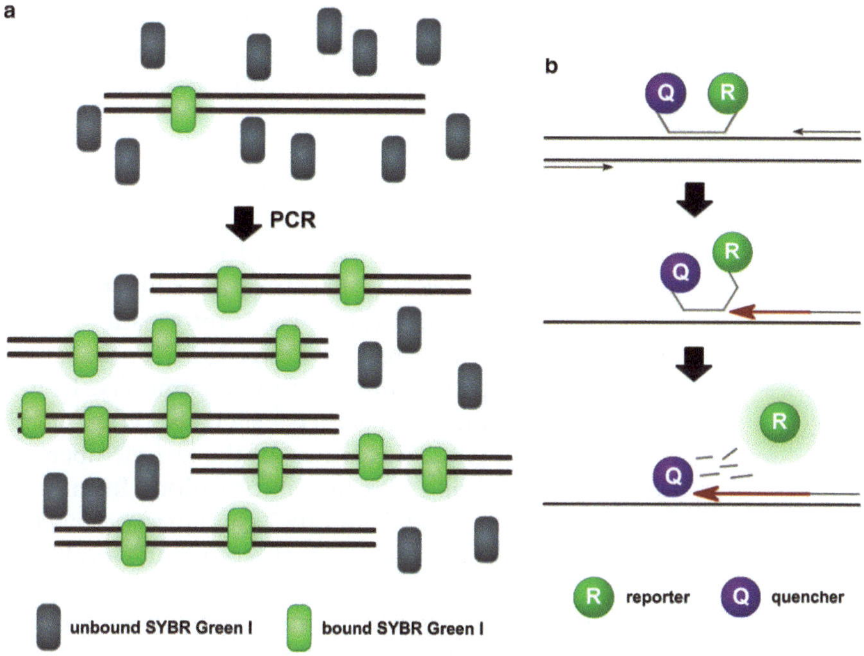

Fig. 2. QPCR fluorescent dye chemistries. (**a**) SYBR Green I. During amplification, SYBR Green I binds newly synthesized double-stranded DNA and yields an increase in fluorescence. Unbound dyes do not fluoresce. (**b**) TaqMan® probe. An oligonucleotide probe labeled with a 5′ reporter fluorophore and 3′ quencher binds target DNA during primer annealing. During primer extension, *Taq* polymerase partially displaces the probe and cleaves the reporter, and the unquenched reporter fluoresces.

separates the quencher from the reporter, resulting in an increase of fluorescence during amplification of the target DNA. Specific attachment of the fluorescent probe within a primer-targeted amplicon makes this technique more specific than DNA binding dye-based detection (see Note 1).

A typical QPCR assay reports the number of gene copies in a given unit of the sample with the following relationship:

$$\text{Number of genecopies per mL} = \frac{\text{Target gene conc. (g target DNA/mL)}^{(1)} \times \text{Avogadro's number } (6.023 \times 10^{23} \text{ copies/mol})}{(\text{plasmid} + \text{amplicon or genome size})(\text{mol bp/mol target})^{(2)} \times 660 \text{ (g DNA/mol bp)}^{(3)}}.$$

Notes for the equation: (1) This value is obtained from NanoDrop Spectrophotometer. (2) The plasmid size is obtained from the cloning kit used, plus the target gene amplicon region. The amplicon size should be known by the user, based on the primers used. (3) The molecular weight of each nucleotide (A, G, C, or T) pair.

For calibration, template DNA having known gene copies (this is calculated based on the concentration of DNA and the formula above using DNA of a pure culture or a plasmid containing the gene of interest) is serially diluted and analyzed in a real-time

PCR thermocycler along with the samples. Most real-time PCR cyclers are equipped with software that automatically generates a calibration curve (a semilog plot of CT values against log copy numbers of the standard).

1.1.3. Using Pyrosequencing and QPCR in Concert

We use pyrosequencing to define the structure of a bacterial community for which we have little or no previous knowledge. We can then identify the key members based on the relative dominance of their 16S rRNA-gene tags, their function (if it can be inferred from their identity), or both. Information obtained with pyrosequencing can be used to design QPCR primers and probes targeting important members of the community. In parallel, we track key community members, such as different groups of *Archaea*, by QPCR. The identity of the key members may come from pyrosequencing or other information. In either case, we combine data obtained from QPCR and pyrosequencing to infer community functions and to form ecological hypotheses.

1.2. Applications of Pyrosequencing and QPCR

We have successfully used pyrosequencing targeting bacterial populations and QPCR targeting *Archaea* together to investigate three complex microbial communities: the human large intestinal microbiota, the anaerobic microbial community in a methanogenic digester receiving pretreated sludge, and the biofilm community that develops on the anode of a microbial electrolysis cell (MEC). We provide summaries of each to illustrate the power of combining both approaches.

1.2.1. Human Large Intestinal Microbiota and Its Relationship to Obesity

Gut microbes play a key role in energy extraction in the human intestines. Targeting the hypervariable V6 region of the bacterial 16S rRNA gene, we used pyrosequencing to produce more than 180,000 sequences from the stools of nine individuals, three in each of the categories of normal weight, morbidly obese, or post-gastric-bypass surgery (7). The length of these sequences, which was around 60 bp (excluding primers), was set by the V6 primers. This length was shown to be a good target for the earlier 454 GS-20 pyrosequencer that Sogin et al. (32) used in their rare-biosphere study, but was shorter than the maximum read length (250 bp) of the 454 GS-FLX model. Our high-throughput pyrosequencing survey showed that the gut has very high species richness, with more than 2,000 bacterial phylotypes, and that obese individuals have markedly distinct bacterial community structures from those found in normal-weight individuals.

Even more importantly, pyrosequencing revealed that phylotypes affiliated with the H_2-producing bacteria within the genus *Prevotella* were highly enriched in obese individuals. This coincided with another important observation, made with QPCR: H_2-oxidizing methanogens also were highly enriched in obese individuals. This unique *Bacteria–Archaea* coexistence only in obese individuals points to a syntrophy that is associated with obesity and explains higher energy uptake by obese individuals. Bacterial fermentation in obese individuals is able to produce more acetate, since the removal of H_2 by methanogens clears the thermodynamic roadblock for fermentation. Acetate is rapidly taken up through the intestinal epithelium of the human host, leading to higher energy uptake and a tendency toward obesity.

1.2.2. Microbial Community's Response to Sludge Pretreatment

In the second example, we investigated how the microbial ecology in full-scale anaerobic digester was altered when the digester's methane production rate was significantly increased by Focused Pulsed (FP) (OpenCEL®) Technology pre-treatment, which make complex biological solids more bioavailable by exposing them to rapid pulses of a very strong electric field (19, 20). Using QPCR, we first observed a shift in methanogens to the acetate-cleaving *Methanosaeta* and away from the H_2-oxidizing *Methanoculleus*. By analyzing 36,797 pyrosequencing tags of around 60 bp from the V6 region of the bacterial 16S rRNA-gene (again, using the GS-FLX model), we found that the bacterial community became more diverse after FP pretreatment and was populated more by phylotypes associated with cellulose fermentation (*Ruminoccoccus*), scavenging of biomass-derived organic carbon (*Chloroflexi*), and homoacetogenesis (*Treponema*). In this example, we used pyrosequencing, along with QPCR, to demonstrate that, as the overall activity of the community was stimulated by addition of more bioavailable organic matter, the bacterial community became more phylogenetically diverse to take advantage of the added input of biodegradable material and in response to the more efficient utilization of acetate by *Methanosaeta*.

1.2.3. The Ecology of MEC Anodes: The Importance of H_2 Scavengers

In the third example, we looked at the ecology of the anode of a MEC (21, 22). When an MEC is working properly, anode-respiring bacteria (ARB) oxidize simple organic substrates at the anode chamber (e.g., acetate) and transfer almost all of the electrons to the anode. Those electrons then move through an electrical circuit to the cathode, where they reduce protons (H^+) to generate hydrogen gas (H_2) as the energy output. Because we observed methane production from the anode in some cases, we suspected that methanogenesis was an undesired sink for electrons that otherwise could go to current. To find out whether this was true, we set up two MECs fed with ethanol, which is fermented to acetate and H_2. In one, we allowed methanogens to develop, while in the second, we suppressed methanogenesis by adding bromoethanesulfonate (BES). After 30 days, we quantified electron recovery in the current and found that 24% more current was generated when methanogenesis was suppressed, and the difference in current could be attributed to methane formation in the other MEC.

By using QPCR with group-specific primers and probes, we found that *Archaea* were present only in the biofilm of the MEC that showed CH_4 formation. Furthermore, H_2-oxidizing methanogens made up nearly 100% of the total *Archaea*. This corresponded well to the observation of CH_4 formation, but it did not explain how the electrons in the H_2 got routed to current when methanogenesis was suppressed.

From earlier work of others and our laboratory (23–27), we knew that ARB are efficient at oxidizing acetate and producing electric current. However, having ARB that can oxidize H_2 to make current was far from established. Another possibility is that the H_2 was oxidized by homoacetogenic bacteria, which made acetate that was then consumed by the regular ARB. So, we used 16S-rRNA-gene-based pyrosequencing to detect general bacteria, which include homoacetogens (21). Since homoacetogens occur across multiple bacterial phyla, we also detected them by cloning the functional gene encoding formyl tetrahydrofolate synthetase (FTHFS), which is specific to homoacetogenesis. The presence of homoacetogens was confirmed by pyrosequencing and further bolstered by the FTHFS-gene-based clone library analysis. When methanogenesis was suppressed, we found homoacetogens mostly belonging to the genus *Acetobacterium*, thus confirming that they were a channel for electron flow from H_2 to current through acetate. Figure 3 illustrates the characteristics

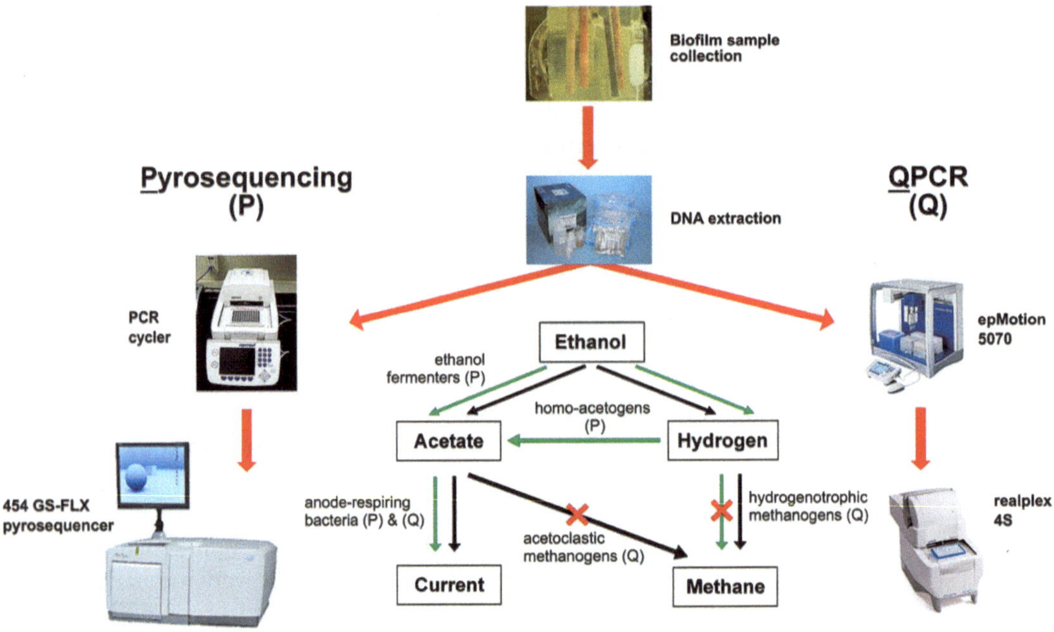

Fig. 3. Using pyrosequencing (P) and QPCR (Q) in parallel to identify anode biofilm community members in an MEC. The *black arrows* indicate electron flow path when methanogenesis was allowed, and the *green arrows* indicate the flow when methanogenesis was inhibited. The *cross sign* (*red*) indicates the pathways that were not observed. The method used to identify key members mediating the observed electron flow is indicated as either P or Q. For example, ethanol fermenters were identified by using P (pyrosequencing), and hydrogenotrophic methanogens were identified with Q (QPCR).

of the two communities that developed when methanogens or homoacetogens consumed H_2 from ethanol fermentation, as well as how we used pyrosequencing and QPCR to gather the evidence for each key microbial pathway.

2. Materials

2.1. General Materials and Equipment

1. Nuclease-free water.
2. Ice.
3. –20°C freezer.
4. UV NanoDrop® Spectrophotometer.
5. Pipette tips – certified RNAse, DNAse and pyrogen free. (Axygen, USA).
6. Pipettes (1,000, 200, 100, 10 μL).
7. Sterile microcentrifuge tubes (1.5 and 2.0 mL, VWR Scientific).
8. UV cross-linker for inactivating nucleases (Stratalinker).
9. PCR purification kit (Qiaquick, Qiagen).
10. Agarose gel (1% w/v).
11. Gel electrophoresis unit.

2.2. Sample Collection and DNA Extraction

1. Kits for DNA extraction (MoBio soil or QIAamp DNA stool kit).
2. Tweezers.
3. Tabletop centrifuge (e.g., Eppendorf 5415D).
4. Top-load balances (e.g., Mettler Toledo).
5. Biomass samples for DNA extraction.

2.3. Pre- and Postpyrosequencing

1. Thermocyclers (i.e., Eppendorf gradient cycler).
2. PCR tubes.
3. PCR reagents and enzymes (i.e., 5 PRIME PCR MasterMix).
4. Sequence processing and phylogeny software tools available at the pyrosequencing pipeline: http://pyro.cme.msu.edu.
5. The “mothur” program for generating rarefaction curves and comparing microbial communities, available at http://www.mothur.org/wiki/.
6. A computer with access to Unix shell commands. This could be a Linux/Solaris computer, a Mac OS X or later, or a Windows machine with Cygwin (http://www.cygwin.com) installed.

2.4. QPCR and Automated QPCR

1. Semiskirted twin tec 96-white-well plate (http://www.eppendorf.com).
2. Optically clear heat sealing film (http://www.eppendorf.com).
3. Heat sealer, 115 V (http://www.eppendorf.com).
4. 5 PRIME RealMaster mix probe with or without ROX (http://www.5prime.com). Alternatively, BioRad IQ supermix with/without SYBR green I can also be used, at the concentrations recommended by the manufacturer (http://www.bio-rad-gene.com).
5. 5 PRIME RealMaster mix with SYBR Green I.
6. Primers (http://www.idtdna.com based on optimum design from Probe Match or other software for probe design).
7. Probes for TaqMan® based assays (also from http://www.idtdna.com with proper selection of the fluorescent reporter dye on the 5′ end and appropriate quencher at the 3′ end).
8. Tabletop centrifuge with a maximum speed of 4,000 rpm.
9. White ice block with a plastic adapter.
10. Automated liquid handler (e.g., epMotion 5075).
    (a) 50-μL pipette tool.
    (b) Compatible 50-μL PCR-grade filter-barrier tips.
    (c) Compatible 1.5-mL tube rack/holder.
11. epMotion 5070.
12. 96-Well thermoblock, prechilled to 0–4°C.

3. Methods

3.1. DNA Extraction

1. First, collect the appropriate volume of solid sample, or pellet the liquid sample to obtain an appropriate mass for the DNA extraction (for example, 0.25–1 g for Power soil DNA extraction kit). For biofilm samples from the anode of microbial electrolysis cells, follow the following instructions:
    (a) Remove the graphite electrode and use a sterile pipette tip to scrape the biofilm from the surface into a microcentrifuge tube.
    (b) Clean the thin layer of biofilm that is attached to the electrode with a known volume of nuclease-free water (around 100 μL).
2. Follow the manufacturer's instructions to extract DNA.
3. Use 2 μL of the extracted DNA on a NanoDrop spectrophotometer to quantify the yield of DNA in the sample and also

to establish DNA quality (260/280 nm absorbance ratio, which is a measure of nucleic acids over proteins; the 260/230 nm ratio indicates the amount of phenolic derivatives from the extraction that is present in the final sample).

4. Store the DNA at −20°C, whichever is available.

3.2. Designing Bar-Coded Primers for 16S rRNA-Gene Tag Pyrosequencing

Because the maximum nucleotide length that can be accurately pyrosequenced with the current 454 GS-FLX model is around 250 bp (9), one needs to design PCR primers to target only a portion of the gene of interest, in this case the 16S rRNA gene, which is typically around 1,500 bp. The 16S rRNA has variable regions named V1 through V9 (28). Specifically, variable regions V2 (29), V3 and V6 (30), V4 (31), and V6 (7, 20, 22, 32, 33) have been used. The following instructions provide an example of designing bar-coded V6 primers.

1. The forward primer has the structure: 454 Adaptor A + barcodes + V6 forward primer. The sequence of 454 adaptor A is "gcctccctcgcgccatcag." The sequence of V6 forward primer is "CAACGCGAAGAACCTTACC."
2. In order to combine multiple samples in a single 454 GS-FLX run and sort out datasets belonging to each sample, the design of unique bar codes of five nucleotides fused to the 5′ end of V6 forward primers and the 3′ end of the 454 adaptor A is necessary. For example, AGACT and AGATC, which differ by two bases, were two of the nine bar codes used in our human microbiome study. Table 1 lists other bar code sequences for the example applications in this chapter.
3. In designing bar codes, one important consideration is to avoid holopolymers, for example a string of five adenines (AAAAA), because 454 pyrosequencing is prone to errors around holopolymers (15, 34), resulting in insertions (AAAAAA) or deletions (AAAA). In practice, even bar codes with two adjacent identical bases, for example ATTGC, should be avoided.
4. Fuse the reverse primer to the 3′ end of 454 adaptor B. In the case of V6 region, the reverse primer is 5-gccttgccagccc gctcagCGACAGCCATGCANCACCT-3.
5. Synthesize the primers, usually by commercial vendors, such as Integrated DNA Technologies (http://www.IDTdna.com). The standard desalt purification of the primers is adequate.

3.3. Pyrosequencing PCR Amplicon Library Preparation

1. Use standard PCR reagents and protocols for generating PCR amplicons.
2. Prepare 100-μL PCRs as follows: 2 mM magnesium, 0.2 mM of total dNTPs (N = A, G, C, T), 0.2 μM of each primer, 4 units of *Taq* polymerase, and around 10 ng of template DNA. A high-fidelity polymerase such as PfuTurbo polymerase

Table 1
Bar codes of five nucleotides (also called MID or multiplexing identifiers) designed for multiplexing samples in one 454 pyrosequencing run

Topic	Sample	Bar code (MID)
Human gut	nw1	AGACT
	nw2	AGATC
	nw3	GACTA
	ob1	GATCA
	ob2	GATAC
	ob3	GACAT
	gb1	GAGCT
	gb2	GAGTC
	gb3	CTAGA
Sludge pretreatment	MCSB	CATGA
	MTSB	CGCGA
MEC anode	MEC1	CTCAG
	MEC2	CTCGA
	MEC3	TCAGC
	MEC4	TCGAC

(Stratagene, La Jolla, CA) should help reduce error rates in the PCR step.

3. Set up PCR temperature programs as follows: 94°C for 2 min, 25 cycles of denaturation at 94°C for 30 s, 57°C annealing for 45 s, and 72°C for 1 min extension, followed by a final extension at 72°C for 2 min.
4. Purify the PCR products with QIAquick (Qiagen) spin columns to remove excess primer dimers and dNTPs.
5. Measure the purified PCR products with a NanoDrop spectrophotometer.

3.4. Pyrosequencing Data Analysis

Translating pyrosequencing reads into taxon names requires two steps. The data analysis starts with removing low-quality tags and trimming off low-quality end bases. The second step is to find a full-length sequence match for the tags in a reference database via a process called alignment. The taxonomic classification (phylum, class, order, etc.) of the full-length match is then used as the pyrosequencing tags. Stand-alone or Web-based data processing pipelines, such as "mothur" (35), RDP pyrosequencing pipeline (36), GAST (37), are available.

1. Remove low-quality sequences using mothur: trim.seqs(fasta = vtags.fasta, minlength = 57, maxlength = 250, maxambig = 0, qfile = user.qual, qaverage = 27, oligos = user.oligo). For example, the "minlength" setting here tells the program to keep sequences that are 57 bp or longer. The reader should refer to the mothur user manual for additional parameters.
2. Classify the pyrosequencing tags using the RDP classifier or by comparing reads against locally maintained reference database. The RDP classifier (http://www.cme.msu.edu/rdp/classifier) accepts sequences that are longer than 50 bases, and it returns phylogenetic information from domain to the genus level. In classifying sequences, it is important to set classification reliability to at least 80% to get accurate taxonomy information (36).
3. One can also use GAST (37). In this method, each sequence is compared through similarity searches, using the program BLASTN (38), against a reference database including 44,011 nonidentical V6 sequences extracted from 119,480 bacterial rRNA genes. Collect the top 150 BLAST hits with alignments longer than 57 bp and align along with the query using the program MUSCLE (39) set to the following parameters: -diags and -maxiters 2: muscle -maxiters 2 -diags -in infile -out muscles.pfam. The alignment serves as the input for the program quickdist (32), which generates a distance matrix: ./quickdist -t muscles.pfam > query.dist. Based on this distance matrix, collect sequence(s) from the reference database having the minimum distance to the original query. The complete perl script of this pipeline is available upon request.

The second task of comparing community structures answers ecological questions related to species richness, evenness, and similarities and differences between two or more communities. This involves sequence alignments, generating distance matrices from alignments, and sequence clustering into OTUs at different distance cutoffs (0, 0.03, 0.06, etc.).

1. Trim off primers using mothur, as mentioned above, or using unix shell command: sed –i 's/^[forward_primer_sequence]//' sequences.fasta. The symbol ^ indicates the forward primer sequence is at the beginning of a new line. Sequences.fasta is the sequence file in fasta format.
2. Align pyrosequencing tags with RDP's inferno aligner, which takes into consideration of the secondary structure of the 16S rRNA molecule.
3. Calculate distance matrices using dist.seqs command in mothur, DNADIST in the PHYLIP software package (40), quickdist (32).

4. Assign sequences to OTUs and compare community structures. We use the mothur program (35): cluster( ). Detailed description of this program exceeds the scope of this review. Readers should consult the original references for instructions.
5. Comparing community structures with OTU-based methods described here helps to answer questions like whether a community is a subset of another or to what degree members from two communities overlap (see Note 2).

3.5. QPCR

1. Design probes and primers, or modify primers and probes available in the probeBase (http://www.microbial-ecology.net/probebase/).
2. Perform all reactions in a very clean room, under dark conditions to avoid photobleaching of the fluorescent reagents and probes.
3. Clean the bench space first with DNAse Away wipes.
4. Treat the microcentrifuge tubes to be used for preparing the megamix under UV light (a process called cross-linking, which renders replication of any free DNA impossible).
5. Place the required reagents, primers, PCR water after thawing in a white ice block with plastic adapter so that they stay around 4°C, but do not freeze them.
6. The following are the steps for setting up a QPCR by hand, while an automated liquid handler can also be programmed to set up the reaction (see details on the QPCR automation section 3.7).
7. Precalculate the requirements of the various reagents needed for the megamix for the total no. of reactions to be set up and an extra reaction per ten reactions as a safety factor for pipetting losses. Include a six-point calibration curve along with your templates for quantification (the serial dilutions could be prepared as per the instructions under Subheading 3.6, and the gene copy numbers are calculated as per the formula illustrated in Subheading 1.1.2). A general tabulation of the quantities for a 20-μL reaction volume (for IQ SYBR Green Supermix as the Mastermix reagent) is shown in Table 2.
8. Plan your orientation and movement while setting up the reaction mix, and minimize handling of the tubes frequently, even if it is with gloved hands.
9. Load the master mix at the end while preparing the QPCR mixture, as it can introduce significant bubbles if not pipetted properly.
10. Vortex the reaction mixture well before starting to pipette 18 μL to each tube.

Table 2
Volumes of reagents for megamix preparation of a SYBR Green I or TaqMan®-based QPCR assay

Ingredient	Concentration (μM)	Final concentration (nM)	Volume per 20-μL reaction (μL)	Total volume for 20 reactions (μL)
PCR grade water	–		6/5.97[a]	120/119.4
Master mix	2×	1×	10	200
Forward primer	10	500	1	20
Reverse primer	10	500	1	20
TaqMan® probe	100	300	0/0.03[b]	0/0.6
		Total megamix	18	360/20 = 18
DNA[c]			2	

[a]Volume of PCR-grade water per reaction is obtained as a difference of megamix (18 μL/reaction) and the remaining reagents per reaction
[b]The probe is either added or not, depending upon the assay
[c]DNA is generally added at the end to each reaction tube after 18 μL of megamix has been added

11. While transferring the reaction mixture to the tubes, it is important that you are consistent in pipetting the same quantity in all the tubes; this is more important than being particular about transferring all the quantity. It is possible that some of the mixture sticks to the end of the pipette tip. As long as you are adding the same quantity to each tube, while not trying to add all of it and introduce bubbles, the results should not be affected.
12. Ideally, use the same pipette tip for loading all the tubes with the reaction mixture, as long as you have not contaminated the pipette tip. Touch only the inner wall of the reaction tube, not anywhere else.
13. Load the template and change pipette tips at each step. Touch the template at the inner wall of the reaction tube as well, and be sure to transfer all the contents of the tube. This will affect the accuracy of the quantification.
14. Seal the 96-white-well plate at the end of the sample transfer with heat transfer film and a heat sealer.
15. Cool the tabletop centrifuge to 4°C and spin the 96-well plate down so that the template and the reaction mix are uniformly mixed.
16. The plate is now ready to be set up in the real-time PCR machine; follow the instructions specific for the unit you are

going to use. The following are the steps for setting up the QPCR in the Eppendorf Realplex 4S unit.

(a) Switch on the power switch of the Realplex unit, which is behind it. It is also advisable to switch it on several minutes before setting up the assay.

(b) Then, click the Realplex software shortcut on the desktop. Open the template file for setting up your assay.

(c) Insert the 96-well plate into the Realplex 4S unit.

(d) Gently wipe the tops of the strip tubes to remove any minute dust particles that might interfere with the fluorescence data collection.

(e) Close the top lid of the Realplex unit and gently lower the lid cover. This is critical for smooth operation of the equipment: If the lid is not lowered, the photomultiplier tubes could not get the light signals from the LEDs.

(f) Do not include the 4°C hold step after the PCR program; this could introduce condensation on the well surface and potentially affect detection from the surface.

The following are the steps for specifying the plate setup to the real-time PCR unit (Realplex 4S, in this case).

(a) Select the dye of interest in the filter 520-nm drop-down tab: both FAM and SYBR Green I should be available. Leave the three other filters blank.

(b) Select the reaction volume (10 μL in this case) and the type of assay (SYBR Green I or TaqMan®) and the appropriate background calibration from the respective drop-down tabs.

(c) Assign the serially diluted standards in triplicate as standard samples (there are four types of samples that the Realplex understands – standard, negative control, positive control, and unknown or sample to be quantified).

(d) Specify the gene copies calculated for each point in the calibration curve with the relevant units at the appropriate location.

(e) Assign the samples from the experiment as unknown.

(f) Always include a negative control (which is the reaction mixture with PCR-grade water instead of DNA) in triplicate, and specify the same in the plate layout.

The following are the steps to insert a PCR program:

(a) On the drop-down menu, select insert program option.

(b) Right click on each of the program steps to add/modify the details as needed.

(c) As the default, the following options should always be left checked on: Impulse step, Turn off heat from lid when block temperature is lower, and TSP heated lid.

(d) If the program will be used more than once, save the program as a template. You can then run the specific run again from this template.

17. Once the program is entered and the plate layout is defined, click the run option (a green button that starts the run). You will be prompted to confirm the background option for the run.
18. Always leave the 96-well plate cover over the wells when the equipment is not in use.
19. After an assay is finished, wait for the block temperature to cool down to room temperature, after which one can remove the strip tubes/96-well plate and close the lid.
20. Exit the Realplex software program and turn off the Realplex 4S equipment's switch behind the instrument.
21. Do not disturb the USB adapter (CAN bus adapter) from behind the computer. Minimize any disturbances to the CAN modulator that connects Realplex with the computer, as this could seriously disrupt the data acquisition process.
22. At the end of an assay, the Realplex 4S equipment (as do most other real-time PCR units) automatically generates a calibration curve for the assay. This is a semilog plot of number of gene copies (log scale) on the *X*-axis against the CT value on the *Y*-axis. The intercept value is the threshold CT value, which corresponds to the least number of copies you can reliably detect with the assay. The slope of the calibration curve represents the CT value separation between each *X*-fold dilution of the standard target.
23. The inverse of the calibration curve's slope is a measure of the amplification efficiency, and it should generally be between 80 and 105%, while the coefficient of determination should be greater than 0.95.

3.6. Automated Preparation of Standard Serial Dilutions for QPCR

1. A fresh preparation of DNA standard is recommended for each QPCR run.
2. Determine the concentration of the DNA standard (see above) using a NanoDrop UV spectrophotometer.
3. Calculate the appropriate dilution factor for normalization of the standard concentration to 10 ng/μL in a final volume of 50 μL.
4. Dilute the DNA standard to 10 ng/μL in nuclease-free water. Label this tube the 10^1 dilution and keep the tube on ice.

5. Label empty, UV-cross-linked, 1.5-mL microcentrifuge tubes 10^0 through the lowest dilution desired.
6. Switch on the computer that operates the epMotion 5075 (processor switch behind the 5075 unit) and then the green switch for the epMotion itself on the left-front end. Activate the icon for epMotion software on the desktop and log in
7. Program a method with the software that accompanies the automated liquid handler. The following is specific for the epMotion 5075:
   (a) Specify placement of PCR-grade tips and 1.5-mL microcentrifuge tube rack on the worktable. Plan a logical flow of the robot head from the tips to the sample rack and then to the waste container.
   (b) Move to the "Method" tab. For each step, specify "TS_50" as the pipette tool and check the box for "filter tips."
   (c) Specify the following steps in the method file, in the following order:
      (i) Transfer 45 μL of nuclease-free water to each empty microcentrifuge tube. Make note of the placement of each tube in the rack. Placing empty tubes at the start of a new row is recommended. Discard the tip at the end of this step.
      (ii) Do not use "multidispense" mode, as this can compromise volumetric accuracy.
      (iii) Using "pipette" mode, transfer 5.0 μL of the DNA standard (10^1) to the 10^0 tube. Set the mixing to ten cycles of 45 μL at 9.0 mm/s. Important: Discard the pipette tip before moving to the next dilution. This improves volumetric accuracy and avoids carryover of any unmixed liquid sticking to the tip.
      (iv) Repeat step (ii) as many times as necessary to achieve the desired number of standard dilutions.
   (d) Save the method file and check for errors.
8. Carefully open empty 1.5-mL microcentrifuge tubes and use tweezers to place tubes at the correct rack position specified in the method file. Do the same for the tube containing nuclease-free water.
9. Place the 1.5-mL microcentrifuge rack at the correct position on the worktable.
10. Ensure that the TS_50 pipette tool is situated in the tool holder position.
11. Check placement of 50-μL PCR-grade tips on the worktable and remove the lid.

12. Close the shutter door and run the method.
13. When the method has completed, open the shutter door and carefully retrieve the 1.5-mL microcentrifuge tube rack.
14. Use tweezers to remove tubes from the rack. Avoid touching the tube interior and close tubes firmly.
15. Place diluted standards on an ice block until ready for use.

3.7. Automated Preparation of QPCRs

The following protocol describes use of an automated liquid handler for preparing individual QPCRs at a final volume of 10 µL in a 96-well plate. Instructions are specific to the epMotion 5070, but are generally applicable to other automated liquid handlers (see Note 3).

1. Switch on the epMotion 5070 and create a new method file.
2. Specify placement of PCR-grade tips, 1.5-mL microcentrifuge tube rack, and 96-well QPCR plate on the worktable. Plan a logical flow of the robot head from the tips to the sample rack and then to the QPCR plate.
3. Plan the method commands first to aliquot a fixed volume of QPCR master mix to each well and then to add appropriate volume of template DNA individually to each well.
4. Using "pipette" mode and checking the appropriate box for "filter tips" throughout, program the following in the method file:
   (a) Transfer 6 µL of QPCR master mix to the appropriate position(s) in the 96-well plate. Exchange tips at the end of a row, column, or group of reactions (6–12 aliquots before discarding the tip is recommended).
      - *Important*: To avoid bubbling and/or spraying of master mix, adjust the liquid type settings to custom values optimized by trial and error. If using 5 PRIME reagents for the master mix, adjust the liquid parameters as follows: aspirate speed, 12.0 mm/s; dispense speed, 3.0 mm/s; delay blow, 3,000 ms; blow speed, 66.0 mm/s; and movement blow, 90%.

   (b) Using the default liquid settings for "Water," individually transfer 4 µL of DNA template to the appropriate position in the 96-well plate. Exchange the pipette tip after each transfer. Do not mix.
5. Save the method file and check for errors.
6. Prepare serial dilutions of DNA standard as described in Subheading 3.6 protocol above.
7. Prepare QPCR master mix(es) as described in the "Quantitative PCR" protocol, steps 1–6 above.
8. Prechill the 96-well plate adapter to 0–4°C.

9. Ensure that the samples of interest are diluted appropriately to fall within end points on the standard curve.
10. Cross-link the QPCR plate under UV for 5 min.
11. Using tweezers, carefully place 1.5-mL microcentrifuge tubes containing standards, unknowns, and master mix(es) in the correct position in the 1.5-mL microcentrifuge tube rack specified in the method file.
12. Place the 1.5-mL microcentrifuge tube rack at the correct position on the worktable.
13. Place the UV-cross-linked, 96-well, QPCR plate into the prechilled thermoblock adapter. Make note of the plate orientation, especially if every well is to contain a reaction.
14. Place the 96-well plate and thermoadapter at the correct position on the worktable.
15. Ensure that 50-μL PCR-grade pipette tips are placed at the correct position on the worktable.
16. Close the shutter door and run the method. Periodically check progress and supply of pipette tips.
17. Continue at step 11 of the "Quantitative PCR" protocol described above.

4. Notes

1. In QPCR, the specificity of the primers and the structure and nature of target region of interest determine the PCR conditions. High ramp rates enable the performance of combined annealing and extension of the target DNA, which saves time between cycles, with minimal impact on amplification efficiency. This strategy of combined annealing and extension works more efficiently for amplicons less than 150 bp, while a three-step cycle consisting of denaturation, annealing, and extension is superior for bigger amplicons.
2. It has been reported that overestimation of species richness can occur due to pyrosequencing errors, such as those around long holopolymers (34). Excising flowgram preclustering has been shown effective to suppress such errors (34), but the algorithm currently requires cluster computing and exceeds the capability of a single laboratory or even a department. An effective remedy is to use an appropriate clustering algorithm such as the average neighbor (41). In addition, programs for detection of chimeric sequences in large dataset will also be helpful. Overall, sequencing errors affect richness estimation

more at the species level than higher taxa such as genus and family (34).

3. As with manual reaction preparation, all steps involving QPCR master mixes should be performed in the dark.

References

1. Rittmann, B. E., Hausner, M., Loffler, F., Love, N. G., Muyzer, G., Okabe, S., et al. (2006) A vista for microbial ecology and environmental biotechnology. *Environ. Sci. Technol.* **40**, 1096–1103.
2. Rittmann, B. E., Krajmalnik-Brown, R., and Halden, R. U. (2008) Pre-genomic, genomic and post-genomic study of microbial communities involved in bioenergy. *Nat. Rev. Microbiol.* **6**, 604–612.
3. Coates, J. D., Michaelidou, U., Bruce, R. A., O'Connor, S. M., Crespi, J. N., and Achenbach, L. A. (1999) Ubiquity and diversity of dissimilatory (per)chlorate-reducing bacteria. *Appl. Environ. Microbiol.* **65**, 5234–5241.
4. Wu, J., Unz, R. F., Zhang, H., and Logan, B. E. (2001) Persistence of perchlorate and the relative numbers of perchlorate- and chlorate-respiring microorganisms in natural waters, soils, and wastewater. *Biorem. J.* **5**, 119–130.
5. Hugenholtz, P., Goebel, B. M., and Pace, N. R. (1998) Impact of culture-independent studies on the emerging phylogenetic view of bacterial diversity. *J. Bacteriol.* **180**, 4765–4774.
6. Amann, R. I., Ludwig, W., and Schleifer, K. H. (1995) Phylogenetic identification and in-situ detection of individual microbial cells without cultivation. *Microbiol. Rev.* **59**, 143–169.
7. Zhang, H., DiBaise, J. K., Zuccolo, A., Kudrna, D., Braidotti, M., Yu, Y., et al. (2009) Human gut microbiota in obesity and after gastric-bypass. *Proc. Natl. Acad. Sci. U. S. A.* **106**, 2365–2370.
8. Eckburg, P. B., Bik, E. M., Bernstein, C. N., Purdom, E., Dethlefsen, L., Sargent, M., et al. (2005) Diversity of the human intestinal microbial flora. *Science* **308**, 1635–1638.
9. Hugenholtz, P., and Tyson, G. W. (2008) Microbiology - Metagenomics. *Nature* **455**, 481–483.
10. Jones, R. T., Robeson, M. S., Lauber, C. L., Hamady, M., Knight, R., and Fierer, N. (2009) A comprehensive survey of soil acidobacterial diversity using pyrosequencing and clone library analyses. *ISME J.* **3**, 442–453.
11. Hamady, M., Walker, J. J., Harris, J. K., Gold, N. J., and Knight, R. (2008) Error-correcting barcoded primers for pyrosequencing hundreds of samples in multiplex. *Nat. Methods* **5**, 235–237.
12. Suzuki, M. T., Taylor, L. T., and DeLong, E. F. (2000) Quantitative analysis of small-subunit rRNA genes in mixed microbial populations via 5′-nuclease assays. *Appl. Environ. Microbiol.* **66**, 4605–4614.
13. Yu, Y., Lee, C., Kim, J., and Hwang, S. (2005) Group-specific primer and probe sets to detect methanogenic communities using quantitative real-time polymerase chain reaction. *Biotechnol. Bioeng.* **89**, 670–679.
14. Ritalahti, K. M., Amos, B. K., Sung, Y., Wu, Q., Koenigsberg, S. S., and Loffler, F. E. (2006) Quantitative PCR targeting 16S rRNA and reductive dehalogenase genes simultaneously monitors multiple *Dehalococcoides* strains. *Appl. Environ. Microbiol.* **72**, 2765–2774.
15. Margulies, M., Egholm, M., Altman, W. E., Attiya, S., Bader, J. S., Bemben, L. A., et al. (2005) Genome sequencing in microfabricated high-density picolitre reactors. *Nature* **437**, 376–380.
16. Gharizadeh, B., Kalantari, M., Garcia, C. A., Johansson, B., and Nyren, P. (2001) Typing of human papillomavirus by pyrosequencing. *Lab. Invest.* **81**, 673–679.
17. Zhang, T., and Fang, H. H. (2006) Applications of real-time polymerase chain reaction for quantification of microorganisms in environmental samples. *Appl. Microbiol. Biotechnol.* **70**, 281–289.
18. Talbot, G., Topp, E., Palin, M. F., and Masse, D. I. (2008) Evaluation of molecular methods used for establishing the interactions and functions of microorganisms in anaerobic bioreactors. *Water Res.* **42**, 513–537.
19. Rittmann, B. E., Lee, H. S., Zhang, H., Alder, J., Banazak, J. E., and Lopez, R. (2008) Full-scale application of Focused-Pulsed pretreatment for improving biosolids digestion and conversion to methane. *Water Sci. Technol.* **58**, 1895–1901.
20. Zhang, H., Banaszak, J. E., Parameswaran, P., Alder, J., Krajmalnik-Brown, R., and

Rittmann, B. E. (2009) Focused-Pulsed sludge pre-treatment increases the bacterial diversity and relative abundance of acetoclastic methanogens in a full-scale anaerobic digester. *Water Res.* **43**, 4517–4526.

21. Parameswaran, P., Zhang, H., Torres, C. I., Rittmann, B. E., and Krajmalnik-Brown, R. (2009) Microbial community structure in a biofilm anode fed with a fermentable substrate: The significance of hydrogen scavengers. *Biotechnol. Bioeng.* In press.
22. Parameswaran, P., Torres, C. I., Lee, H. S., Krajmalnik-Brown, R., and Rittmann, B. E. (2009) Syntrophic interactions among anode respiring bacteria (ARB) and non-ARB in a biofilm anode: electron balances. *Biotechnol. Bioeng.* **103**, 513–523.
23. Liu, H., Cheng, S., and Logan, B. E. (2005) Production of electricity from acetate or butyrate using a single-chamber microbial fuel cell. *Environ. Sci. Technol.* **39**, 658–662.
24. Jung, S. and Regan, J. M. (2007) Comparison of anode bacterial communities and performance in microbial fuel cells with different electron donors. *Appl. Microbiol. Biotechnol.* 77, 393–402.
25. Torres, C. I., Marcus, A. K., Parameswaran, P., and Rittmann, B. E. (2008) Kinetic experiments for evaluating the Nernst-Monod model for anode-respiring bacteria (ARB) in a biofilm anode. *Environ. Sci. Technol.* **42**, 6593–6597.
26. Lee, H. S., Parameswaran, P., Kato-Marcus, A., Torres, C. I., and Rittmann, B. E. (2008) Evaluation of energy-conversion efficiencies in microbial fuel cells (MFCs) utilizing fermentable and non-fermentable substrates. *Water Res.* **42**, 1501–1510.
27. Logan, B. E. (2009) Exoelectrogenic bacteria that power microbial fuel cells. *Nat. Rev. Microbiol.* 7, 375–381.
28. Neefs, J. M., Vandepeer, Y., Hendriks, L., and Dewachter, R. (1990) Compilation of small ribosomal subunit RNA sequences. *Nucleic Acids Res.* **18**, 2237–2317.
29. Turnbaugh, P. J., Hamady, M., Yatsunenko, T., Cantarel, B. L., Duncan, A., Ley, R. E., et al. (2009) A core gut microbiome in obese and lean twins. *Nature* **457**, 480–484.
30. Dethlefsen, L., Huse, S., Sogin, M. L., and Relman, D. A. (2008) The pervasive effects of an antibiotic on the human gut microbiota, as revealed by deep 16S rRNA sequencing. *PLoS Biol.* **6**, e280.
31. Claesson, M. J., O'Sullivan, O., Wang, Q., Nikkila, J., Marchesi, J. R., Smidt, H., et al. (2009) Comparative analysis of pyrosequencing and a phylogenetic microarray for exploring microbial community structures in the human distal intestine. *PLoS One* **4**, e6669.
32. Sogin, M. L., Morrison, H. G., Huber, J. A., Mark Welch, D., Huse, S. M., Neal, P. R., et al. (2006) Microbial diversity in the deep sea and the underexplored "rare biosphere". *Proc. Natl. Acad. Sci. U. S. A.* **103**, 12115–12120.
33. Huber, J. A., Mark Welch, D., Morrison, H. G., Huse, S. M., Neal, P. R., Butterfield, D. A., et al. (2007) Microbial population structures in the deep marine biosphere. *Science* **318**, 97–100.
34. Quince, C., Lanzen, A., Curtis, T. P., Davenport, R. J., Hall, N., Head, I. M., et al. (2009) Accurate determination of microbial diversity from 454 pyrosequencing data. *Nat. Methods* **6**, 639–641.
35. Schloss, P. D., Westcott, S. L., Ryabin, T., Hall, J. R., Hartmann, M., Hollister, E. B., et al. (2009) Introducing mothur: open source, platform-independent, community-supported software for describing and comparing microbial communities. *Appl. Environ. Microbiol.*, doi:AEM.01541–01509.
36. Wang, Q., Garrity, G. M., Tiedje, J. M., and Cole, J. R. (2007) Naive Bayesian classifier for rapid assignment of rRNA sequences into the new bacterial taxonomy. *Appl. Environ. Microbiol.* 73, 5261–5267.
37. Huse, S. M., Dethlefsen, L., Huber, J. A., Mark Welch, D., Relman, D. A., and Sogin, M. L. (2008) Exploring microbial diversity and taxonomy using SSU rRNA hypervariable tag sequencing. *PLoS Genet.* **4**, e1000255.
38. Altschul, S. F., Madden, T. L., Schaffer, A. A., Zhang, J. H., Zhang, Z., Miller, W., et al. (1997) Gapped BLAST and PSI-BLAST: a new generation of protein database search programs. *Nucleic Acids Res.* **25**, 3389–3402.
39. Edgar, R. C. (2004) MUSCLE: multiple sequence alignment with high accuracy and high throughput. *Nucleic Acids Res.* **32**, 1792–1797.
40. Felsenstein, J. (2004) PHYLIP (Phylogeny Inference Package) version 3.6. Department of Genome Sciences, University of Washington, Seattle. *Distributed by the author.*
41. Schloss, P. D. and Handelsman, J. (2005) Introducing DOTUR, a computer program for defining operational taxonomic units and estimating species richness. *Appl. Environ. Microbiol.* 71, 1501–1506.

Chapter 9

Tag-Encoded FLX Amplicon Pyrosequencing for the Elucidation of Microbial and Functional Gene Diversity in Any Environment

Yan Sun, Randall D. Wolcott, and Scot E. Dowd

Abstract

Comprehensive evaluation of microbial diversity in almost any environment is now possible. Questions such as "Does the addition of fiber to the diet of humans change the gastrointestinal microbiota?" can now be answered easily and inexpensively. Tag-encoded FLX-amplicon pyrosequencing (TEFAP) has been utilized to evaluate bacterial, archaeal, fungal, algal, as well as functional genes. Using the new tag-encoded FLX amplicon pyrosequencing (bTEFAP) approach, we have evaluated the microbial diversity using a more cost-effective and largely reproducible method that would allow us to sequence the ribosomal RNA genes of microorganisms (hereafter focused on bacteria), without the need for the inherent bias of culture methods. These developments have ushered in a new age of microbial ecology studies, and we have utilized this technology to evaluate the microbiome in a wide range of systems in almost any conceivable environment.

Key words: Bacterial diversity, FLX, Amplicon, Pyrosequencing, Bar code, 454

1. Introduction

The evaluation of bacterial diversity in the world around us has gone from the biochemical, physiological, and microscopic methods begun in the days of Koch. Culture-based methods were champions in the 1900s and up until the late 1980s, providing us the ability to begin to understand bacteria. Indeed, they are still useful in the advanced characterization of bacteria that are able to be isolated and cultured, yet they have fallen short in their ability to evaluate fastidious and viable nonculturable organisms. In other words, we do not often have the ability to grow in the laboratory most of the bacteria we now know exist in the various

Young Min Kwon and Steven C. Ricke (eds.), *High-Throughput Next Generation Sequencing: Methods and Applications*, Methods in Molecular Biology, vol. 733, DOI 10.1007/978-1-61779-089-8_9, © Springer Science+Business Media, LLC 2011

environments around us. Estimates in most environments as to the percentage of bacteria able to be cultured often range from 1 to 10%. Thus, we cannot easily culture in the laboratory up to 90% of bacteria from most environments. This realization has led to the development and use of molecular methods such as cloning and Sanger sequencing, DGGE, and other molecular methods (1). In early 1997, we began investigating the use of next-generation sequencing as a method to provide deep resolution of the diversity in microbial environments. Our goal was to develop cost-effective and largely reproducible methods that would allow us to sequence the ribosomal RNA genes of bacteria without the need for the inherent bias of culture methods, to reduce the labor and cost involved in the previous gold standard of ribosomal sequencing, cloning, selecting, and sequencing (1). Our early studies showed that pyrosequencing definitely had the potential to perform complex and high-resolution analysis of diverse microbial populations, but the cost was still a limiting factor. Months later, we developed bacterial tag-encoded FLX amplicon pyrosequencing (bTEFAP). This is a method where a unique tag is encoded within the amplicons. This tag is unique for each individual sample and is used to identify any given sequence related to which sample it originated from. As we could label samples, we could then combine samples after labeling and mix them together into a single pyrosequencing run. At the time, the cost of one sample in our proof-of-concept studies was in the range of $5,000; however, because we could put 100s of samples, each individually tagged with a short unique piece of incorporated DNA, we could reduce the per sample cost. Today, years later, we have greatly matured this technology, and it is now commercially available through Research and Testing Laboratory (Lubbock, TX) where the cost per sample now is in the $110–$150 range, which includes up to 3,000 individual sequences for each sample.

These developments have ushered in a new age of microbial ecology studies, and we have utilized this technology to evaluate the microbiome in a wide range of systems including cattle feces, pig cecum, chicken houses, air samples, ant gardens, water samples, soil samples, and chronic wounds (2–10). The method shows incredible promise for the evaluation of microbial diversity in almost any conceivable environment. The following methods represent the in-depth description of bTEFAP data acquisition.

2. Materials

2.1. DNA Extraction

1. TissueLyser and TissueLyser tube rack assemblies (QIAGEN, Valencia, CA).
2. QIAamp DNA mini kit and RLT lysis buffer (QIAGEN).

3. Glass beads of 0.1 mm (Scientific Industries, Bohemia, NY).
4. Steel beads of 5 mm (QIAGEN).

2.2. PCR

1. Nanodrop spectrophotometer (Nyxor Biotech, France).
2. MyCycler ThermoCycler (BIO-RAD, Hercules, CA).
3. HotStarTaq Plus Master Mix Kit (QIAGEN).
4. Lonza FlashGel System (Lonza Group Ltd, Switzerland).

2.3. bTEFAP FLX Pyrosequencing

1. Genome Sequencer FLX System (Roche, Branford, CT).
2. PCR enclosure for PCR reagent preparation (Fisher Scientific, Pittsburgh, PA).
3. Vented hood for second PCR and emulsion breaking.
4. Centrifuge and rotor with swinging buckets, 15- and 50 ml tube adaptors for swinging bucket rotor (Beckman Coulter, Brea, CA).
5. BioAnalyzer and DNA and RNA LabChip (Agilent Technologies, Santa Clara, CA).
6. Particle counter (Beckman Coulter).
7. Full set of micropipettes (2–1,000 μl).
8. Platinum HiFi *Taq* DNA polymerase (Invitrogen, Carlsbad, CA).

3. Methods

3.1. DNA Extraction

Samples were collected and placed in 2 ml screw cap microcentrifuge tubes and were resuspended in 500 μl of TE buffer. A sterile 5-mm steel bead and 500 μl of glass beads were then added for complete bacterial lyses. The tubes were put on a TissueLyser and run at 30 Hz for 5 min. The samples were centrifuged briefly and 100 μl of supernatant was aliquoted to each new microcentrifuge tube after which 350 μl of RLT buffer (with β-ME, 10 μl per 1 ml buffer RLT) and 250 μl of 100% ethanol were added. After pulse vortex, this solution was added to a DNA spin column, and DNA recovery protocol was followed as instructed in the QIAamp DNA Mini Kit menu. The extracted DNA was stored at −20°C.

3.2. Partial Ribosomal DNA Amplification

All DNA samples were adjusted to 100 ng/μl. A 100 ng (1 μl) aliquot of each sampled DNA was used for a 20-μl PCR. The 16S universal eubacterial primers 530F (5′-GTG CCA GCM GCN GCG G) and 1100R (5′-GGG TTN CGN TCG TTG) were used for amplifying the 600 bp region of 16S rRNA genes. HotStarTaq Plus Master Mix Kit was used for PCR under the following conditions: 94°C for 3 min followed by 30 cycles of 94°C for 30 s;

60°C for 40 s and 72°C for 1 min; and a final elongation step at 72°C for 5 min (see Note 1). A secondary PCR was performed for FLX amplicon sequencing under the same condition by using designed special fusion primers with different tag sequences, namely, LinkerA-Tags-530F and LinkerB-1100R (Table 1). The use of a secondary PCR prevents amplification of any potential bias that might be caused by inclusion of tag and linkers during initial template amplification reactions (see Note 2).

3.3. The Emulsion-Based Clonal Amplication (emPCR)

3.3.1. Small Fragment Removal and Quantification of Pooled PCR Products

1. The second PCR products were checked by using a Lonza FlashGel System and then pooled into a microcentrifuge tube.
2. Add 35 µl of the AMPure beads to a new 1.5-ml tube and then add 50 µl of the pooled DNA into the 1.5-ml tube with AMPure beads. Incubate for 5 min at room temperature.
3. Place the tubes in a magnetic particle collector (MPC) to pellet the beads against the wall of the tube (about 1 min).
4. Remove the supernatant and wash the beads twice with 500 µl 70% Ethanol. Remove all the supernatant.
5. Place the tubes in a 37°C incubator for 5 min with the lids open to dry the beads.
6. Add 50 µl of buffer AE into the tube. Tap the tube gently to mix and put the tube back in an MPC for 1 min.

Table 1
Primer sequences utilized for secondary PCR (The letters shaded *gray* are bar codes)

Name	Primer sequence (5′–3′)
530-454-F1	GCCTCCCTCGCGCCATCAGCGCCCACATAGTGCCAGCMGCNGCGG
530-454-F2	GCCTCCCTCGCGCCATCAGCCTTTGCCGAGTGCCAGCMGCNGCGG
530F-454-F3	GCCTCCCTCGCGCCATCAGCGAACGAGTTGTGCCAGCMGCNGCGG
530F-454-F4	GCCTCCCTCGCGCCATCAGCGAGGCAGATGTGCCAGCMGCNGCGG
530F-454-F5	GCCTCCCTCGCGCCATCAGCGAGGTACACGTGCCAGCMGCNGCGG
530F-454-F6	GCCTCCCTCGCGCCATCAGCGAGGTTCTAGTGCCAGCMGCNGCGG
530F-454-F7	GCCTCCCTCGCGCCATCAGCGAGTACGCTGTGCCAGCMGCNGCGG
530F-454-F8	GCCTCCCTCGCGCCATCAGCGAGTATGATGTGCCAGCMGCNGCGG
530F-454-F9	GCCTCCCTCGCGCCATCAGCGAGTGAATGGTGCCAGCMGCNGCGG
530F-454-F10	GCCTCCCTCGCGCCATCAGCGAGTGGATAGTGCCAGCMGCNGCGG
LinkerB-1100R	GCCTTGCCAGCCCGCTCAGGGGTTNCGNTCGTTR

7. Remove 40–45 μl of the eluted bead-purified pool DNA to a new 1.5-ml tube.
8. Run 1 μl of the bead-purified pool DNA on a BioAnalyzer DNA 1000 LabChip to assess the size and quantity of the DNA sample.
9. Given the library concentration (in ng/μl) determined, calculate the equivalence in molecules/μl, using the following equation:

$$\text{Molecules} / \mu\text{l} = \frac{(\text{Sample conc; ng} / \mu\text{l}) \times (6.022 \times 10^{23})}{(328.3 \times 10^{9}) \times (\text{avg.fragment length; nt})},$$

where 6.022×10^{23} is Avogadro's number (molecules/mole), and 328.3 is the average molecular weight of nucleotides, in g/mol.

3.3.2. Preparation of the Live Amplification Mix

Allow the frozen kit components of two GS emPCR Kits II to fully thaw (with the exception of the enzyme tubes which should be left at −15 to −25°C). After thawing, vortex the reagents for 5 s. Prepare the Live Amplification Mix based on the number of emulsion reactions you are making. Store it at +2 to +8°C.

3.3.3. Washing the DNA Capture Beads

1. Transfer 19.2 million DNA Capture Beads from the stock tube to a 2-ml tube (for 32 reactions; the stock is 10,000 beads/μl, so use 1.920 ml). These beads will immobilize the amplified DNA.
2. Pellet the beads in a benchtop minifuger and remove the supernatant.
3. Wash the beads twice with 1 ml of 1× Capture Bead Wash buffer and vortex for 5 s to resuspend the beads.
4. Remove most of the supernatant without disturbing the bead pellet. Discard the supernatant.

3.3.4. Addition and Annealing of the DNA Fragments to the Capture Beads

1. Obtain an aliquot of the quantitated DNA to be amplified.
2. To the tube of washed Capture Beads, add the correct amount of the DNA to provide optimal amplification. Vortex for 10 s to make sure that the library is evenly distributed throughout the beads.
3. Determine the exact, total volume you have by measuring with a pipette. This should be around 40 μl per emulsion. To make sure the beads remain resuspended while you are aliquoting, vortex or pipette the beads up and down several times during the process.

3.3.5. Emulsification

1. In the Emulsion enclosure, vortex 32 tubes of the Emulsion Oil for 10 s. Add 240 μl of Mock Amplification Mix to each tube of Emulsion Oil.

2. Raise the TissueLyser safety shield and remove the tube rack assemblies by unscrewing the clamp holding them in place. (There are two tube racks, each with a capacity of eight tubes/outermost row). Remove the lid of the TissueLyser tube racks and place the tubes containing the Mock Amplification Mix and Emulsion Oil in the outermost rows in the racks. Run the TissueLyser for 25/s for 5 min.
3. While the TissueLyser is running, add 160 μl of the Live Amplification Mix to each tube of annealed DNA beads.
4. When the TissueLyser stops shaking, remove the TissueLyser tube racks as above. Pipette the bead mixtures up and down three times to ensure proper mixing, and add each to one of the emulsion tubes. Place the tubes back into the TissueLyser tube racks (outermost row only), and insert the racks again securely into the TissueLyser.
5. Set the TissueLyser to 15/s for 5 min and start shaking. This will create an emulsion with aqueous-phase micelles of the appropriate size to contain single beads with amplification mix.

3.3.6. Amplification

1. When the emulsification is complete, remove the tube racks from the TissueLyser, and open the emulsion tubes. Dispense the emulsions into 96-well plates, 200 μl per well.
2. Check for the presence of air bubbles at the bottom of each well, as bubbles can cause breakage of the emulsion. If there are air bubbles, tilt the plate to dislodge them.
3. Cover the plate with 8-strip snap cap lids. In either case, make sure that all the wells are properly sealed.
4. Place all the plates containing the emulsified amplification reactions in thermocyclers. Set up and launch the amplification program.

3.3.7. Bead Recovery

1. Prepare a set of six to eight 10-ml syringes by screwing a 16 gauge blunt, flat tip needle directly onto the end of each syringe. Assemble Swin-Lok filter holders, each with a nylon filter.
2. Draw about 5 ml of emulsion from the 96-well plates into each syringe, but try not to draw in excess air. Draw isopropanol into each syringe up to the 10ml mark and set the syringes aside.
3. Add 100 μl of isopropanol into each well and pipette up and down a few times. Draw the emulsion–isopropanol mix from each well into the syringes. Add 100 μl of isopropyl alcohol to the first column of eight wells, wash the wells thoroughly, collect the remaining beads in the same set of tips, and add this into the second column of wells. Perform similar washes until all the beads from eight columns are collected together. Collect this material in syringes.

4. Remove the blunt needle from the syringe, attach one of the assembled Swin-Lok filter units to the syringe, and attach the blunt needle to the Swin-Lok filter unit. Mix the isopropanol with the emulsified beads by vigorous shaking to obtain a homogenous mix. Gently squirt out the contents of the syringe through the Swin-Lok filter unit, into a waste jar containing bleach. The DNA beads will be retained by the filter, while the emulsion oil is washed away with the isopropanol.
5. Expel the contents into waste and remove the Swin-Lok filter unit from the syringe. Repeat isopropanol wash three times and then wash the beads using bead wash buffer three times. Finally, wash the beads with Enhancing Fluid three times.
6. Disassemble the Swin-Lok filter units and place the filters and the "upstream" component from both Swin-Lok assemblies into a 50-ml tube containing 35 ml of 1× Enhancing Fluid.
7. Place the cover on the 50-ml tube and shake the tube vigorously five to ten times to dislodge the beads from the filter and Swin-Lok unit components. Using plastic tweezers, remove the filters and the plastic Swin-Lok assembly components from the tube, and cap the tube.
8. Pellet the beads in a centrifuge at 3,000 rpm (1847 × *g*) for 5 min. Carefully remove the supernatant until the remaining volume left in the tube (combined bead pellet and liquid) is about 5 ml; make sure to not remove any of your pelleted beads. Resuspend the bead in the remaining supernatant. Transfer 1 ml of the beads suspension into a 1.5-ml microcentrifuge tube, using a micropipettor. Retain the pipet tip for further transfers, to minimize bead loss. Pellet the beads by centrifugation in a minifuger as before (spin 10 s, rotate the tube 180°, and spin again 10 s). Repeat until all the bead suspension from the 50-ml tube has been transferred and sequentially pelleted in the 1.5-ml tube.
9. Add 1 ml of 1× Enhancing Fluid into the 50-ml tube, gently swirl to collect any residual beads, and add this bead slurry to the bead pellet in the 1.5-ml tube. Pellet the beads by centrifugation in a minifuge as before. Remove the supernatant without disturbing the beads pellet. Resuspend the beads in each tube in 500 μl of 1× Enhancing Fluid.

3.3.8. DNA Library Bead Enrichment

1. Vortex the tube of Enrichment Beads for 1 min to resuspend its contents completely.
2. Place 320 μl of Enrichment Beads in each of two 1.5-ml microcentrifuge tubes and add 500 μl of 1× Enhancing Fluid to each tube.
3. Vortex the diluted Enrichment Beads for 5 s.

4. Using a Magnetic Particle Collector (MPC), pellet the paramagnetic Enrichment Beads against the side of the tube.
5. Remove and discard the supernatant from each tube, taking care not to draw off any Enrichment Beads.
6. Remove the tubes from the MPC and add 300 μl of 1× Enhancing Fluid into each tube.
7. Vortex for 3 s to resuspend the beads.
8. Add one of the tubes of the washed Enrichment Beads (from Subheading 3.3.8.7) into each of the tubes of amplified DNA beads (from Subheading 3.3.7.9). Each of the two beads tubes should now contain about 800 μl. Gently pipette up and down three times to mix the DNA and Enrichment Beads. Rotate on a LabQuake tube roller at ambient temperature for 10 min.
9. Place the two bead tubes in the MPC and wait for 2 min to pellet the paramagnetic Enrichment Beads against the side of the microcentrifuge tubes. Invert the MPC several times to collect any beads that may be lodged in the caps. Carefully remove the supernatant from each of the tubes, taking care not to draw off any Enrichment Beads. Remove the tube from the MPC and gently add 1 ml of 1× Enhancing Fluid to the beads.
10. Repeat step 9 three times (*do not* resuspend in 1× Enhancing Fluid after the fourth wash).
11. Remove the tubes from the MPC and resuspend each bead pellet in 700 μl of Melt Solution. Vortex for 5 s, and put the tubes back into the MPC to pellet the Enrichment Beads. Transfer the SUPERNATANTS, containing enriched sstDNA beads, to two separate 1.5-ml microcentrifuge tubes.
12. Repeat step 11 for better sstDNA bead recovery, pooling the two melts (total 1,400 μl in each tube). Pellet the enriched sstDNA beads by centrifugation as before (spin 10 s, rotate the tube 180°, and spin again 10 s). Remove and discard the supernatants, and wash three times with 1 ml of 1× Annealing Buffer. Remove all the supernatant without disturbing the pellet. Resuspend the pellet with 100 μl of 1× Annealing Buffer.
13. Vortex and transfer the enriched sstDNA bead suspension from each 1.5-ml tube, equally into two 0.2-ml tubes. This will make a total of four 0.2-ml tubes for all the enriched beads. Rinse the 1.5-ml tubes each of which contained the collected enriched sstDNA beads with 100 μl of 1× Annealing Buffer, and add 50 μl of this rinse to each of the four 0.2-ml tubes.

3.3.9. Sequencing Primer Annealing

1. Pellet the enriched DNA beads as before. Remove the supernatants. Add 24 μl of Sequencing Primer to each of the tubes. Vortex to mix.
2. Place the tubes into the thermocycler and run the sequencing primer annealing program. When the sequencing primer annealing program is finished, remove the four tubes from the thermocycler. Pool the beads from the four 0.2-ml tubes into a single 1.5-ml microcentrifuge tube. Wash the 0.2-ml tube with 100 μl of 1× Annealing Buffer and add the rinse to the 1.5-ml tube. Spin down the tubes to collect the beads.
3. Wash the pellet with 500 μl of 1× Annealing Buffer. Resuspend the pellet with 200 μl of 1× Annealing Buffer. The beads can be counted in a Coulter Counter, and then stored at 2–8°C for at least 1 month before use.

3.3.10. Count the Beads Using a Coulter Counter

1. Add some Isoton II into a clean Accuvette cup and flush the Coulter Counter.
2. Add 10 ml of Isoton II into a new Accuvette cup. Add 3 μl of the samples after tapping the tube. Swirl the container to mix and make sure that no bubbles are gone.
3. Click Setup and push Start. Record all readings.
4. Flush the Coulter Counter after using it.

3.4. GS FLX Sequencing

3.4.1 Prewash Run

1. Close the previous run and relog in. Open the exterior fluidics door and raise the sipper manifold completely. Slide out the reagent cassette. Remove the reagent bottles and tray and pour fluids down the drain. Tip the reagent cassette into the sink to drain out the waste.
2. Remove the sipper tubes from the sipper manifold. Change gloves and place new sipper tubes on the sipper manifold by turning them clockwise, the four long tubes go to the left and the 11 short tubes are located on the right.
3. Place the prewash tube holder on top of the reagent cassette. Place the prewash tubes in the holder. Fill the tubes to the top with prewash buffer. Slide the prewash cassette into the fluidics area, lower the sipper tubes carefully, checking to make sure that each tube falls into prewash tube, and close the exterior fluidics door.
4. Launch the prewash run. Double-click the instrument icon. Click Start, and then Prewash Run.

3.4.2. Pico Titer Plate Device Preparation

1. Preparation of the bead buffer. Prepare BB2 (Bead Buffer 2) by adding 34 μl of Apyrase solution to the 200 ml of Bead Buffer, swirl to mix, and keep on ice. Prepare two tubes for large volume or one tube for small volume of BB3 or Bead Buffer 3 by mixing 930 μl of BB2 with 20 μl of Bead Buffer Additive. Mix and keep on ice.

2. Preparation of the PTP and Bead Deposition Device. Place PTP on top of the notches in the tray. Submerge PTP with BB2. Let the PTP soak for at least 10 min.
3. Cleaning of the Bead Deposition Device and Bead Loading Gasket. Assembly of the Bead Deposition Device with the PTP and Gasket. Remove the PTP from the tray and wipe the backside of the PTP with a KimWipe. Place the PTP on the Bead Deposition Device and make sure the notch lines up with the notch on the PTP. Place the appropriate gasket on top of the PTP. Secure the Bead Deposition Device by pulling the two latches up at the same time.
4. Wetting of the PTP Device. Fill each loading region with the appropriate volume of BB2. Place the Bead Deposition Device and the counterweight into the centrifuge. Centrifuge the Bead Deposition Device for 10 min at 2,850 rpm (1890 × *g*). Leave the BB2 on the PTP until ready to proceed.

3.4.3. Preparation of the Beads

1. Prepare the DNA beads

   Add the appropriate number of DNA beads and control beads to a clean tube. Centrifuge for 1 min at 10,000 rpm (9600 × *g*) and rotate 180° and repeat. Leave only the appropriate amount of buffer left with the beads. Prepare the DBIM (DNA Bead Incubation Mix). Transfer the correct amount of DBIM to the region's beads. Place the tubes on the rotator for 30 min.

2. Prepare the packing beads

   Wash the packing beads three times in BB2. Add the original amount of BB2 back to the beads after the third wash. Resuspend and keep on ice. Transfer the appropriate volume of Packing beads and DBIM to a 2-ml tube. Place the packing beads on the rotator and incubate them at room temperature.

3. Prepare the enzyme beads

   Vortex to resuspend the enzyme beads and place on MPC (Magnetic Particle Collector). Wash the beads three times with BB2 using the MPC. Add the appropriate volume of BB2 to resuspend the beads. In each of the two for large volume or one for small volume 2-ml tubes, combine the amount of BB2 and enzyme beads. Vortex and place on ice.

3.4.4. Deposition of Beads onto PTP

1. Deposition of the first bead layer (DNA beads)

   Preparation of the first layer beads. Remove the tubes with the DNA beads from the rotator when 30 min has finished. Add the appropriate volume of BB2 and keep on ice. Deposition of the First Layer. Remove as much of the BB2 in the Bead Deposition Device with a pipette. Vortex to resuspend the first layer of beads. Draw the appropriate amount of first layer beads and load them into the Bead Deposition Device with one smooth movement. Repeat for all regions on the PTP. Leave the PTP on the bench top for at least 10 min; do not centrifuge.

2. Deposition of the second bead layer (packing beads)

   Dilution of the packing beads for the second bead layer. Remove the tubes from the lab rotator. Gently remove the beads from each region on the PTP and place the liquid in new tubes. Collect the supernatant by centrifuging the tubes. Carefully remove the supernatant from each region and add the appropriate amount to the packing beads. Deposition of the second layer. Vortex the packing beads. Draw in the appropriate amount of packing beads for the regions used. Pipette up and down three times to resuspend beads and load them into the regions on the PTP. Centrifuge the PTP in the Bead Deposition Device for 10 min at 2,860 rpm (1678 × *g*).

3. Deposition of the third bead layer (enzyme beads)

   Remove gently the supernatant of the second layer and discard. Vortex the enzyme beads. Add the appropriate amount of the enzyme beads depending on the number of regions used. Pipette up and down three times to mix and load the regions of the PTP. Centrifuge for 10 min at 2,860 rpm (1678 × *g*).

3.4.5. The Sequencing Run

1. Load the sequencing reagents into the instrument

   Remove the prewash cassette and clean the fluidics area deck. Click OK when the prewash run is complete. Open the exterior fluidics door and raise the sipper manifold. Slide out the prewash cassette. Remove the prewash tubes and tube holder. Empty the waste and wipe the reagent cassette with a paper towel. Prepare and load the sequencing reagents cassette. Thaw the reagents from the Kit's Sequencing enzyme tray and keep them on ice. Add the proper amount of 1 M DTT to each bottle of Buffer CB and place the bottles of Buffer CB in the reagent cassette. Place the sequencing reagents tray in the reagent cassette when thawed. Add the reagent supplements to tube 11 by adding the appropriate amount of Apyrase to the tube in position 11. Add the dATP reagent to tube 10 by adding the appropriate amount of dATP to the tube in position 10. Secure the lids on all reagent tubes and invert the whole tray several times. Remove the lids from the reagent tray and the four bottles. Load the reagent cassette into the instrument. Lower sipper tubes and close the exterior fluidics door.

2. Clean the PTP cartridge and the camera faceplate.

   Unlock the camera door by selecting "Unlock Camera Door." Remove the PTP by pressing on the PTP frame spring latch. Close the frame and remove the PTP cartridge seal. Wet a Kim Wipe with 50% Ethanol and wipe the surface of the cartridge.

Allow to dry. Gently wipe the camera faceplate with a Zeiss moistened cleaning tissue. Wipe the surface of the PTP cartridge with a KimWipe moistened with 10% Tween-20.

3. Load and set the run script and other run parameters

   Click File, sequencing run. In the sequencing run field, do the following: Under the Scripts field, select the appropriate kit and check whether short reads or long reads want to be generated. Enter the run name in the Run Name Field. Enter the bar code of the plate in the PTP bar code field. Enter the user and user group in the user and user group fields. Select backup under backup class. Click Next and choose the PTP layout currently being used. Click Next. A run name confirmation will appear and if correct click yes. If incorrect, click back to change the information that is incorrect.

4. Insert the PTP Device and launch the sequencing run

   Close the cartridge frame if needed. Install the cartridge seal by placing the ridges face up in the cartridge ridge. If necessary, press the seal gently with a gloved hand. Press the PTP spring latch to lift the PTP frame. After centrifugation of the third layer, gently draw out and discard the supernatant. Remove the PTP from the Bead Deposition Device by: Rotate the latches down on the Bead Deposition Device. Carefully remove the top of the device. Lift off the gasket gently. Remove the PTP but be careful to only handle it by the edges. Slide the PTP into the frame making sure to align the notch properly. Close the PTP frame. Wipe the backside of the PTP with a KimWipe. Close the camera door. Click the finish button on the requirements window. This will start the sequencing run. When the sequencing run is complete, a window will alert the user.

4. Notes

1. Contamination of bacteria 16S is a major issue. Using two different rooms for the primary and secondary PCR will help to reduce the cross-contamination between an amplified PCR product and DNA that still needs to be run. In addition, all areas in those rooms must be decontaminated before any work is completed to reduce the amount of bacteria that is present in the rooms. Any bacteria that might be present in the air can cause contamination.
2. If the PCR product has multiple bands present, then an agarose gel needs to be run. Once the gel separates the bands, the specific target band can be cut out from the gel. The precise band can be purified, and the PCR product can be used in the downstream application.

References

1. Dowd, S.E., Sun,Y., Secor, P.R., Rhoads, D.D., Wolcott, B.M., et al. (2008b) Survey of bacterial diversity in chronic wounds using Pyrosequencing, DGGE, and full ribosome shotgun sequencing. *BMC Microbiol.* **8**, 43.
2. Bailey, M.T., Walton, J.C., Dowd, S.E., Weil, Z.M., and Nelson, R.J. (2009) Photoperiod Modulates Gut Bacteria Composition in Male Siberian Hamsters (*Phodopus sungorus*). *Brain Behav. Immun.* **24**, 577–584.
3. Callaway, T.R., Dowd, S.E., Wolcott, R.D., Sun, Y., McReynolds, J.L., et al. (2009) Evaluation of the bacterial diversity in cecal contents of laying hens fed various molting diets by using bacterial tag-encoded FLX amplicon pyrosequencing. *Poult. Sci.* **88**, 298–302.
4. Dowd, S.E., Wolcott, R.D., Sun, Y., McKeehan, T., Smirh, E., et al. (2008d) Polymicrobial nature of chronic diabetic foot ulcer biofilm infections determined using bacterial tag encoded FLX amplicon pyrosequencing (bTEFAP). *PLoS One* **3**, e3326.
5. Dowd, S.E., Sun, Y., Wolcott, R.D., Domingo, A., and Carroll, J.A. (2008c) Bacterial tag-encoded FLX amplicon pyrosequencing (bTEFAP) for microbiome studies: bacterial diversity in the ileum of newly weaned Salmonella-infected pigs. *Foodborne Pathog. Dis.* **5**, 459–472.
6. Dowd, S.E., Callaway, T.R., Wolcott, R.D., Sun, Y., McKeenhan, T., et al. (2008a) Evaluation of the bacterial diversity in the feces of cattle using 16S rDNA bacterial tag-encoded FLX amplicon pyrosequencing (bTEFAP). *BMC Microbiol.* **8**, 125.
7. Sen, R., Ishak, H.D., Estrada, D., Dowd, S.E., Hong, E., et al. (2009) Generalized antifungal activity and 454-screening of Pseudonocardia and Amycolatopsis bacteria in nests of fungus-growing ants. *Proc. Natl. Acad. Sci. U.S.A* **106**, 17805–17810.
8. Suchodolski, J.S., Dowd, S.E., Westermarck, E., Steiner, J.M., Wolcott, R.D., et al. (2009) The effect of the macrolide antibiotic tylosin on microbial diversity in the canine small intestine as demonstrated by massive parallel 16S rRNA gene sequencing. *BMC Microbiol.* **9**, 210.
9. Wolcott, R.D., Gontcharova, V., Sun, Y., and Dowd, S.E. (2009a) Evaluation of the bacterial diversity among and within individual venous leg ulcers using bacterial tag-encoded FLX and titanium amplicon pyrosequencing and metagenomic approaches. *BMC Microbiol.* **9**, 226.
10. Wolcott, R.D., Gontcharova, V., Sun, Y., Zischakau, A., and Dowd, S.E. (2009b) Bacterial diversity in surgical site infections: not just aerobic cocci any more. *J. Wound Care* **18**, 317–323.

Chapter 10

Pyrosequencing of Chaperonin-60 (*cpn60*) Amplicons as a Means of Determining Microbial Community Composition

John Schellenberg, Matthew G. Links, Janet E. Hill, Sean M. Hemmingsen, Geoffrey A. Peters, and Tim J. Dumonceaux

Abstract

The chaperonin-60 universal target (*cpn60* UT) is generated from a set of PCR primers and provides a universally conserved, phylogenetically informative sequence signature for determining the composition of microbial communities by DNA sequencing. Pyrosequencing of *cpn60* UT amplicons is emerging as a next-generation tool for providing unprecedented sequencing depth and resolution of microbial communities in individual samples. Owing to the increase in sequencing depth, the dynamic range across which the presence and abundance of individual species can be sampled experimentally also increases, significantly improving our ability to investigate microbial community richness and diversity. The flexible format of the pyrosequencing reaction setup combined with the ability to pool samples through the use of multiplexing IDs makes the generation of microbial profiles based on the *cpn60* UT both feasible and cost-effective. We describe here the methods we have developed for determining microbial community profiles by pyrosequencing of *cpn60* UT amplicons, from generating amplicons to sequencing and data analysis.

Key words: Pyrosequencing, Chaperonin-60, *cpn60*, Microbial communities, Metagenomic profiling

1. Introduction

Microbial communities play an important and often underappreciated role in a diverse array of phenomena, including human and animal health, biomolecule conversion, and pollutant degradation. The taxonomic composition of microbial communities and how it changes with respect to various environmental perturbations is currently a rapidly expanding field of study. Traditionally, the identification of microbes present in complex communities has required identification of cultured organisms to assess species

Young Min Kwon and Steven C. Ricke (eds.), *High-Throughput Next Generation Sequencing: Methods and Applications*, Methods in Molecular Biology, vol. 733, DOI 10.1007/978-1-61779-089-8_10, © Springer Science+Business Media, LLC 2011

richness. However, culture-based methods of isolating organisms greatly limit the taxonomic breadth of organisms that can be identified. Culture-independent tools to examine microbial community richness and diversity are therefore critical. In order to assess species diversity for microbial communities in a culture-independent fashion common sequenced-based targets are exploited to provide a molecular signature for a given taxonomic group (e.g., genus or species). To date, the most widely used sequence targets are those based on ribosomal RNA-encoding genes (1, 2). In order to address some of the recognized limitations of this target, we have developed a set of tools using the chaperonin-60 (*cpn60*) gene (3) for microbial identification.

Type I chaperonins comprise a family of proteins found in prokaryotes and in the mitochondria and plastids of eukaryotes. Cpn60 (also known as GroEL or Hsp60) plays an important role in the folding of cellular proteins and is present in nearly all prokaryotic and eukaryotic cells (3). Degenerate universal PCR primers that amplify a fragment of *cpn60* from virtually any organism of interest, the *cpn60* universal target (*cpn60* UT), have been developed (4) and it was quickly appreciated that the sequence of the *cpn60* UT is highly discriminatory among bacterial isolates (5, 6). Moreover, the interspecies variability appears to be largely uniform across the length of the 549–567 bp *cpn60* UT (Links and Hill unpublished). To aid in the assignment of microbial identities based on the *cpn60* UT, the chaperonin-60 database (cpnDB) was developed and is now available as a broad, Web-based resource for microbial ecology (http://cpndb.cbr.nrc.ca/) (7).

Determining the composition of microbial communities using this tool begins with the generation of *cpn60* UT amplicons from a DNA extract derived from a microbial community. In one approach, a clone library is generated from *cpn60* UT amplification products. Sequencing individual clones from such a library results in a phylogenetic profile of the microbial community by the determination of the "nearest neighbor" of each experimentally derived clone to reference sequences contained in the cpnDB (8–11). While this traditional sequencing approach provides a wealth of information, a repository of cloned *cpn60* UT fragments, and a microbial profile for a given sample or group of samples, the sequencing depth that is achievable is resource-limited due to the manual nature of the procedure. Thus, rarer organisms from the community or those which cause reduced efficiency in cloning will be underrepresented or missing from the microbial profile.

Recently, pyrosequencing of *cpn60* amplicons generated from microbial communities has emerged as a suitable "Next-Generation" technology for deriving profiles of microbial communities (12). Pyrosequencing avoids the need for cloning the PCR product generated from the microbial community, thus

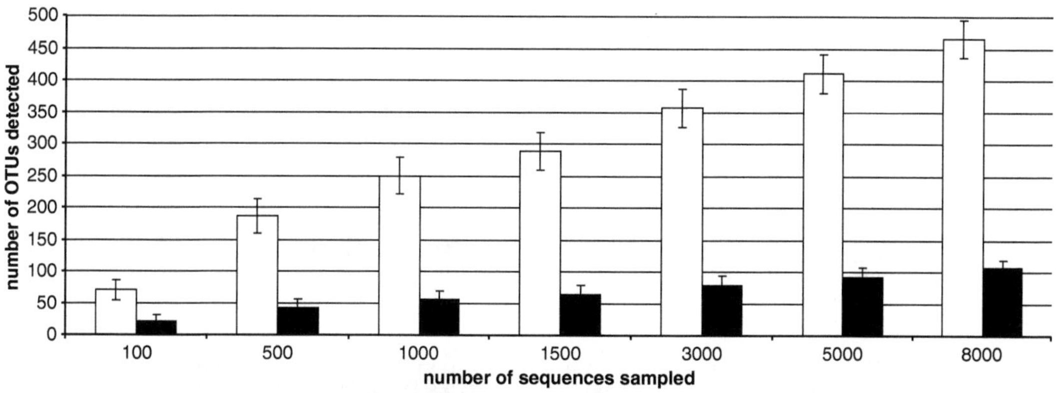

Fig. 1. Comparison of the number of operational taxonomic units (OTUs) generated by *cpn60* UT (*white bars*) primers and 16S rRNA-encoding gene- (*black bars*) targeted primers for those sequences identified as *Lactobacillus* spp. from vaginal microbial communities. Identical samples were used as template for *cpn60* UT primers or for 16S-targeted primers (16) and amplicons were analyzed by pyrosequencing as described in the text. Sequences were compared to one another within their respective datasets using the nearest-neighbor algorithm of DOTUR (17) and the number of OTUs was determined at various sampling depths for each dataset. A difference cutoff of 3% was used (i.e., to be considered to be derived from the same OTU, two sequences had to be at least 97% identical to one another). Note that the very large sample sizes of the matched datasets necessitated the analysis of only those reads that mapped to a single genus at a time. This demonstrates the ability of *cpn60* to provide more resolution of *Lactobacillus* spp. as compared to 16S rRNA for these samples. The data presented here derived from the subset of data identified as *Lactobacillus* spp.; these were found to be represented with nearly identical frequencies in the *cpn60* and 16S datasets (12).

eliminating one possible level of bias in the community composition that is determined. Although the length of pyrosequencing reads do not span the entire *cpn60* UT (GS-FLX Roche), we have found that reads >150 nucleotides are sufficient to determine the identity of the organism from which it was derived (12). Further, Roche/454's titanium format routinely delivers >400 bp reads. With pyrosequencing, the achievable depth of sequencing from a given sample increases significantly compared to large Sanger sequencing projects (12). Comparing matched samples amplified using 16S-targeted universal primers and *cpn60* UT primers found that the *cpn60* UT provides far greater taxonomic resolution (Fig. 1 and (12)), resulting in a deep, precise, and reproducible microbial profile from a given community (12). Finally, the use of multiplexing IDs (MIDs) enables the parallel sequencing of multiple libraries within a single physical region of a pyrosequencing picotiter plate (PTP), greatly reducing the per-sample cost and offering a means to generate and compare deep microbial profiles for large numbers of samples in a single sequencing reaction. Alternatively, when multiple samples are prepared from similar microbial communities (differing, for example, by individual or across multiple time points), data can be pooled to determine the total complement of organisms that can be present in a given microbial community under the conditions analyzed (Fig. 2).

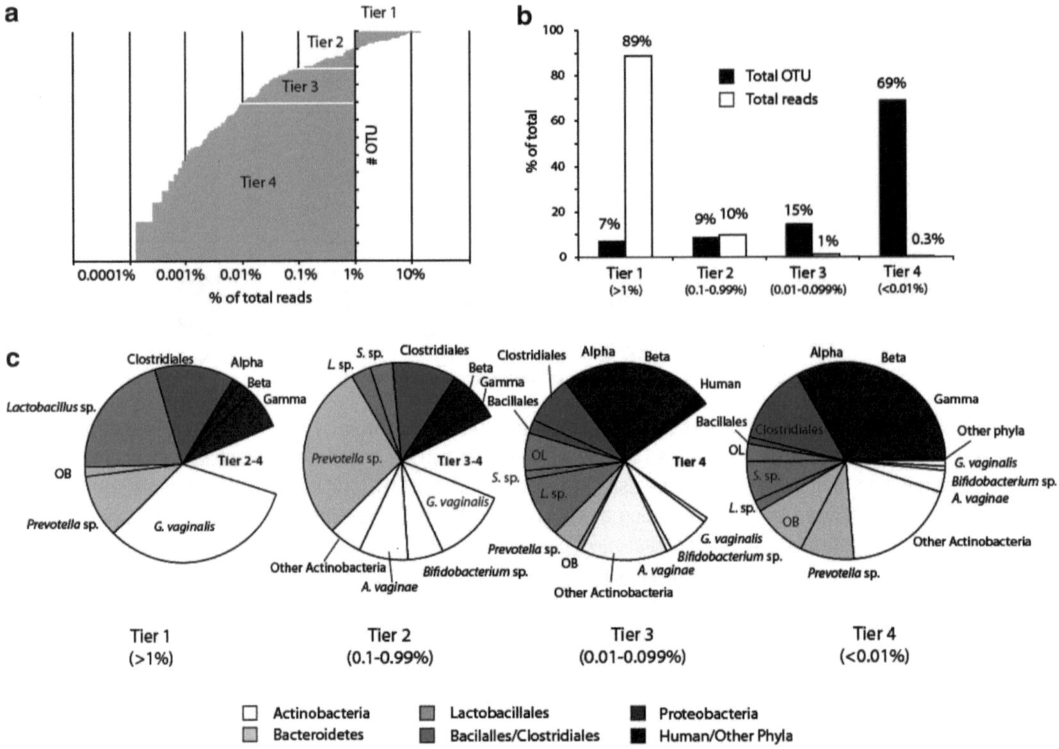

Fig. 2. Pooling data from 60 individual vaginal microbial communities reveals the "universe of possibilities," or all organisms likely to be present in these communities. This analysis also reveals an inverse relationship between the abundance and the diversity of operational taxonomic units (OTU). (**a**) All OTU (291 taxa derived from 798,756 reads from 60 vaginal samples) sorted by abundance (% of total reads). Tier 1 is defined as all taxa with ≥1% of total reads, Tier 2 as taxa with between 0.01 and 0.99% of all reads, Tier 3 as taxa with 0.01–0.099% of total reads, and Tier 4 as taxa with ≤0.01%. (**b**) Proportion of total reads and proportion of total OTU for each tier of abundance, showing increased taxonomic diversity in lower tiers of abundance. (**c**) Tiers of abundance by OTU and phylogenetic group. Slices are coded by phylogenetic group (L. sp.: *Lactobacillus* spp.; S. spp.: *Streptococcus* spp.; OL: other Lactobacillales; OB: other Bacillales/Clostridiales). This demonstrates that each group is well-represented at all tiers of abundance.

The purpose of this chapter is to describe the methods by which microbial profiles are generated by pyrosequencing of *cpn60* UT amplicons from a microbial community of interest. While the methods are based on our recently published description using Roche's GS-FLX pyrosequencing chemistry (12), the method has already been applied to samples sequenced using Roche/454's titanium chemistry and has achieved average read lengths of 400 bps (Hemmingsen, unpublished).

2. Materials

2.1. cpn60 Amplicon Generation

1. Programmable thermocycler with a 96-well heating block capable of achieving annealing temperature gradients (e.g., Bio-Rad C1000).

2. PCR primers recognizing the *cpn60* UT, 100, 25, or 10 μM working stocks. These primers can be modified at the 5′ end to contain emulsion PCR (emPCR)/sequencing primer sequences and/or unique multiplexing IDs (MIDs) (Table 1). See Note 1 for information regarding the choice of modified or unmodified primers and the use of MIDs.
3. Thermostable DNA polymerase for PCR. The polymerase is chosen based on personal preference as well as empirical experience. A "hot start" polymerase is essential. We have used both AmpliTaq Gold (Applied Biosystems) and Platinum Taq (Invitrogen) with success.
4. 96-Well PCR plates or strip tubes compatible with the thermocycler.
5. Removable or puncturable sealing tape or foil for 96-well PCR plates.
6. Two workstations reserved for PCR work fitted with ultraviolet bulbs (e.g., PCR Cleanspot, Coy Laboratory Products). See Note 2 for cautionary steps to be taken to avoid contamination.
7. Amicon Ultra 0.5-mL microconcentrators with 30 kDa molecular weight cutoff (Millipore).
8. Rotary vacuum concentrator (optional).
9. AMPure resin (Beckman Coulter) (optional).
10. Magnetic separator that accommodates 1.5- and 2-ml Eppendorf tubes (optional).

Table 1
Sequences of the *cpn60* amplification primers. Note that there are two upstream and two downstream primers that are mixed in a 3:1 molar ratio to maximize community representation (15)

Primer name	Primer sequence (5′-3′)[a, b]
H279	(emPCR primer A)(MID tag)- GAIIIIGCIGGIGAYGGIACIACIAC
H280	(emPCR primer B)(MID tag)- YKIYKITCICCRAAICCIGGIGCYTT
H1612	(emPCR primer A)(MID tag)- GAIIIIGCIGGYGACGGYACSACSAC
H1613	(emPCR primer B)(MID tag)- CGRCGRTCRCCGAAGCCSGGIGCCTT

[a]Note that the 5′ ends of the amplification primers may be modified to facilitate MID tagging and emPCR/pyrosequencing, as described in Note 1. Typically, both of the sequences indicated in parentheses or only the MID tags are added to the 5′ ends of the primers. For reactions using the titanium format, we recommend adding only the sequences of the MID tags (available from Roche) to the 5′ ends of the amplification primers
[b]I = inosine; Y = C or T; R = A or G; K = T or G; S = C or G

11. Gel purification system of choice.
12. Quant-iT DNA quantification kit (Invitrogen) or analogous fluorescence-based DNA quantification system and fluorometer (e.g., Qubit fluorometer, Invitrogen).

2.2. Pyrosequencing Reaction Setup

This method was originally optimized for the GS FLX sequencing platform (454 Life Sciences/Roche) but is amenable to adaptation to the Titanium sequencing chemistry.

1. GS FLX sequencer (Roche) (other next-generation sequencing platforms may be sufficient but have not been tested).
2. PicoGreen dye kit or other suitable method for accurately quantitating double stranded DNA, along with a fluorometer capable of reading PicoGreen dye (e.g., Beckman Coulter DTX 880 Multimode Detector).
3. GS emPCR kit I, II, or III for the GS FLX, or other suitable amplicon emPCR kit for the Titanium chemistry, with all appropriate accessories, such as thermocycler, emulsion oil, TissueLyser (Qiagen), 96-well PCR plates, heat sealing film for 96-well PCR plates.
4. GS emulsion bead recovery reagent kit; with required accessories such as syringes, vacuum pump, isopropanol, centrifuges, Eppendorf tubes, etc.
5. Coulter counter model Z1 (Beckman Coulter) or other appropriate bead counting device, with required accessories.
6. GS sequencing kits with all required accessories, such as sipper tubes, PTP, sequencing buffers, prewash tubes, prewash buffer, etc.

2.3. Data Analysis

Computational hardware requirements will vary depending on the number of libraries being studied and the depth at which the amplicons are sequenced. A Titanium run on a pyrosequencer configured with two physical regions will generate SFF files of approximately 2 GB per region. It is therefore reasonable to estimate that 4–5 GB of disk space per full run on a Roche/454 Genome Sequencer will be required to store the SFF data. We commonly use Linux servers running CentOS (www.centos.org) for performing the analyses. For warehousing of multiple projects, we use APED (http://aped.sourceforge.net) to store and track the sequencing data. Tools for working with SFF files (splitting by MID and extracting FASTA sequence data) are available from Roche/454 (http://www.454.com). Mapping experimental sequences against a reference collection can be done using a watered_BLAST approach (12) using PERL scripts maintained as part of the APED software package.

3. Methods

3.1. cpn60 Amplicon Generation

Procedures to extract microbial community DNA to be used as a template for the *cpn60* UT amplification are not described. This procedure is highly sample-dependent, and each project will have its own challenges to be met. See Notes 3 and 4 for information regarding template preparation from microbial communities. The procedure described below assumes that template DNA extracted from the microbial community of interest is already available.

1. In the "white" PCR workstation (see Note 2), prepare a primer-free PCR mastermix sufficient for the desired number of reactions. It is best to prepare sufficient mastermix for your entire project at once, with some allotment for failed reactions. Set up the mastermix according to Table 2. This is a suggested setup for Platinum Taq DNA polymerase; for other polymerases and/or other mastermix volumes, the values may need to be adjusted. Dispense this mastermix into aliquots and store at −20°C.
2. Also in the white workstation, prepare the various sets of PCR primer mixes according to Table 2. The table is calculated for 100 μM primer stocks; for other concentrations or other volumes of mastermix, adjust the volumes accordingly. If applicable, be highly attentive to matching the appropriate MID tags in the primer mixes. Prepare a large volume of each primer mix that can be used for multiple samples and store it in aliquots at −20°C.
3. To analyze up to four individual samples in one 96-well PCR block, prepare individual MID-tagged mastermixes in the white PCR workstation according to Table 2. Thermocyclers will vary as to the orientation of the gradient: "horizontal" (row of 12 PCRs) or "vertical" (column of 8 PCRs). For a horizontal gradient, one sample occupies two rows; for a vertical gradient, one sample occupies three columns (24 PCRs per sample in either case). Either 96-well PCR plates or strip tubes can be conveniently used.
4. Distribute 48 μl of this mastermix into each of 24 PCR plate wells, oriented across two rows (horizontal gradient) or down three columns (vertical gradient). Up to four such mastermixes with different tags can be accommodated in one 96-well block. Each mastermix should have sufficient leftover volume to prepare one "no-template" control for each; distribute 48 μl of each into a separate capped labelled tube. These tubes will be run separately from the total plate to provide a control for template contamination.

Table 2
Suggested reaction setup for generating amplicon for pyrosequencing from a microbial community DNA template

Mastermix (Mmx) 1: buffer, Mg, and dNTP (prepare ahead of time and store at −20°C)				
Component	*μl/reaction*	*μl/100 reactions*	*μl/300 reactions*	*Final concentration in PCR*
10× PCR buffer	5	500	1,500	1×
50 mM $MgCl_2$	2.5	250	750	2.5 mM
10 mM dNTP	1	100	300	0.2 mM
Water	34.3	3,430	10,290	–
Total	42.8	4,280	12,840	–

Mastermix (Mmx) 2: primer mixtures (prepare ahead of time and store at −20°C)				
Primer	*[Stock], μM*	*μl/reaction*	*μl/480 reactions*	*Final concentration in PCR*
H279	100	0.05	24	100 nM
H280	100	0.05	24	100 nM
H1612	100	0.15	72	300 nM
H1613	100	0.15	72	300 nM
Water	–	4.6	2,208	–
Total	–	5	2,400	–

Mastermix 3: mixture for distribution into plates/tubes and template addition		
Component	*μl/reaction*	*μl/26 reactions (2 rows or 3 columns)*
Mmx1 (buffer, Mg, dNTP)	42.8	1,105
Mmx2 (primers)	5	130
Platinum Taq DNA polymerase, 5 U/μl	0.2	13
Total	48	1,248

5. In the “gray” PCR workstation (see Note 2), add 2 μl template DNA to the reactions as required. Be extremely attentive to match the MID tag in each group of wells to the appropriate template, as a mix-up here will be very difficult to rectify and may result in the assignment of the wrong microbial profile to your samples. Also, add 2 μl of PCR-grade water to the no-template controls and cycle them in a separate thermocycler using standard *cpn60* amplification conditions (see step 6) with a T_a of 42°C.
6. Seal the plate with an appropriate PCR lid (or cap the tubes) and then place it in the thermocycler. Cycle the reactions under the following conditions:

   95°C 5 min (1×)

95°C 30 s; 42–60°C 30 s; 72°C 30 s (40×)

72°C 2 min (1×)

Note that the exact structure of the thermal gradient will vary with the thermocycler used, and for some thermocyclers the ramp rate is adjustable. See Note 5 regarding PCR optimization.

7. When the PCRs have been completed, pool 12 PCRs from each sample into a 1.5- or 2-ml Eppendorf tube (total of ~1,200 μl per sample).
8. Concentration and purification of the PCR products prior to gel purification can be accomplished by several means, two of which are described here.
   (a) To purify and concentrate using AMPure resin, proceed as described below. If a robotic liquid handling system for purifying PCRs by AMPure is available to you, see Note 6.
      - Divide the pooled amplicon into two 1.5-ml Eppendorf tubes. Add 1,080 μl of AMPure resin to each tube and mix well with a pipettor.
      - Seal the tubes and rotate for at least 5 min and then let stand for at least 5 min.
      - Place the tubes on the magnet for 10 min.
      - Remove the supernatant and discard. The DNA is now bound to the AMPure resin.
      - Wash twice (keep the tubes on the magnet for the washes) with 1.5 ml freshly prepared 70% ethanol. Add ethanol and let the samples stand on the magnet for 5 min and then remove the wash and repeat. Vacuum aspirate the last wash to get the resin as dry as possible.
      - Dry the beads in the flow hood for 20 min with the tubes still in the magnet (see Note 7).
      - Elute each sample with 120 μl of water or 10 mM Tris–HCL pH 8.5 by removing the tube from the magnet and washing the resin down into the tube and then pipetting up and down ten times. Let it stand for a minute or two to increase yields. Place the tube back on the magnet for 10 min and then remove the eluent containing the DNA into a fresh tube. This will leave two tubes of ~120 μl pooled PCR product for each sample.

   (b) To purify and concentrate using Amicon Ultra-0.5 microconcentrators, proceed as follows:

- Prerinse filter by adding 200 μl water and centrifuging at 14,000 × g for 4 min.
- Add material to be concentrated (up to 500 μl). Centrifuge for 10 min at 14,000 × g. The retained volume should be around 20 μl.
- Continue adding pooled PCR product to the cartridge until the entire 1.2 ml has been added. Centrifuge as above at each step, removing the flow-through from the bottom of the tube as necessary.
- Add 180 μl of TE buffer to give a final volume of ~200 μl. Invert the filtration cartridge into a new tube and centrifuge for 5 min at 2,500 × g to recover DNA (see Note 8).

9. Add loading buffer to the purified, concentrated amplicon. Use this solution to load 3–4 lanes of a 1% agarose gel. Run the gel with ethidium bromide and cut out the ~600 bp *cpn60* UT amplicon (~620 bp if MID tags are used). Minimize exposure of the amplicon to ultraviolet light while cutting out the bands. Purify the amplicon from the gel slice using your method of choice. Elute each sample in a final volume of 30–50 μl per well and pool all wells from the same sample.
10. Determine the concentration and yield of amplicon using PicoGreen. See Note 9 regarding sample yield and concentration.

3.2. Pyrosequencing Reaction Setup

1. Pool the amplicons of interest in equimolar ratios, using concentrations as determined by PicoGreen (see Note 9). The number of pools to prepare is determined by data requirements, the gasket type, and PTP being used for the sequencing run.
2. If required, ligate emPCR/sequencing adaptors to MID-tagged amplicons, following the instructions in the Roche manual [Amplicon Library Preparation Method Manual – GS FLX Titanium Series (October 2009)]. Determine the yield of ligated pooled amplicon using PicoGreen.
3. Calculate the number of molecules per microliter of the amplicon pools with a molar calculator, using the average length of the amplicon (620 bp using a typical amplicon size of 550 bp and including MID-tagged primers) and the concentration of each pool determined after ligation. It is useful to note that the pooled amplicons are double-stranded DNA, whereas typical GS FLX libraries are comprised of single-stranded DNA.
4. Calculate the number of molecules required to carry out titration emPCRs for each pool, typically using a range of 0.5, 1.0, 2.0, and 4.0 copies per bead (cpb) as a useful starting

range. The number of beads used in a single emulsion differs between sequencing chemistries, with 600,000 used for the GS FLX and 2,400,000 used for Titanium.

5. Using TE buffer, serially dilute the amplicon pool to an appropriate stock such that a volume between 1 and 10 μl is required to deliver the desired template amount for each titration point.
6. Carry out the emPCR setup and amplification for single tube emulsions as recommended by the GS emPCR Kit User's Manual for the kit being used (or other appropriate amplification protocol). The seven main points of the process include preparation of the live amplification mix, DNA library capture, emulsification, amplification, bead recovery, DNA library bead enrichment, and sequencing primer annealing. The Titanium version of the GS amplicon emPCR kit (the equivalent of the GS FLX emPCR kit II) is due to be released by the end of 2009.
7. Count the number of DNA beads recovered after enrichment using the Coulter counter or suitable alternative; we typically use 2 μl of beads in 10 ml of Isoton with counts done in triplicate. Determine the bead recovery by dividing the number of beads recovered post-enrichment by the number of DNA capture beads used in the emPCR. Recoveries in the 5–15% range have been found to give good sequencing metrics for pyrosequencing. If recoveries are too low, insufficient beads are obtained to efficiently fill the region of interest without performing an excessive number of reactions, although the beads would likely be of good quality. When the recovery is too high, the beads tend to give poor-quality reads due to the fact that more than one molecule is captured per bead ("mixed" reads). If no recoveries fall within the 5–15% recovery range and no usable ratio can be extrapolated from the recoveries observed, a second titration series with appropriately adjusted ratios should be carried out for the amplicon pool.
8. Set up a sufficient number of emulsions at the optimal cpb ratio determined via titrations for each of the amplicon pools to be sequenced. For a typical two-region gasket on the GS FLX ~900,000 beads are loaded, but for the same region with Titanium chemistry 2,000,000 beads are suggested. The number of emulsions required for each particular amplicon pool will vary depending on the gasket type and sequencing chemistry being used. In a single 1/16th region on the GS FLX enough beads may be recovered from a single emPCR to fulfill the bead loading requirements; for the Titanium, large volume (LV) cups of emulsion oil can be used if two-region gaskets are being run.

9. Carry out the emPCR setup, amplification, and enrichment processes as recommended by the GS emPCR kit user manual for the kit being used (or other appropriate amplification protocol). Determine the bead recovery by counting on the Coulter counter, ensuring that sufficient beads are recovered to optimally utilize the appropriate gasket and PTP combination.

If bead recovery is sufficient, carry out the standard pyrosequencing protocol on the beads using the GS FLX sequencer according to the GS FLX (or Titanium) Sequencing Method Manual with no further modifications.

3.3. In-Silico Separation of Pyrosequencing Data Via MIDs

Roche/454's pyrosequencing pipeline will produce one Standard Flowgram Format (SFF) file per physical region being sequenced. In order to extract all of the read data associated with a specific MID, you will need to obtain the appropriate SFF software from Roche/454 (http://www.ncbi.nlm.nih.gov/Traces/trace.cgi?cmd=show&f=formats&m=doc&s=format). Roche/454's software includes a utility called sfffile, which can be used to extract MID specific read data from an SFF file. The following command is an example of how on a Linux computer the read data associated with MID1 could be extracted to a new file.

```
sfffile –o path/to/new/file MID1@Region01.sff
```

If you are using a novel set of MID sequences, you will need to create a MID configuration file which defines the MID sequences as well as the number of base-pair errors allowed for them. The software available from Roche/454 includes a definition for their default set of supported MIDs (GSMIDs, which was originally released for the GS-FLX chemistry) in a file called MIDConfig.parse. Within a MID configuration file, a given MID would be defined similar to

```
mid = "NML1", "ACACAC", 0;
```

This defines a MID called NML1 having the sequence ACACAC and allows for 0 bp errors in its sequence.

Once separated by MID the pyrosequencing data can be extracted into FASTA format for additional processing.

```
sfffile –s /path/to/SFFfile
```

Once extracted as FASTA format, additional tools can be used for further analyses such as mapping to a reference database via wateredBLAST (12). To manage large collections of *cpn60* UT libraries we commonly use APED (http://aped.sourceforge.net) to warehouse and analyze data from complex experiments.

4. Notes

1. The pyrosequencing protocol is sufficiently flexible to allow for a number of different approaches to multiplexing. The *cpn60* UT primers can be used unmodified, which generates amplicons with ends that must be modified by the ligation of MID tags and/or emPCR/sequencing primer sequences. This approach is not generally recommended, however, since the ligation process is somewhat labor-intensive, particularly when MIDs are used and each sample must be ligated in separate reactions, and it requires more input template DNA. An alternative is to incorporate both the emPCR/sequencing primer sequences and the MID tag directly into the PCR primers themselves. We have used this approach successfully in the past (12) and its main advantage is the lack of requirement for any ligation reaction and therefore much less input DNA is required for the sequencing reaction. Another result of this approach is that all sequences are generated from the same end of the *cpn60* UT, since the sequencing primer landing site is incorporated into the amplification primer sequence. With the advent of the Titanium chemistry, however, this approach is somewhat less desirable due to the increased length of the emPCR/sequencing primer and MID sequences, which makes for very long amplification primers. As a compromise, then, we have opted to use *cpn60* UT primers that are modified at the 5′ ends only with the MID tag sequences available from Roche (Technical Bulletin no. 005-2009). In this case, ligation to the emPCR/sequencing primer sequences is still required, but due to the fact that all amplicons have the unique MIDs already incorporated into the PCR products, one ligation reaction can be performed for a number of pooled MID-tagged samples. This will generate sequencing information from both ends of the *cpn60* UT, and this should be kept in mind when analyzing the data.
2. The PCR workstations are ideally located in different rooms but should at a minimum be as distant from one another as is feasible. One workstation ("white") is reserved for mastermix preparation and is rigorously free from both template DNA and especially amplicon. The other workstation ("gray") is reserved for the addition of template DNA to the mastermix that is prepared in the white workstation. Gloves and lab coats must be changed when moving from the white workstation to the gray workstation, and both workstations should have a set of pipettors and aerosol-resistant tips reserved for use in that particular workstation. Both workstations should be pre- and posttreated with ultraviolet light for at least 10 min, and for ongoing work weekly cleaning of the workstation and

pipettors with DNA Away (Molecular BioProducts) is highly recommended.

3. Attention must be paid to lysing the entire complement of organisms found within a sample, without regard to cell-wall structure and degree of difficulty of lysis. A particularly thorough procedure is described by Apajalahti et al. (13), but a number of commercially available kits or manual procedures could be used for DNA extraction. It may be useful to assess the efficacy of cell lysis by microscopy or by spiking samples with particular target cells and determining the recovery of genomic DNA in the extract using quantitative PCR (14).
4. The input material varies widely according to the particular application, and the quality of the output depends on the quality of the input. It is critically important to check the final extract for the presence of inhibitors of PCR, since this can lead to a decrease in PCR product yield, and in serious cases, in no PCR product at all. The simplest way to determine if PCR inhibition is a problem is to analyze a small dilution series of the raw extract; the yield of PCR product will initially increase with dilution if inhibition is a concern. PCR product yield can be measured accurately using qPCR targeting an organism known to be present in the extract (14). Alternatively, the yield of PCR product can be estimated by visually examining an ethidium-bromide-stained gel of PCR products generated from the dilution series using the *cpn60* universal primers. The dilution that gives the maximally intense band with 2 μl of template DNA should be chosen for the preparation of PCR product.
5. If the thermocycler used enables the adjustment of ramp rates, an optimized ramp rate can improve the yield of PCR products from certain templates. In general, we have found that slower ramp rates tend to improve yields in most cases.
6. Some labs have an automated system for doing the AMPure reactions in 96-well formats so that each PCR can be purified independently and automatically. In this case, pool all 24 AMPure-purified reactions into a single 1.5-ml Eppendorf tube and concentrate the solution down to a final volume of ~50 μl using a rotary vacuum concentrator with heat. Load the concentrated solution into two wells of a 1.2% agarose-TBE gel for gel purification.
7. If the beads are not dried sufficiently, there will be residual ethanol in the eluate that may cause your sample to float out of the well when you try to load it in the gel. Conversely, overdrying the beads may lead to a decrease in yields. If this is a problem, it might be useful to concentrate the pooled AMPure eluate using a rotary vacuum concentrator with heat; this should evaporate all residual ethanol efficiently.

8. With this approach, you could in theory concentrate the PCR product to the extent that you could load all of it in a single lane. Be cautious not to overload the gel with PCR product, as this can decrease the efficiency of DNA separation by gel electrophoresis. The actual final volume of PCR product and the number of lanes utilized will depend on the yield of PCR product from the 24 PCRs and will vary from sample to sample.
9. The key is to generate sufficient gel purified PCR product for highly accurate quantification to facilitate the setup of the pyrosequencing reaction. Accurate quantification enables the proper determination of the optimal bead to template ratio (copies per bead), which generates the maximum number of valid pyrosequencing reads. When there is no need for ligation, as is the case when both the MID tag and the emPCR/sequencing primer sequences are incorporated into the *cpn60* amplification primers, the amplicon is normally diluted significantly prior to bead coupling. When a ligation reaction is required, such as when unmodified *cpn60* amplification primers or primers modified only with the MID tag are used, there is a need for more – a total of 1,000 ng of amplicon at a concentration of 50 ng/μl (20 μl). When multiplexing using MID-tagged samples, the actual amount of each amplicon that is required will depend on the number of MID tags that are pooled. For example, in a pooled sample with 10 MID tags, 100 ng of each of the 10 amplicons is pooled in a total volume of 20 μl. If the pooled volume exceeds 20 μl, the sample may be easily concentrated using Amicon Ultra 0.5-mL ultrafiltration membranes (30 kDa molecular weight cutoff).

References

1. Hamady, M., and Knight, R. (2009) Microbial community profiling for human microbiome projects: Tools, techniques, and challenges, *Genome Res.* **19**, 1141–1152.
2. Tuohy, K. M., Gougoulias, C., Shen, Q., Walton, G., Fava, F., and Ramnani, P. (2009) Studying the human gut microbiota in the trans-omics era - Focus on metagenomics and metabonomics, *Current Pharmaceutical Design* **15**, 1415–1427.
3. Hemmingsen, S. M., Woolford, C., van der Vies, S. M., Tilly, K., Dennis, D. T., Georgopoulos, C. P., Hendrix, R. W., and Ellis, R. J. (1988) Homologous plant and bacterial proteins chaperone oligomeric protein assembly, *Nature* **333**, 330–334.
4. Goh, S. H., Potter, S., Wood, J. O., Hemmingsen, S. M., Reynolds, R. P., and Chow, A. W. (1996) HSP60 gene sequences as universal targets for microbial species identification: studies with coagulase-negative staphylococci, *J. Clin. Microbiol.* **34**, 818–823.
5. Goh, S. H., Facklam, R. R., Chang, M., Hill, J. E., Tyrrell, G. J., Burns, E. C., Chan, D., He, C., Rahim, T., Shaw, C., and Hemmingsen, S. M. (2000) Identification of Enterococcus species and phenotypically similar Lactococcus and Vagococcus species by reverse checkerboard hybridization to chaperonin 60 gene sequences, *J. Clin. Microbiol.* **38**, 3953–3959.
6. Hill, J. E., Paccagnella, A., Law, K., Melito, P. L., Woodward, D. L., Price, L., Leung, A. H., Ng, L. K., Hemmingsen, S. M., and Goh, S. H. (2006) Identification of Campylobacter spp. and discrimination from Helicobacter and Arcobacter spp. by direct sequencing of PCR-amplified cpn60 sequences and comparison to cpnDB, a chaperonin reference sequence database, *J Med Microbiol* **55**, 393–399.

7. Hill, J. E., Penny, S. L., Crowell, K. G., Goh, S. H., and Hemmingsen, S. M. (2004) cpnDB: A Chaperonin Sequence Database, *Genome Res.* **14**, 1669–1675.
8. Dumonceaux, T. J., Hill, J. E., Hemmingsen, S. M., and Van Kessel, A. G. (2006) Characterization of intestinal microbiota and response to dietary virginiamycin supplementation in the broiler chicken, *Appl. Environ. Microbiol.* 72, 2815–2823.
9. Dumonceaux, T. J., Hill, J. E., Pelletier, C., Paice, M. G., Van Kessel, A. G., and Hemmingsen, S. M. (2006) Molecular characterization of microbial communities in Canadian pulp and paper activated sludge and quantification of a novel Thiothrix eikelboomii -like bulking filament, *Can J Microbiol* **52**, 494–500.
10. Hill, J. E., Hemmingsen, S. M., Goldade, B. G., Dumonceaux, T. J., Klassen, J., Zijlstra, R. T., Goh, S. H., and Van Kessel, A. G. (2005) Comparison of ileum microflora of pigs fed corn-, wheat-, or barley-based diets by chaperonin-60 sequencing and quantitative PCR, *Appl.Environ.Microbiol.* **71**, 867–875.
11. Hill, J. E., Seipp, R. P., Betts, M., Hawkins, L., Van Kessel, A. G., Crosby, W. L., and Hemmingsen, S. M. (2002) Extensive profiling of a complex microbial community by high-throughput sequencing, *Appl. Environ. Microbiol.* **68**, 3055–3066.
12. Schellenberg, J., Links, M. G., Hill, J. E., Dumonceaux, T. J., Peters, G. A., Tyler, S. D., Ball, T. B., Severini, A., and Plummer, F. A. (2009) Pyrosequencing of the chaperonin-60 universal target as a tool for determining microbial community composition, *Appl Environ Microbiol* **75**, 2889–2898.
13. Apajalahti, J. H., Sarkilahti, L. K., Maki, B. R., Heikkinen, J. P., Nurminen, P. H., and Holben, W. E. (1998) Effective recovery of bacterial DNA and percent-guanine-plus-cytosine-based analysis of community structure in the gastrointestinal tract of broiler chickens, *Appl. Environ. Microbiol.* **64**, 4084–4088.
14. Dumonceaux, T. J., Hill, J. E., Briggs, S. A., Amoako, K. K., Hemmingsen, S. M., and Van Kessel, A. G. (2006) Enumeration of specific bacterial populations in complex intestinal communities using quantitative PCR based on the chaperonin-60 target, *J.Microbiol. Methods* **64**, 46–62.
15. Hill, J. E., Town, J. R., and Hemmingsen, S. M. (2005) Improved template representation in cpn60 polymerase chain reaction (PCR) product libraries generated from complex templates by application of a specific mixture of PCR primers, *Environ. Microbiol.* **8**, 741–746.
16. Spear, G. T., Sikaroodi, M., Zariffard, M. R., Landay, A. L., French, A. L., and Gillevet, P. M. (2008) Comparison of the diversity of the vaginal microbiota in HIV-infected and HIV-uninfected women with or without bacterial vaginosis, *J Infect Dis* **198**, 1131–1140.
17. Schloss, P. D., and Handelsman, J. (2005) Introducing DOTUR, a computer program for defining operational taxonomic units and estimating species richness, *Appl. Environ. Microbiol.* **71**, 1501–1506.

Chapter 11

Prescreening of Microbial Populations for the Assessment of Sequencing Potential

Irene B. Hanning and Steven C. Ricke

Abstract

Next-generation sequencing (NGS) is a powerful tool that can be utilized to profile and compare microbial populations. By amplifying a target gene present in all bacteria and subsequently sequencing amplicons, the bacteria genera present in the populations can be identified and compared. In some scenarios, little to no difference may exist among microbial populations being compared in which case a prescreening method would be practical to determine which microbial populations would be suitable for further analysis by NGS. Denaturing density-gradient electrophoresis (DGGE) is relatively cheaper than NGS and the data comparing microbial populations are ready to be viewed immediately after electrophoresis. DGGE follows essentially the same initial methodology as NGS by targeting and amplifying the 16S rRNA gene. However, as opposed to sequencing amplicons, DGGE amplicons are analyzed by electrophoresis. By prescreening microbial populations with DGGE, more efficient use of NGS methods can be accomplished. In this chapter, we outline the protocol for DGGE targeting the same gene (16S rRNA) that would be targeted for NGS to compare and determine differences in microbial populations from a wide range of ecosystems.

Key words: DGGE, Microbial, Populations, Screening, 16S rRNA, Comparison

1. Introduction

An estimated 99% of microorganisms are unable to be cultured; however, molecular methods made it possible to visualize complex microbial communities, independent of their culturability and without prior knowledge on the complexity and diversity of the ecosystem (1). There are multiple methods that can be used to profile microbial populations including DGGE, terminal restriction length polymorphism (T-RFLP), and pyrosequencing (2–4). A recent study utilized culturing, pyrosequencing, and DGGE to characterize anaerobic bacteria populations in a wound. The study reported that culturing provided poor recovery;

Young Min Kwon and Steven C. Ricke (eds.), *High-Throughput Next Generation Sequencing: Methods and Applications*, Methods in Molecular Biology, vol. 733, DOI 10.1007/978-1-61779-089-8_11, © Springer Science+Business Media, LLC 2011

however, the molecular methods were able to characterize the populations more fully (5). In addition, the DGGE and pyrosequencing techniques provided quantitative data and determined diversity of bacterial species.

DGGE is presently the most commonly used molecular method for determining community diversity (2). When comparisons of microbial populations are based on one target gene, both DGGE and NGS typically utilize the 16S rRNA gene (2, 6, 7). This gene is present in all bacteria and because the gene is essential for survival, it is relatively well conserved. However, there are target areas within the gene that are sufficiently unique that differences among genera can be detected. Initially, this gene would be targeted and amplified using conventional PCR techniques. Because the PCR product size is the same regardless of the bacteria being used, separation of the sequences cannot be accomplished by conventional gel electrophoretic techniques. The sequences, however, are different, and therefore PCR amplified products can be separated using a denaturing gradient gel.

The gradient gel has a denaturant (typically urea) with a lower concentration at the top and increases gradually with the highest desired concentration at the bottom. PCR products are loaded into the gel and during electrophoresis these products are denatured based on sequence composition. The forward PCR primer utilized for DGGE contains a 40 bp G-C sequence called a G-C clamp that ensures that the DNA will remain partially double stranded so that the PCR products will not migrate completely through the gel (Fig. 1). However, the remainder of the PCR product (the 16S rRNA sequence) will denature into single strands dependent on the G-C and A-T content. In this way, the PCR products form a "wishbone" shape with the G-C clamp remaining double stranded, but the 16S rRNA portion of the sequence becoming single stranded. The PCR products with 16S rRNA sequences having a higher G-C content will migrate further than those sequences with a higher A-T content. After electrophoresis, gels are stained and visualized on a transilluminator and the banding patterns produced by the bacterial populations can be visually compared (Fig. 2).

In this chapter, we outline the protocol for DGGE targeting the 16S rRNA gene to compare and determine differences in microbial populations from a wide range of microbial communities.

2. Materials

2.1. DNA Extraction

1. Acid-Washed Zirconium Beads: 200 μm (OPS Diagnostics, Lebanon, NJ).
2. 2-ml Screw cap microcentrifuge tubes with O-ring caps (Thermo Scientific, Dubuque, IL).

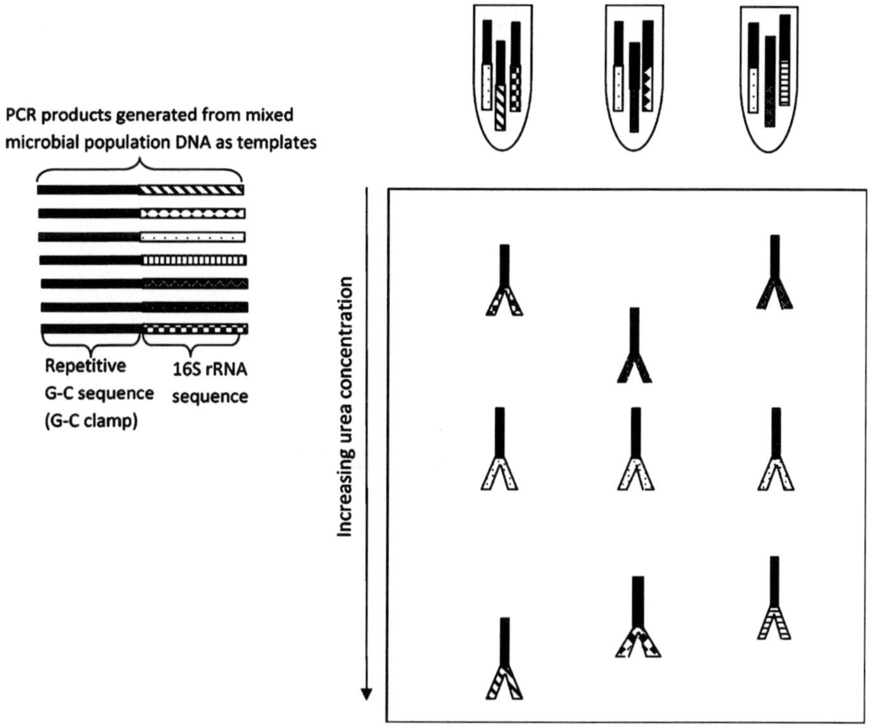

Fig. 1. A diagram of the principle of denaturing gel electrophoresis (DGGE). In this diagram, the G-C sequence at the 5′ end of the PCR product (black bar) is identical in all PCR products, but the 16S rRNA sequence is different (as depicted by different patterns). As the PCR products migrate through the denaturing gel, the G-C clamp remains double stranded, while the 16S rRNA sequence will denature dependent on the A-T and G-C ratio.

3. 25 Phenol:24 chloroform:1 isoamyl alcohol, pH 8.0.
4. Sodium Buffer: 200 mM NaCl, 200 mM Tris–HCl, 20 mM EDTA.
5. 20% Sodium dodecyl sulfate (SDS).
6. 100% Isopropanol (2-propanol).
7. 100% Ethanol.
8. Sodium acetate: 3 M NaAC, pH 5.2.
9. Bead beater: FastPrep®-24 System (MP Biomedicals, LLP, Solon, OH).

2.2. PCR Amplification of Target Gene

1. DNA isolated from microbial community.
2. PCR mix: JumpStart Taq ReadyMix (St. Louis, MO) (see Note 1).
3. Nuclease-free water.
4. PCR primers (see Note 2).

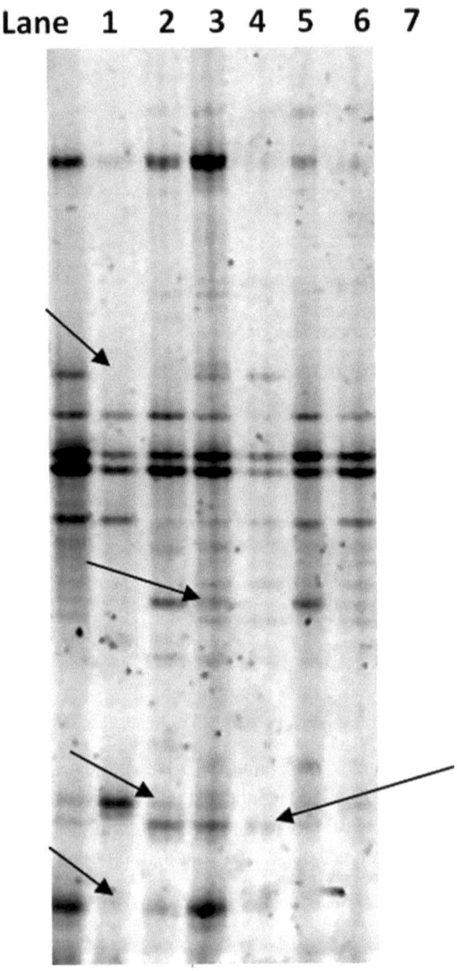

Fig. 2. A typical banding pattern after electrophoresis of PCR products using DNA obtained from rat intestinal samples each lane representing a different feed treatment (*lanes* 1–7). In many cases, only unique members of the population are necessary to distinguish, and therefore it is not necessary to identify all microflora present in the samples. The *arrows* indicate unique banding patterns. Using the DGGE gel below as an example of this, unique bands of interest (*arrows*) could be excised, purified, and sequenced. Because the profiles of the banding patterns in these seven lanes are similar, DGGE can be used as a prescreening method for NGS, which allows the choice to be made to sequence all or only the samples with banding patterns of particular interest for most efficient use of NGS.

5. Sterile DNA, RNA and nuclease free PCR tubes: 0.2 ml.
6. Inhibitor neutralizer: bovine serum albumin (BSA) 10 mg/ml (New England Biolabs, Ipswich, MA).

2.3. Denaturing Gradient Electrophoresis

1. Gradient maker: GM-100 (C.B.S. Scientific, Delmar, CA).
2. 40% Acrylamide and bis-acrylamide solution: 37.5:1 (Bio-Rad, Hercules, CA).

3. Vacuum filter: 150-ml receiver, 0.45-µm filter (VWR Int., Philadelphia, PA).
4. Formamide: Deionized (Sigma Chemical Company, St. Louis, MO).
5. Urea (Sigma Chemical Company, St. Louis, MO).
6. *N,N,N',N'*-Tetramethylethylenediamine (TEMED) (Sigma Chemical Company, St. Louis, MO).
7. Electrophoresis Buffer (50×): Tris–acetate–EDTA buffer (TAE), 2 M Tris–acetate, 0.05 M EDTA.
8. Ammonium persulfate: 10% (see Note 3) (Sigma Chemical Company, St. Louis, MO).
9. Vertical Electrophoresis Unit: D-Code Universal Mutation detection system with two glass plates (18×18 cm and 18×15 cm), spacers (1 mm thick), two clamps (19 cm in length), an assembly stand, an electrophoresis core, electrophoresis buffer tank, lid (Bio-Rad, Hercules, CA).
10. Plumbers grease (PlumbShop, Southfield, MI).
11. Pasteur pipettes (VWR Int., Philadelphia, PA).
12. Bibulous paper (VWR Int., Philadelphia, PA).
13. 100-ml Beaker.
14. 15-ml Centrifuge tubes (VWR Int., Philadelphia, PA).
15. Loading dye: 0.02% bromophenol blue; 0.02% xylene cyanol; 70% glycerol; water to volume.
16. Large stirring plate.
17. Small stirring bean.
18. Flow pump: variable speed (VWR Int., Philadelphia, PA).

2.4. Band Purification

1. Small scalpel or razor blade.
2. Microcentrifugation devices: 0.2 µM, Spin-X tubes (Corning, NY).
3. 1.7-ml Centrifuge tubes (VWR Int., Philadelphia, PA).
4. 0.5-ml Centrifuge tubes (VWR Int., Philadelphia, PA).
5. 21 Gauge needle.
6. 7.5 M NH_4OAc.
7. DNA carrier: Glycogen, 20 mg/ml (EMD Biosciences, San Diego, CA).
8. 100% Ethyl alcohol (EtOH).
9. 70% Ethyl alcohol (EtOH).
10. Tris–EDTA buffer (TE; 10 mM Tris–HCl; 1 mM EDTA; pH 8.0).

3. Methods

3.1. DNA Extraction

In general, no specialized DNA extraction is required for DGGE. However, clean samples are essential for successful amplification and improvements on original extraction methods have been developed. A bead beating method that can produce high quality and a high yield of DNA is described in the next section (8). Several preparatory techniques and precautions can also improve DNA recovery and purity. Keeping alcohols in the freezer and buffers in the refrigerator will promote DNA precipitation. Because the primers utilized in this protocol will amplify any bacteria, great caution should be taken to avoid contamination. All materials being used should be exposed to UV light for at least 30 min prior to use. All DNA preparations should be prepared under an enclosure that has been exposed to UV light for at least 30 min prior to use.

1. In a 2-ml screw cap tube, add 500 μl of acid-washed zirconium beads, 500 μl of 25 phenol:24 chloroform:1 isoamyl alcohol buffered to pH 8.0, 500 μl of sodium buffer, 200 μl of 20% SDS, and approximately 200 mg of sample.
2. Bead beat in FastPrep®-24 system on highest speed for 2 min.
3. Centrifuge at 8,000 × *g* for 5 min.
4. Remove the aqueous layer into a clean 1.7-ml microcentrifuge tube and add 500 μl of 25 phenol:24 chloroform:1 isoamyl alcohol buffered to pH 8.0.
5. Centrifuge at 4°C at 14,000 × *g* for 5 min.
6. Remove the aqueous layer into a clean 1.7-ml microcentrifuge tube and add 600 μl of isopropanol and 60 μl of NH_4OAc.
7. Incubate at −20°C for at least 15 min (see Note 4).
8. Centrifuge at 4°C at 14,000 × *g* for 20 min.
9. Pour off supernatant and wash the pellet twice with cold 100% ethanol.
10. Allow the pellet to dry for 2–5 min and resuspend in 50 μl of TE (see Note 5).

3.2. PCR Amplification of Target Gene

Because the primers utilized in this protocol will amplify any bacteria, great caution should be taken to avoid contamination. All materials being used should be exposed to UV light for at least 30 min prior to use, with the exception of the DNA and primers. All PCR preparations should be prepared under a PCR enclosure that has been exposed to UV light for at least 30 min prior to use.

1. In a PCR tube, mix 25-μl total reaction volume consisting of 12.5 μl of Jump start ready mix, 1 μl of forward primer final

concentration 50 pmol, 1 μl of reverse primer final concentration 50 pmol, DNA (approximately 250 ng/μl), nuclease free water to volume (see Note 6).

2. Thermal cycler conditions are as follows (see Note 7):
   (a) 94.9°C for 2 min.
   (b) 94.0°C for 1 min.
   (c) 67.0°C for 45S (decreasing by 0.5°C per cycle).
   (d) Extension at 72°C for 2 min.
   (e) Repeat steps 2–4 for 17 cycles.
   (f) 94.0°C for 1 min.
   (g) 58.0°C for 45 s.
   (h) Repeat steps 6 and 7 for 12 cycles.
   (i) Extension at 72.0°C for 7 min.
   (j) Hold at 4°C.

3.3. Pouring the Denaturing Gradient Gel

3.3.1. Preparing the Gel Solutions

The percent of urea utilized in the gel solutions may vary depending on the G-C content of the amplicons. Some researchers utilize either a wider or narrower gradient for this reason. However, the preparation of 0 and 100% denaturing solutions allows researchers to save time by making only one set of denaturing solutions and adjusting solutions to the desired final concentrations.

1. In two separate glass beakers, mix 20 ml of 40% acrylamide/bis (37.5:1) and 2 ml of 50× TAE.
2. For the 0% gradient solution (low gel), fill to 100 ml with deionized sterile water and put aside.
3. For the 100% solution (high gel), add 40 ml of deionized formamide, 42 g of urea and bring to 100-ml total volume with deionized sterile water.
4. Mix until dissolved with a stir bar on a stir plate. After dissolving, degas for 10 min by letting the container to sit uncovered.
5. For both solutions, filter-sterilize using a 0.45-μm filter and protect the container from light by covering in aluminum foil. Solutions should be used within 1 month (see Note 8).

3.3.2. Assembling the Casting Plates

1. Put a thin film of grease on each spacer. Change gloves to avoid getting grease all over the glass plates.
2. Lay the large glass plate down and place the greased spacers on the sides with the grease to the outside edges.
3. Place the smaller glass plate on top of the spacers.
4. Insert the sides of the plates into the clamp with the arrows of the clamp facing you.

5. Stand the assembly up on the lab bench for alignment. Loosen the clamps slightly and insert the alignment card to align the spacers and at the same time make sure the bottoms of the glass plates are flush.
6. If aligned properly, tighten the clamps.
7. Insert the assembly into the stand for pouring the gel. Turn the clamps on either side of the stand 180° to tighten the assembly into the stand.

3.3.3. Pouring the Gel

1. Put a small stir bean into the high side of the gradient maker.
2. Make sure both valves are closed.
3. Hook the tubing to the luer valve from the gradient maker through the pump and to the assembly.
4. Hook a pipette tip to the tubing and colored tape to the assembly.
5. Add 11.5 ml of high and 11.5 ml of low denaturing gel into separate 15-ml centrifuge tubes.
6. Add 81 μl of ammonium persulfate (APS) and 4.5 μl of TEMED to each solution and invert to mix.
7. In the high side of the gradient maker, pour in the high gel solution and open valve 2 slightly to allow some of the gel to flow into the low chamber and then close the valve. Remove any gel in the low side with a Pasteur pipette and put back into the high side (see Note 9).
8. Start the stirring bean on the high side and pour the low gel into the low side.
9. Turn on the pump to a speed of approximately 5 ml per minute.
10. Open the luer valve and then the channel valve that is between the low and high side.
11. You should be able to see the gel traveling along the tubing and an "oily" appearance in the high side as low gel is mixing with the high gel.
12. Tip the gradient maker slightly by putting a tip under the low side to help facilitate emptying of the low side.
13. Once the gel has been poured, overlay the top of the gel with 3–4 ml of 1× TAE to get an even layer.
14. After the gel is poured, remove the tubing from the assembly and add water to both sides of the gradient maker. Flush the tubing and gradient maker with water into an empty beaker.
15. Allow 1 h or more for polymerization.
16. Pour off the 1× TAE layer and dry mostly with bibulous paper.

17. Add 5 ml of 0% denaturing gel into a 15-ml centrifuge tube with 66.7 μl of APS and 6 μl of TEMED and invert to mix (cap solution).
18. Put comb between the glass plates of the assembly and add the cap solution with a Pasteur pipette until filled.
19. Allow at least 15 min to polymerize.
20. Take out the comb and flush the wells with a Pasteur pipette (see Note 10).

3.4. Electrophoresis of PCR Products Through Denaturing Gradient Gel

3.4.1. Preparing the Buffer and Tank and Loading the Gel

1. Place the buffer tank on top of a large stirring plate and add a large stirring bar at the bottom of the tank.
2. Pour 7 l 1× TAE into the gel tank (see Note 11), put core into the tank and put on the lid. Set the heater to 59°C.
3. Turn on the large stir bar at the bottom of the tank.
4. Snap the gel onto the core, making sure the gel is on the front.
5. Replace the core into the tank and put on the lid. Continue heating the buffer until 59°C is reached.
6. While the buffer gets heated, mix PCR products with loading buffer in a 1:1 ratio.
7. Remove the lid to load samples (see Note 12), acting quickly to avoid losing too much heat.
8. Replace the lid. Make sure that heater, stir rod, stir bar, and pump are all operating. Turn on the power supply to 60 V (see Note 13).
9. Perform the electrophoresis for 17 h.

3.4.2. Staining and Viewing the Gel

1. Dilute 25 μl of SYBR green in 500 ml of 1× TAE (see Note 14).
2. Take gel from core and place in a clean plastic container.
3. Move spacers at top slightly outward and push up with thumbs to separate glass plates (see Note 15).
4. At this point, it is practical to mark a corner of the gel to help determine which side is left and which is right. The left corner of the gel can be clipped for this purpose.
5. Once top plate is removed, slide the gel from the bottom plate into the plastic container.
6. Add the SYBR green stain and place the plastic container on a shaker or rocker with gentle motion for 40 min.
7. Destain for 10 min in 1× TAE.
8. View gel using a UV transilluminator.

3.5. Band Purification for Sequencing Purposes

The bands of interest can be sequenced after the banding pattern has been created using a modified protocol described by Wang et al. (9). After bands are purified, they can be sequenced using the same primers that were used for amplification without the G-C clamp. Once again, because the primers utilized in these protocols amplify any bacteria present in the sample, caution should be taken to avoid contamination. All materials should be exposed to UV light 30 min prior to use.

1. Cut the gel piece containing the band of interest.
2. Place each gel piece into a 0.5-ml microfuge tube.
3. Pierce bottom of each 0.5-ml tube with a 21 gauge needle.
4. Place the 0.5-ml tubes in 2.0-ml microfuge tubes and spin at full speed for 5 min to disrupt the gel.
5. Discard the 0.5-ml tubes and add 300 μl of TE to the 2.0-ml tubes.
6. Vortex the tubes and place at 65°C for 15 min.
7. Transfer contents of each tube to a Spin-X tube.
8. Centrifuge Spin-X tubes at full speed for 5 min.
9. Remove the filtered supernatant and place it in a clean and sterile 1.7-μl microcentrifuge tube and add:

7.5 M NH4OAc	133 μl
Glycogen	3 μl
100% EtOH	1,000 μl

10. Vortex tubes and place at −70°C for at least 10 min, but they can be incubated overnight.
11. Spin tubes for 20 min at full speed.
12. Wash the pellet twice with cold 70% EtOH.
13. Resuspend the pellet in 5.0 μl of TE.

4. Notes

1. It is not necessary to utilize the JumpStart ReadyMix Taq polymerase. Other PCR premixes may work, but if PCR fails utilizing a different PCR mix may improve the results.
2. Several primer sets are available each having slight variations depending on the microbial community being analyzed. Please refer to the papers of Hanning et al. and Ercolini (2, 10) for some suggestions and explanations of differences in targets.

3. Prepare the ammonium sulfate fresh for each gel.
4. Incubate for at least 20 min, but samples may be incubated as long as overnight.
5. Samples can be placed at 4°C overnight if pellet does not go into solution immediately. If pellet does not go into solution after overnight at 4°C, add additional TE (1 μl) until pellet goes into solution.
6. It is not essential to add BSA to the PCR reaction. If the DNA was obtained from a fecal, water, or soil sample, BSA may improve the reaction by binding inhibitors. If BSA is used, replace 1 μl of water with 1 μl of BSA.
7. Other thermal cycler conditions have been reported. However, a touchdown PCR cycle greatly reduces the formation of primer dimers, which are inherent in the use of a large G-C sequence.
8. Most failures with DGGE can be attributed to using old and expired reagents. Using fresh reagents will improve the rate of success.
9. If this step is neglected, no gel will flow from the high side to the low side.
10. Neglecting to flush out the wells will leave gel in the wells and thus reduce the total amount of PCR product that can be loaded.
11. TAE can be prewarmed in a water bath set at 60°C. TAE can also be warmed in the electrophoresis chamber, but the amount of time to warm the buffer is much longer.
12. Each well will hold a total volume of about 40 μl. The amount loaded will vary depending on the expected outcome. If many bands are expected, a larger volume should be loaded so that fainter bands can be visualized.
13. Make sure that samples do not get to the gradient portion of the gel until temperature is at 59°C. If temperature has dropped significantly (45–55°C), run the power supply at 20 V until 59°C is reached and then increase to 60 V.
14. Since SYBR green adheres to glass, dye solution must be made in a plastic container, and gels must be removed from glass casting assembly to stain.
15. A wash bottle of TAE can be used to help slide the gel off the glass.

Acknowledgments

This book chapter was funded by an Institute of Food Science and Engineering grant through the University of Arkansas.

References

1. Huys, G., Vanhoutte, T., and Vandamme, P. (2008) Application of sequence-dependent electrophoresis fingerprinting in exploring biodiversity and population dynamics of human intestinal microbiota: what can be revealed? *Interdiscip Perspect. Infect. Dis.* **2008**, 597–603
2. Ercolini, D. (2004) PCR-DGGE fingerprinting: novel strategies for detection of microbes in food. *J. Microbiol. Methods* **56**, 297–314.
3. Deng, W., Xi.,D., Mao, H., and Wanapat, M.. (2008) The use of molecular techniques based on ribosomal RNA and DNA for rumen microbial ecosystem studies: a review. *Mol. Biol. Rep.* **35**, 265–74.
4. Petrosino, J.F., Highlander, S., Luna, R.A., Gibbs, R.A., and Versalovic, J. (2009) Metagenomic pyrosequencing and microbial identification. *Clin. Chem.* **55**, 856–66.
5. Dowd, S., Sun, Y., Secor, P., Rhoads, D., Wolcott, B., James, G., and Wolcott, R. (2008) Survey of bacterial diversity in chronic wounds using pyrosequencing, DGGE, and full ribosome shotgun sequencing. BMC Microbiol. **8**, 1-5. Available at http://www.biomedcentral.com/1471–2180/8/43
6. Muyzer, G., De Waal, E., and Uitierlinden, A. (1993) Profiling of complex microbial populations by denaturing gradient gel electrophoresis analysis of polymerase chain reaction-amplified genes coding for 16S rRNA. *Appl. Environ. Microbial.* **59**, 695–700.
7. Singh, J., Behal, A., Singla, N., Joshi, A., Birbian, N., Singh, S., Bali, V., and Batra, N. (2009) Metagenomics: Concept, methodology, ecological inference and recent advances. *Biotechnol. J.* **4**, 480–94.
8. Turnbaugh P., Hamady, M., Yatsunenko, T., Cantarel, B.L., Duncan, A., Ley, R.E., Sogin, M.L., Jones, W.J., Roe, B.A., Affourtit, J.P., Egholm, M., Henrissat B., Heath, A.C., Knight, R. and Gordon, J.L. (2009) A core gut microbiome in obese and lean twins. *Nature* **457**, 480–484.
9. Wang, T., Vogelstein, B., Kinzler, K.W., and Velculescu, V.E.. (2002) Digital Karyotyping Detailed Protocol. Version 1.0A; December 2, 2002. The Howard Hughes Medical Institute and The Sidney Kimmel Comprehensive Cancer Center, The Johns Hopkins University Medical Institutions Available at: http://www.digitalkaryotyping.org/Digital%20Karyotyping%20Protocol%20v%201.0A.pdf
10. Hanning, I., Jarquin, R., and Slavik, M. (2008) *Campylobacter jejuni* as a secondary colonizer of poultry biofilms. *J. Appl. Microbiol.* **105**, 1199–1208.

Part IV

Metagenomics

Chapter 12

Metagenomics

Jack A. Gilbert, Bonnie Laverock, Ben Temperton, Simon Thomas, Martin Muhling, and Margaret Hughes

Abstract

Metagenomics has evolved over the last 3 decades from the analysis of single genes and their apparent diversity in an ecosystem to the provision of complex genetic information relating to whole ecosystems. Metagenomics is a vast subject area in terms of methodology, which encompasses a suite of molecular technologies employed to investigate genomic information from all members of a microbial community. However, the relatively recent developments in high-throughput sequencing platforms have meant that metagenomic can be performed simply by extracting DNA and sequencing it. Here, we outline explicit methodologies for the extraction of metagenomic DNA from marine and sediments/soil environmental samples, the subsequent production and sequencing of large-insert metagenomic libraries, and also shotgun pyrosequencing considerations. We also provide relevant advice on bioinformatic analyses of the complex metagenomic datasets. We hope that the information provided here will be useful to establish the techniques in most reasonably equipped molecular biology laboratories.

Key words: Metagenomics, Pyrosequencing

1. Introduction

Culture remains to be the most powerful method in microbiology. However, it is limited by the fact that the vast majority of microorganisms are averse to isolation and culture using common culturing methods. However, the genetic information that codes for the metabolic and catabolic activities of the microorganisms is now accessible to scientific investigations using new methods in molecular biology. This area of research, which aims at the functional or sequence-based analysis of all microbial genomes present in an environmental sample, is commonly referred to as microbial metagenomics. When combined with high-throughput sequence analysis, metagenomics becomes an extremely powerful

Young Min Kwon and Steven C. Ricke (eds.), *High-Throughput Next Generation Sequencing: Methods and Applications*, Methods in Molecular Biology, vol. 733, DOI 10.1007/978-1-61779-089-8_12, © Springer Science+Business Media, LLC 2011

tool for investigating microbial communities, as it can provide a detailed profile of the functions and diversity of the community. Traditionally, metagenomics approaches such as shotgun clone library construction were sequenced one clone at a time using Sanger sequencing, requiring considerable sequencing cost and manpower for a limited number of reads (e.g., 1,000 Sanger sequences would cost ~£4,000) consequently placing it out of reach of most laboratories. Notable exceptions to this rule were the Global Ocean Survey (GOS) (1) and the HOT Fosmid sequencing project (2). The invention of high-throughput sequencing platforms such as pyrosequencing and Illumina has democratized sequencing capability so that 600,000 pyrosequencing reads or 30,000,000 Illumina reads can now be sequenced for around £4,000. Whereas, previously, ecological studies such as GOS were only feasible with multimillion dollar private contracts, the sequencing revolution means that it is now possible for every laboratory to perform similar scale experiments on their chosen habitats. This data explosion comes with its own difficulties of an exponential increase in data storage requirements and computer processing, but such issues are outside the focus of this chapter.

Pyrosequencing technology currently produces the longest reads of all the viable second-generation platforms, at approximately 450 bp and likely to increase up to 800 bp by mid-2010 (Roche communication). These long read lengths provide valuable information for annotating genetic fragments, as greater read length increases the likelihood of identifying the sequence using available databases (3). As a result, we focus this methodology on producing DNA for pyrosequencing analysis (although many of the strategies are readily applicable for use with the Illumina or SOLiD platforms). Metagenomic pyrosequencing is particularly useful for generating genetic "fingerprints" for examining how microbial communities exposed to different gradients of changes (e.g., temperature, time, salinity, etc.) respond, by changing the relative proportions of taxa and/or functional characteristics within the community. This provides a route to start investigating the holy grail of microbial ecology – absolute understanding of a microbial ecosystem. It should also be noted in this context that due to the fact that this approach does not require any cloning, it typically avoids the biases associated with the cloning of (e.g., toxic) genes in heterologous hosts (4).

In this chapter, we outline several key DNA extraction techniques for aquatic and soil-derived environmental samples. Additionally, we highlight methods for amplification of genetic material prior to pyrosequencing from samples with limited biomass. We would like to note that methods used for the isolation of DNA and RNA from aquatic and soil/sediment environments mainly differ from each other by the approaches taken to lyse microbial cells and on how to recover and purify the fraction of genomic DNA from the lysate. The methods we highlight here are tried and tested for producing DNA of sufficient quality and quantity for metagenomic pyrosequencing.

2. Materials

Standard chemicals should be of high grade and can be purchased from Sigma-Aldrich.

2.1. Sampling Microbial Cells

1. 140 mm diameter, 1.6-μm pore size GF/A filters (Whatman, USA).
2. Sterivex filter cartridges – 0.22-μm pore size filters (Millipore, UK).
3. Peristaltic pump capable of holding 16-mm (inner diameter) Tygon LFL tubing.
4. 142-mm (diameter) filter rig (Millipore, UK).
5. Liquid nitrogen.

2.2. DNA Extraction from Soil

1. 710–1,180 μm acid-washed glass beads (Sigma-Aldrich).
2. 0.1 M Sodium phosphate buffer: 0.1 M monobasic NaH_2PO_4, 0.1 M dibasic Na_2HPO_4, pH 8.
3. Phenol–chloroform–isoamyl alcohol (25:24:1), pH 8 and chloroform– isoamyl alcohol (24:1), pH 8.
4. 100% Molecular-grade ethanol.
5. Sodium acetate, pH 5.2.
6. Sterile water.

2.3. DNA Extraction from Filter-Isolated Water Samples

1. SET buffer: 40 mM EDTA, 50 mM Tris–HCl, pH 9, 0.75 M sucrose.
2. 9 mg/ml Lysozyme in 10 mM Tris/HCl, pH 8.
3. 10% (w/v) Sodium dodecyl sulphate solution.
4. Proteinase K (20 mg/ml) in 20 mM Tris–HCl buffer, pH 8.
5. MaxTrack Gel lock tubes (Qiagen, USA).
6. 7.5 M Ammonium acetate.
7. Phenol–chloroform–isoamyl alcohol (25:24:1), pH 8 and chloroform: isoamyl alcohol (24:1), pH 8.
8. 100% Molecular-grade ethanol.
9. Sterile water.

2.4. Pyrosequencing

1. AMPure 60 ml kit (p/n 000130 Agencourt).
2. RNA 6000 picochip kit (p/n 5067-1513 Agilent).
3. DNA 7500 LabChip kit (p/n 5067-1506 Agilent).
4. MinElute PCR purification kit (p/n 28004 Qiagen).
5. RiboGreen RNA Quantitation kit (p/n R-11490 Invitrogen).
6. Quant-iT Picogreen DNA reagent (p/n P7581 Invitrogen).
7. 3 M Sodium acetate buffer.
8. 10 N Sodium hydroxide.
9. Isopropanol (reagent grade).
10. Ethanol (reagent and molecular biology grades).
11. GS Titanium General library prep kit (p/n 05233747001 Roche).
12. GS Titanium LV emPCR kit (p/n 05233542001 Roche).
13. GS Titanium SV emPCR kit (p/n 05233615001 Roche).
14. GS Titanium emPCR breaking kit (p/n 05233658001 Roche).
15. GS Titanium emPCR filters (p/n 05233674001 Roche).
16. GS Titanium Sequencing kit (p/n 05233526001 Roche).
17. No stick 1.5-ml tubes (p/n 2410 Alpha Labs).
18. Nitrogen gas.
19. DynaMag Z (p/n 12321D Invitrogen).
20. Rubber stoppers.

3. Methods

3.1. DNA Extraction from Soil or Sediment (Method Adapted from (5)) (see Note 1)

1. Weigh 0.5 g of glass beads into a 2-ml Eppendorf tube.
2. Into this tube, weigh 0.5 g of soil or sediment that has previously been homogenized by stirring with a sterile spatula.
3. Add 0.5 ml of 0.1 M sodium phosphate buffer pH 8 (use this to wash the tube with sample).
4. Add 0.5 ml of phenol–chloroform–isoamyl alcohol into the tube.
5. Bead-beat at 8,000 × g for 30 s, put on ice for 30 s, and repeat for another 30 s.
6. Centrifuge for 5 min at 16,000 × g (4°C).
7. Transfer supernatant to a clean 2-ml tube.
8. Add equal volume (~500 ml) of chloroform–isoamyl alcohol and vortex to mix.
9. Centrifuge for 5 min at 16,000 × g (4°C).

10. Transfer supernatant to a new 2-ml tube.
11. Add 2× volume of *ice-cold* 100% ethanol and 1/10 volume sodium acetate and mix (for RNA extraction, use 2.5× the volume of ethanol).
12. Precipitate for at least 1 h in freezer (can leave overnight to increase yield of DNA).
13. Centrifuge for 30 min at 16,000 ×*g*.
14. Pipette off and discard liquid, being careful with the pellet.
15. Wash with 200 μl of 70% ethanol; vortex to mix.
16. Centrifuge for 10 min at 16,000 ×*g*.
17. Repeat last two steps and remove remaining ethanol with a P10 pipette, being careful with the pellet.
18. Air-dry for 10 min. Resuspend pellets in sterile water.

3.2. DNA Extraction from Seawater (see Note 1)

1. Filter 10–15 L of seawater through a 140-mm diameter, 1.6-μm GF/A filter (Whatman), to reduce eukaryotic cell abundance and maximize the proportion of prokaryotic cells.
2. Apply filtrate directly to a 0.22-μm Sterivex filter (Millipore).
3. Following filtration, pump Sterivex cartridge dry and quick-freeze in liquid nitrogen and store at −80°C until DNA isolation.
4. Thaw Sterivex cartridge on ice.
5. Add 1.6 ml of SET lysis buffer directly onto the top of Sterivex using a 2.5-ml syringe with a 25G 5/8″ needle.
6. Add 180 μl of fresh lysozyme and seal the sterivex (Blu-Tack works well).
7. Incubate at 37°C for 30 min under continuous rotation in a Hybaid oven.
8. Add 200 μl of SDS (10% w/v) and 55 μl of fresh proteinase K (20 mg/ml).
9. Incubate at 55°C for 2 h with continuous rotation in a Hybaid oven.
10. Withdraw lysate from Sterivex using a 5-ml syringe.
11. Add 1 ml of fresh SET buffer to Sterivex and rotate to rinse.
12. Withdraw SET buffer into the same 5-ml syringe.
13. Add lysate into a 15-ml Maxtract tube (Qiagen) containing 2 ml of phenol–chloroform–isoamyl alcohol (25:24:1), pH 8.
14. Shake gently until mixed. Then, centrifuge at 1,500 ×*g* for 5 min.
15. Add an additional 2 ml of phenol–chloroform–isoamyl alcohol (25:24:1). Shake gently until mixed and centrifuge at 1,500 ×*g* for 5 min.
16. Add 2 ml of chloroform–isoamyl alcohol (24:1). Shake gently until mixed and centrifuge at 1,500 ×*g* for 5 min.

17. Decant aqueous phase to a sterile 20-ml centrifuge tube and add 0.5 V of 7.5 M ammonium acetate. Mix briefly and then add 2.5 V of pure ethanol.
18. Mix and leave at −20°C for >1 h (overnight is fine).
19. Centrifuge at 10,000×*g* for 30 min at 4°C and decant ethanol.
20. Add 2 ml of 80% ethanol and rinse the tube, then centrifuge at 10,000×*g* for 20 min at 4°C and decant ethanol.
21. Repeat above washing step.
22. Decant ethanol and leave inverted for 15 min in fume hood (provides air flow).
23. Suspend invisible pellet in 200 μl of sterile water. Leave on ice for approximately 1 h with frequent finger-tapping to rinse tube walls.
24. DNA can be quantified by visualizing using agarose gel electrophoresis (see Sambrook et al., 2000 (6)) or by using any standard DNA quantification techniques (e.g., NanoDrop spectrophotometer).
25. For pyrosequencing, it is necessary to have ~5 μg of DNA, which this technique will produce in excess. Ideally, the DNA should be at a concentration of approximately 500 ng/μl. If DNA needs concentration, please follow precipitation steps 17–23 above.

3.3. Pyrosequencing

1. The DNA needs to be accurately measured by fluorescence using a Quant-iT PicoGreen assay. The ideal amount is 3–5 μg for a fragment library, but it is possible to use less with success. It also needs to be of a reasonable molecular weight to maximize random shearing.
2. The DNA in a volume of 100 μl is mixed with 500 μl of nebulization buffer supplied with the library kit. The DNA is sheared by placing it in a nebulizer vessel and connecting the latter to nitrogen gas at 30 psi for 1 min.
3. The DNA is recovered from the nebulization chamber and cleaned with Qiagen MinElute kit following the instructions supplied by the kit, using 2.5 ml of PB buffer and eluting the DNA in 100 μl of elution buffer.
4. At this stage, the small fragments are removed using AMPure beads. A calibrated amount is added to the DNA so that material of less than 300 bp is left in solution. The higher MW DNA binds to the beads, is washed with 70% ethanol, is dried, and is recovered by eluting in 10 mM Tris–HCl pH 7.5.
5. The DNA is checked (1-μl aliquot) for size distribution on a DNA 7500 LabChip using an Agilent Bioanalyser.

6. Further manipulations are all described in the library kit. The ends of the DNA are polished, the adapters (with or without bar codes) added, fragments containing adapters selected on Dynal beads, and the ends filled. The single stranded library is recovered by melting from the Dynal beads with NaOH (0.125 N) as per the manufacturer's instructions and cleaned up with a MinElute column.
7. The library is assessed by running on a RNA picochip and the amount determined with RiboGreen assay.
8. A predetermined amount of the library is used to set up an emulsion PCR reaction using either the large- or small-volume kit from Roche. This involves binding the DNA to capture beads that are specific for one of the adapters.

3.4. Some Thoughts on Bioinformatics (see Note 2)

The output from a standard 454-pyroseqeuncing run on a GS-flx platform is a sequence quality file (.qual), a binary output file (.sff), and a fasta file (.fasta). Each of these files can be used for specific analysis; however, for the majority of users the fasta file and quality file are the most informative. Many pyrosequencing centres will automatically remove low-quality sequences. The majority of the remaining analysis can be performed using the fasta file. Annotation (the assignment of a functional and/or taxonomic classification to a read) can be performed using several different pipelines. Typically, such pipelines compare the read against a database of sequences for which the classification is known and a match is called if certain score parameters are met. Several methods exist to compare sequences to a reference database, and a review of the currently available methods is outside the scope of this chapter. Briefly, the two most commonly used methods are traditional pairwise analysis using tools such as the Basic Local Alignment Search Tool (BLAST) (7) and analysis using hidden Markov model (HMM) protein profiles, such as HMMER (http://hmmer.wustl.edu). BLAST aligns two sequences and then scores the alignment by rewarding nucleotide or amino-acid matches and penalizing mismatches. Gaps are inserted into a sequence to improve a match if the penalty of opening a gap is less than the reward for the improved match. Statistical tests are then performed to evaluate how likely such a match is to occur at random. An excellent review of BLAST can be found in the book "BLAST" (8). Analysis using HMM protein profiles (also known as Position Specific Score Matrices, PSSM) works on a similar basis, with one notable difference. In programs such as BLAST, the reward or penalty of a match is independent of its position in the sequence; thus, a match for a given amino acid does not take into account whether its neighbors also match. HMM protein profiles are statistical models of multiple sequence alignments and therefore the match/mismatch score values vary depending upon the high or low region of conservation, in which it occurs. Subsequently, HMM protein profiling is more sensitive than BLAST for identifying distant homologues and orthologues of a particular gene and is used in the construction of the Pfam (http://pfam.sanger.ac.uk/) and Conserved Domain (http://www.ncbi.nlm.nih.gov/cdd) public databases (see Note 2).

It is worth noting that use of such annotation pipelines is not trivial as score parameters and subsequent matches are a function of the degree of sequence conservation for a given assignment. For example, parameters used to identify reads associated with the highly conserved 16S rRNA gene would be far more stringent than those used to identify highly diverse functional proteins. Furthermore, the computational power required to run these pipelines currently exceeds that available on desktop computers, requiring dedicated servers with significant storage capacity, processing power, and memory. Fortunately, several online resources exist to annotate sequence data, such as MG-RAST (http://metagenomics.nmpdr.org/) for metagenomic sequences and VAMPS (http://vamps.mbl.edu/index.php) or RDPII (http://rdp.cme.msu.edu) for 16S rRNA reads. Users can upload their datasets to these resources and then visualize the functional and taxonomic data once analysis is complete, providing significant cost-saving to running the analysis in-house, at the cost of a loss of fine-granularity control over the pipeline implementation.

Regardless of whether sequence data is annotated in-house or via a Web service, it is recommended that sequence data is pretreated prior to annotation to remove potential biases associated with each sequencing technology. 454 pyrosequencing data should first be filtered for artificially replicated sequences (9) and then treated for base-calling errors resulting from long homopolymers, using software such as PyroNoise (10). Finally, to compare the abundance of specific genes in two or more datasets, it is critical to remove any effects of differential sampling effort. This can be achieved by randomly resampling the sequences in each sample to the number of reads in the smallest dataset. Postannotation, it is also worth accounting for the effects of possible differences in the average genome size of a sample on the probability of sampling a particular gene.

4. Notes

1. Sampling must be done in as sterile a manner as possible; for water sampling, the use of Sterivex cartridges ensures sterility by preventing handling of the filter membrane. However, it is important to remember to make sure that all tubing used in the analysis is kept as clean as possible (e.g., using 10% HCl wash). For soil and sediment, samples are usually frozen

immediately in liquid nitrogen and subsequently stored at the appropriate temperature. When extracting DNA, thaw samples thoroughly on ice and homogenize by stirring before weighing out the desired amount. Depending on soil or sediment type, the sample can be bead-beaten for slightly longer to increase DNA yield, but it is advisable to check the DNA for shearing before proceeding with downstream applications. For functional gene studies with less-robust primers, it may be necessary to further clean DNA to completely remove inhibitors from the sample.

2. It is essential to use metagenomic pyrosequencing in an appropriate manner so as to be able to test your specific hypothesis. To be able to test a hypothesis, it is essential you use statistical analysis; hence, to make sure that your interpretation is valid, you must consult a statistician or a researcher with experience in statistically relevant experimental design. Here, I outline a recent experiment we have performed to determine the relative impact of three environmental variables on bacterial community function using a mesocosm experiment and pyrosequencing metagenomics. This experiment is an excellent example of the power of statistical design in facilitating downstream processing of data.

    We wanted to determine the relative importance of the impact of temperature, salinity, and phosphate concentration on the microbial metagenome from a surface water sample. We isolated 800 L of surface water and divided this into forty 20 L acid-washed carboys, which were sealed with a gas permeable membrane. We then applied a three-way crossed replicated analysis. We exposed each carboy to two levels for each variable, producing eight treatments, which were replicated five times each (Table 1).

    To determine the original community metagenomic profile, five replicate 1 L samples of the original water were filtered onto 0.22-μm Sterivex filters and stored at −80°C. The experiment was run for 4 weeks, at the end of which 1 L of water was collected in triplicate from each carboy and processed in the same way. DNA was extracted using the methodology above from all Sterivex and then pyrosequenced as described. The resulting fasta file was annotated against the MG-RAST database, and the relative abundance of specific functions was determined in each sample. Statistical analysis of this data was performed using nonparametric multivariate techniques for community composition and univariate analysis of variance tests for diversity measures. The benefit of the three-way crossed analysis is that it enables us to determine if a particular combination of the factors (an interaction) causes an even greater shift in the diversity of the community.

Table 1
Representation of the eight treatment conditions for statistical experimental design

High temperature	High salinity	High phosphate concentration
		Low phosphate concentration
	Low salinity	High phosphate concentration
		Low phosphate concentration
Low temperature	High salinity	High phosphate concentration
		Low phosphate concentration
	Low salinity	High phosphate concentration
		Low phosphate concentration

Univariate tests of diversity indices use higher-way ANOVA, but carried out with distribution-free, permutation-based (PERMANOVA) routines (11). Additionally, following functional characterization of the communities and production of an abundance matrix of operational taxonomic units against experimental condition, community similarity between samples was represented by calculating a Bray–Curtis similarity matrix. Nonmetric multidimensional scaling was used to visualize the relationship between the experimental factors and formally tested using a combination of permutation-based PERMANOVA and fully nonparametric ANOSIM tests (12). Essentially, the experiment was designed as a simple three-way, fixed-factor, fully crossed design. The PERMANOVA tests determine whether main effect differences exist between the levels of a particular factor (e.g., high/low temperature) and/or whether there is evidence of these interacting with other factors in the design (e.g., effects only seen for high temperature with high phosphate concentration, not with low phosphate concentration etc). PERMANOVA is important for the multivariate compositional data, where it is applied to test for main effects and interactions. The robustness of these results for particular main effects (not interactions) can be assessed by the fully nonparametric ANOSIM tests.

Acknowledgments

The author would like to thank Neil Hall from the NERC/ University of Liverpool Advanced Genomics Facility.

References

1. Rusch DB, Halpern AL, Sutton G, Heidelberg KB, Williamson S, et al. (2007) The Sorcerer II Global Ocean Sampling expedition: northwest Atlantic through eastern tropical Pacific. *PLoS Biol.* **5**, 398–431
2. DeLong EF, Preston CM, Mincer T, Rich V, Hallam SJ, et al. (2006) Community genomics among stratified microbial assemblages in the ocean's interior. *Science* **311**, 496–503.
3. Gilbert JA, Field D, Huang Y, Edwards R, Li W, et al. (2008) Detection of Large Numbers of Novel Sequences in the Metatranscriptomes of Complex Marine Microbial Communities. *PLoS ONE* 3(8): e3042. doi:10.1371/journal.pone.0003042
4. Temperton B, Field D, Oliver A, Tiwari B, Joint I, Gilbert JA (2009) Bias in assessments of marine microbial biodiversity in fosmid libraries as evaluated by pyrosequencing. *ISME J.* **3**, 792–6.
5. Smith CJ, Nedwell DB, Dong LF, Osborn AM (2007) Diversity and abundance of nitrate reductase genes (*narG* and *napA*), nitrite reductase genes (*nirS* and *nrfA*), and their transcripts in estuarine sediments. *Appl. Environ. Microbiol.* **73**, 3612–3622.
6. Sambrook J, Fritsch EF, Maniatis T (2000) Molecular Cloning: a laboratory manual. 3rd ed. N.Y., Cold Spring Harbor Laboratory, Cold Spring Harbor Laboratory Press.
7. Altschul, SF., Gish W, Miller W, Myers EW, Lipman DJ (1990) Basic local alignment search tool. *J. Mol. Biol.* **215**, 403–10.
8. Korf I, Yandell M, Bedell J (2003) BLAST. O'Reilly Media.
9. Gomez-Alvarez V, Teal TK, Schmidt TM, (2009) Systematic artifacts in metagenomes from complex microbial communities. *ISME* **3**, 1314–1317.
10. Quince C, Lanzan A, Curtis TP, Davenport RJ, Hall N, Head IM, Read IF, Sloan WT (2009) Accurate determination of microbial diversity from 454 pyrosequencing data. *Nature Methods* **6**, 639–641.
11. Anderson MJ (2001) A new method for non-parametric multivariate analysis of variance. *Austral Ecology* **26**, 32–46
12. Clarke KR (1993) Non-parametric multivariate analyses of changes in community structure. Australian Journal of Ecology **18**, 117–143.

Chapter 13

Metagenomic Analysis of Intestinal Microbiomes in Chickens

Taejoong Kim and Egbert Mundt

Abstract

The digestive tract of animals contains a very large numbers of microorganisms with a high diversity. Traditionally, characterization of these microbial communities has relied on the ability to clonally culture each microorganism. With significant improvements in nucleotide sequencing technologies to economically obtain billions of bases, the study of genetic material recovered directly from environmental samples is becoming increasingly affordable. The investigation of microorganisms as a community regardless of their ability to be cultured has become reality. Using the metagenomic approach for analysis of chicken intestinal homogenates, we were able to greatly enhance the understanding of communities of microorganism in healthy and Runting Stunting Syndrome-infected chickens. In particular, comparative analysis of metagenomes from infected and noninfected chickens resulted in the identification of microorganisms as pathogen candidates. In this chapter, we demonstrate step-by-step how tools for comparative metagenomic analysis can facilitate the resolution of complex, multifactor-involved diseases.

Key words: Uncultivable microorganism, Intestine, Microbiomes, Bioinformatics

1. Introduction

Many microorganisms have been isolated and subsequently characterized using conventional approaches such as adaptation into cell culture (viruses) or cultivation on selective media (bacteria, fungi) followed by biological and genetic analysis of each isolate . However, noncultivable microorganisms have been neglected in identification because of difficulties in isolating them as clonal cultures. Electron microscopy has been used as a tool to identify viruses, but it is necessary to first purify and concentrate them with a variety of protocols followed by a time-consuming and imprecise process. With the emergence and steady progression of next-generation sequencing technologies (1–4), many microorganisms have been sequenced not only as clonal cultures but

Young Min Kwon and Steven C. Ricke (eds.), *High-Throughput Next Generation Sequencing: Methods and Applications*, Methods in Molecular Biology, vol. 733, DOI 10.1007/978-1-61779-089-8_13,

also as a microbial community (microbiomes) from human and animal samples (5–7). Most microorganisms involved in enteric disease are not cultivable in vitro or can only be cultivated using sophisticated methods; thus, direct sequencing of nucleic acids from the samples as microbiomes can be an ideal approach to identify and analyze disease-associated microorganisms (5, 8).

In this chapter, the bioinformatics procedure for comparative analysis of metagenomes obtained through the next-generation sequencing from the intestinal microbiomes of Runting Stunting Syndrome (RSS)-infected (9, 10) and uninfected control chicken is described. RSS, an economically important poultry enteric disease, is characterized by diarrhea, cystic enteropathy in small intestine, and growth retardation of chicken. Some virus species have been suggested as the causative pathogens, though the etiological agents remain to be identified (10, 11). The identification of RSS pathogens is the primary aim of the current comparative analysis of metagenomes from RSS-infected and uninfected birds. Individual sequence reads obtained from the Genome Sequencer FLX system (Roche 454® Life Science, Indianapolis, IN) or assembled contigs were screened against the National Center for Biotechnology Information (NCBI) database. The Basic Local Alignment Search Tool (BLAST) results of multiple metagenomes were compared using Metagenome Analyzer (MEGAN) program.

2. Materials

The procedure for isolating and preparing the metagenome is a critical step; thus, a comprehensive plan to produce the metagenome needs to be fully determined in advance. A recent report, the laboratory procedure to generate viral metagenomes (12), is strongly recommended reading for those researchers who plan to conduct viral metagenomics research.

2.1. Intestinal Metagenomes from RSS-Infected and Uninfected Control Chicken (Fig. 1)

These metagenomes were generated from small-intestine homogenates of RSS-infected and uninfected control chickens. The sediments were prepared from the homogenates via centrifugation, filtration, and ultracentrifugation. The nucleic acids were purified from pellets obtained after ultracentrifugation through a 10% sucrose cushion by commercially available nucleic acid (DNA or RNA) isolation kits. The RNAs were transcribed into cDNAs and subsequently amplified by sequence-specific primers. The obtained DNA was amplified with a random hexamer approach. The sequencing reaction was conducted using standard methods for the Genome Sequencer FLX system at the 454 Life Science facility (454 Life Science, Branford, CT, USA).

1. Sample collection from RSS-infected and uninfected control chicken

↓

2. Preparation of samples (e.g. homogenization, centrifugation, filtration, ultracentrifugation)

↓

3. Nucleic acids isolation: DNA and RNA

↓

4. cDNA transcription and amplification of nucleic acids if necessary

↓

5. Sequencing

↓

6. Preparation of the sequence reads to analyze if necessary (e.g. assembly)

↓

7. BLAST analysis (e.g. tBLASTx, BLASTn)

↓

8. MEGAN analysis (e.g. taxonomical analysis, comparative analysis)

Fig. 1. Flowchart of comparative analysis of metagenomes from RSS-infected and uninfected control chicken.

2.2. BLAST Analysis and Comparison of Metagenomes

1. The Linux cluster (rcluster) at the Research Computing Center (RCC), University of Georgia (http://rcc.uga.edu), which is comprised computer nodes with Intel and AMD single-, dual-, and quad-core processors, was used to conduct BLAST analyses. The RCCBatchBlast bioinformatics software (http://rcc.uga.edu/software/app/rccbatchblast/) is available at the RCC.
2. MEGAN software, a computer program used for metagenome analysis, is available at http://www-ab.informatik.unituebingen.de/software/megan (see Note 1).
3. The DNASTAR® Lasergene® v8.0 package including SeqBuilder and SeqMan Pro software (DNASTAR, Inc., Madison, WI, USA) was used for the analysis of the 454 sequence reads (see Note 2).

3. Methods

With improvement in next-generation sequencing technologies in combination with the availability of appropriate amplifications methods of nucleic acids, metagenomics data can be generated from a variety of samples. Several bioinformatics analysis tools of metagenomes have been developed (13–15), and recently a comparative analysis of multiple metagenomes using a user-friendly graphical interface has been established (16–18). Comparative analysis is very practical to identify specific nucleotide sequences present in only a specific metagenome compared to others and also allows the quantitative comparison of multiple metagenomes. The methods described in this chapter are for comparative analysis of metagenomes from RSS-infected and uninfected chicken. With appropriate modifications depending on the availability of bioinformatics resources, the described methods can be applied to most metagenomic analyses.

3.1. Contig Assembly

The sequence read files can be converted as .fas or .fasta formatted files, if necessary, and submitted for analysis against the NCBI NT/NR database. To conduct rapid BLAST analysis, the sequence reads can be assembled as contigs with appropriate thresholds of sequence assembly software such as SeqMan Pro, GS De Novo Assembler (see Note 2); that is, single sequence reads will be analyzed for the presence of very similar or identical sequences between a 5′ sequence of one read with the 3′sequence of a second read. If the overlap is of sufficient length to distinguish it from being a repeat in the sequence, the two sequences must be contiguous (19). This leads to longer continuous sequences or contigs.

1. The sequence reads were assembled as contigs using SeqMan Pro software from DNASTAR® Lasergene®. The current instructions assume the use of SeqMan Pro software of DNASTAR® Lasergene® v8.0 package (DNASTAR, Inc.).
2. After launching SeqMan Pro, click "Add Sequences" to add the sequence outfile (as .fasta or .fas file), click "Project," and select the "Parameter" option. Now, choose "Assembling" from the Seqman parameters window and use the "Pro Assembler" option to assemble the sequence data. It is possible to change in each section the parameters as follows (e.g.,):

   Match size: 35, Minimum match percentage: 99, Match spacing: 150 [default], Minimum sequence length: 40, Gap penalty: 0.00 [default], Gap length penalty: 0.70 [default], Maximum mismatch end bases: 0 [default]. It needs to be mentioned that these parameters are flexible and can be chosen on an empirical basis. In general, the higher the match size and the minimum match percentage are chosen, the more stringent contigs will be assembled. On the other hand, a high stringency might exclude similar but not identical sequences to be assembled into contigs. In other words, these parameters need to be adjusted to the goal of the project.
3. Click "OK" and close the Parameter window, then return to the unassembled sequences window. Click "Assemble." After finishing the assembly, choose the contigs that have a minimum of four sequences in the contig and save them as a single SeqMan file. The number of four sequences has been chosen based on our experience with performed analysis. The saved SeqMan file can be converted to a .fas or .fasta file and subsequently used for analysis against the NCBI NR/NT database.

3.2. BLAST Analysis

The major objective of the current project was the identification of unique sequences in the RSS metagenomes by employing different BLAST analyses and a subsequent subtractive approach of metagenomes of uninfected chickens from RSS-infected chickens. Primarily, tBLASTx analysis was conducted to compare the six-frame translations of nucleotide query sequences against the six-frame translations of NCBI nucleotide sequence database. This approach will identify similar amino-acid sequences, since, due to the degeneration of the genetic code, the probability to identify similar amino acid sequences from the same virus families is higher.

1. The instructions assume the use of the Linux cluster at the University of Georgia RCC. Detailed information can be obtained at the RCC Web site (http://rcc.uga.edu) (see Note 3).
2. Open the rcluster connection to the RCC of UGA using the researcher's personal computer. The RCC staff developed a RCCBatchBlast software which is a semiautomatic pipeline to run NCBI BLAST analysis at the RCC rcluster. In this software, large query files with multiple query sequences were divided into small input files as commanded and run a blastall analysis against the designated NCBI database. After completion of the blastall analysis, the RCCBatchBlast results are reviewed by the rccbatchblast-check software. If the review results are correct, the BLAST results from all divided query sequences will be merged into a single outfile.
3. Here is the example Linux command line to analyze using RCCBatchBlast and rccbatchblast-check software.

   rccbatchblast –i Query.fasta –d/db/ncbiblast-latest/nt –p tblastx –size 50 –e 0.001 –queue r1-10d

   Then, check the submitted job status, and if it is complete, check the completed jobs with the following command:

   perl/usr/local/bin/rccbatchblast-check.pl Query.fasta Query.tblastx.out

   If the tBLASTx results of divided query sequences are successful, the researcher will obtain a single outfile named as *Query.tblastx.out* that contains the tBLASTx results of the query sequences to the NCBI NT database.
4. Similarly, BLASTn analysis that compares a nucleotide query sequence against a nucleotide sequence database also was conducted using RCCBatchBlast software. Here is an example Linux command line to BLASTn analysis:

   rccbatchblast –i Query.fasta –d/db/ncbiblast-latest/nt –p blastn –size 50 –e 0.001 –queue r1-10d

   Also, it is necessary to verify the job and create a single outfile with the following command:

   perl/usr/local/bin/rccbatchblast-check.pl Query.fasta Query.blastn.out

Arguments for tBLASTx and BLASTn of the above are as follows:

– *I*: input file

– *d*: database to be searched

– *p*: BLAST program (i.e., BLASTn, BLASTp, BLASTx, tBLASTn, tBLASTx)

– *size*: determine the size of the split unit of the original query file (The researcher can modify as demand such as 50, 100, or 200.)

– *e*: the cut-off expectation (E) value of BLAST analysis (The researcher can modify the cut-off E value as demand such as 10, 1 or 0.001)

– *queue*: queue-name that the researcher's analysis will be submitted (The researcher can select the available queue, e.g., a dual-core or quad-core and 24-h analysis or 10-day analysis.)

3.3. Comparison of BLAST Analysis Results of Metagenomes by MEGAN

Among several bioinformatics tools for metagenomic data analysis, the Metagenome Analyzer (MEGAN) program provides a comparative analysis of multiple metagenomes through a user-friendly interface and conventional computing power such as a desktop PC (16–18). Currently, MEGAN v3.8 is available at its Web site (http://www-ab.informatik.uni-tuebingen.de/software/megan) and it also provides COG (Clusters of Orthologous Groups of proteins) analysis, which is the functional content of metagenomes using NCBI-NR sequence annotated by COG identifiers as well as GO (Gene Ontology) analysis, which is a comparative analysis of the functional content of metagenome datasets based on gene ontology.

1. The instructions assume the use of MEGAN (v 3.8) software with a Windows-based computer. After launching MEGAN v3.8, click "Add Sequences" to add the sequence outfile (as .fasta or .fas file), Go to "File," then click the "Import blast" button, which will result in the appearance of the READS file window. Select the file (the result file of BLAST analysis described in Subheading 3.2) to import for MEGAN analysis; that is, the researcher needs to specify the reads file (the query file for BLAST analysis with .fasta format), and the researcher provides the MEGAN file name (the outfile of MEGAN analysis). In the "Files" tab, check the researcher's selection of each files (BLAST file, READS file, MEGAN file) and then in the lowest common ancestor (LCA) parameters (params) tab, specify the researcher's thresholds for the minimum support, minimum score, minimum score/length, top percent, and win score option, and then click "Apply" and the READS files window will close (see Note 4).

The thresholds of LCA parameters for current analysis were as follows: minimum support: 5, minimum score: 35, minimum score/length: 0 [default], top percent: 10 [default] and win score: 0 [default].

2. After specifying the thresholds, the automatic import will start shortly. The "Question: Which BLAST format?" window will appear if the researcher do not indicate the BLAST format (e.g., BLASTn, BLASTp, BLASTx). Specify the appropriate BLAST format, and the program will start the parsing and assigning process of the BLAST analysis. The MEGAN program assigns the sequence reads to taxa using the LCA algorithm and then displays the induced taxonomy (Fig. 2). MEGAN provides for the user different ways to explore the results with many different taxonomical levels from Kingdom to Species, and moreover, it supports the statistical analysis of the each reads assigned into specific taxonomy analysis (16–18).

3. For comparative analysis of the multiple metagenomes, each metagenome needs to be analyzed using the MEGAN program with the appropriate LCA algorithm thresholds as described above. Next, click the "Compare" button in the File tab, select the datasets to compare, and then click "OK." The compared MEGAN file will be generated automatically (Fig. 3), and the comparative analysis can also be modified as different chart types with statistical analysis and heat maps for different methods of comparison.

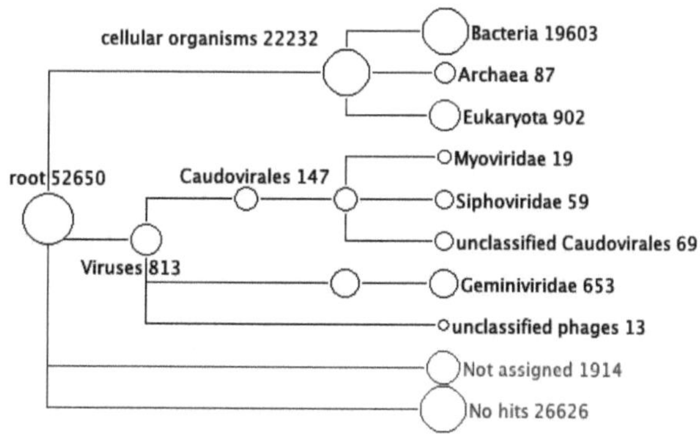

Fig. 2. MEGAN analysis of metagenome from uninfected control chicken. A total of 52,650 sequence reads were analyzed with tBLASTx and assigned to taxa using the MEGAN software. The size of the *circle* indicates the number of nodes (e.g., bacteria 19,603 reads among total 52,650 reads) belonging to each organism.

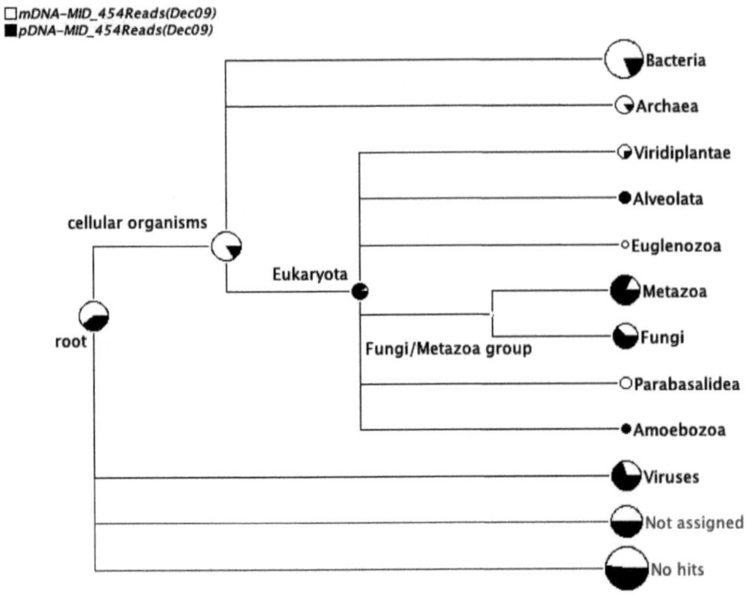

Fig. 3. Comparative MEGAN analysis of metagenomes from RSS-infected and uninfected control chickens. The mDNA-MID_454 Reads are the sequence reads from uninfected control, and pDNA-MID_454 Reads are the sequence reads from RSS-infected chicken. The pie chart indicates the proportion of the reads from each metagenome (i.e., RSS-diseased [*black*] vs. uninfected chicken [*white*]).

4. Notes

1. Besides MEGAN software, a number of bioinformatics tools have been created to analyze metagenome datasets. Also, several Web sites have been developed with tools that perform metagenomics data analysis:

   http://metagenomics.theseed.org (Meta Genome Rapid Annotation using Subsystem Technology)

   http://camera.calit2.net/index.php (Community Cyberinfrastructure for Advanced Marine Microbial Ecology Research and Analysis)

   http://img.jgi.doe.gov/cgi-bin/m/main.cgi (Integrated Microbial Genomes with Microbiome Samples)

2. Currently, DNASTAR® Lasergene® v8.1 package is available as expanded analysis capability for both conventional and the next-generation sequencing applications, multisite gateway cloning support, expanded single-nucleotide polymorphisms (SNP) reporting and management and easy integration with GenVision (http://www.dnastar.com/). Especially, Seq Man

N Gen2.0 was designed to rapidly assemble the sequence data from the next-generation sequencing according to the manufacturer, and it will help to analyze the data with new features as explained at the company's Web site (http://www.dnastar.com/t-products-seqman-ngen.aspx). Also, 454 Life Sciences launched new improved GS data analysis software, GS De Novo assembler that assembles the genomes up to 3 GB in size (http://www.454.com/products-solutions/analysis tools/gs-de-novo-assembler.asp). Since the assembly of contig highly relies on the homology between the individual sequences reads, the researcher need to set the appropriate thresholds for each project, and it might be necessary to validate the assembled contigs with different methods. Particularly, the thresholds described in Subheading 3.1 needs to be modified by researchers depending on the goal of each project.

3. The researcher needs the basic Linux programming language skills to use the UGA-RCC Linux cluster. The researcher also can use Web-based bioinformatics tools for BLAST analysis of the datasets. The Blast2Go, a tool for functional annotation of (novel) sequences and the analysis of annotation data is available at http://www.blast2go.org/. The BLAST results from blast2go software can be exported as .fasta or .fas files and analyzed using MEGAN or other metagenomics analysis software.
4. The principles of LCA algorithm were described in detail (15, 16). To minimize false positives, the minimum support option can be stringent such as 10 or more and minimum score as 50 or more. If the hit does not meet the fixed thresholds, it will be categorized as "Not Assigned." The researcher can modify the thresholds for LCA algorithm as required. In addition, the researcher can adjust the taxonomical level of MEGAN analysis tree as necessary and display or remove the node level as well as taxon names/numbers of the reads in the MEGAN analysis tree.

Acknowledgments

The authors thank Dr. Rebecca Thurber in the Department of Biological Sciences at Florida International University for technical assistances in analysis of chicken intestine metagenome and Dr. Hans Cheng in USDA-ARS Avian Diseases and Oncology Laboratory for his advice and discussion.

References

1. Margulies, M., Egholm, M., Altman, W.E., Attiya, S., Bader, J.S., Bemben, L.A., et al. (2005) Genome sequencing in microfabricated high-density picolitre reactors. *Nature* **437**, 376–380.
2. Shendure, J., Porreca, G.J., Reppas, N.B., Lin, X., McCutcheon, J.P., Rosenbaum, A.M., et al. (2005) Accurate multiplex polony sequencing of an evolved bacterial genome. *Science* **309**, 1728–1732.
3. Soni, G.V., and Meller, A. (2007) Progress toward ultrafast DNA sequencing using solid-state nanopores. *Clinical Chemistry* **53**, 1996–2001.
4. Healy, K. (2007) Nanopore-based single-molecule DNA analysis. *Nanomedicine* **2**, 459–481.
5. Breitbart, M., Hewson, I., Felts, B., Mahaffy, J.M., Nultson, J., Salamon, P., et al. (2003) Metagenomic analyses of an uncultured viral community from human feces. *Journal of Bacteriology* **185**, 6220–6223.
6. Zhang, T., Breitbart, M., Lee, W.H., Run, J.Q., Wei, C.L. Soh, S.W., et al. (2006) RNA viral community in human feces: prevalence of plant pathogenic viruses. *PLoS Biology* **4**, e3.
7. Qu, A., Brulc, J.M., Wilson, M.K., Law, B.F., Theoret, J.R., Joens, L.A., et al. (2009) Comparative metagenomics reveals host specific metavirulomes and horizontal gene transfer elements in the chicken cecum microbiome. *PLoS ONE* **3**, e2945
8. Frank, D.N., and Pace, N.R. (2008) Gastrointestinal microbiology enters the metagenomics era. *Current Opinion in Gastroenterology* **24**, 4–10.
9. Reece, R.L., Hooper, P.T., Tate, S.H., Beddome, V.D., Forsyth, W.M., Scott, P.C., and Barr, D.A. (1984) Field, clinical and pathological observations of a runting and stunting syndrome in broilers. *Veterinary Record* **115**, 483–485.
10. Smart, I.J., Barr, D.A., Reece, R.L., Forsyth, W.M., and Ewing, I. (1988) Experimental reproduction of the runting-stunting syndrome of broiler chickens. *Avian Pathology* **17**, 617–627.
11. Goodwin, M.A., Davis, J.F., McNulty, M.S., Brown, J., and Player, E.C. (1993) Enteritis (so-called runting stunting syndrome) in Georgia broiler chicks. *Avian Diseases* **37**, 451–458.
12. Thurber, R.V., Haynes, M., Breitbart, M., Wegley, L., and Rohwer, F. (2009) Laboratory procedure to generate viral metagenomes. *Nature Protocols* **4**, 470–483.
13. Markowitz, V.M., Ivanova, N., Palaniappan, K., Szeto, E., Korezeniewski, F., Lykidis, A., et al. (2006) An experimental metagenome data management and analysis system. *Bioinformatics* **22**, e359–367.
14. Rodriquez-Brito, B., Rhower, F., and Edwards, R. (2006) An application of statistics to comparative metagenomics. *BMC Bioinformatics* 7, 162.
15. Schloss, P.D. and Handelsman, J. (2008) A statistical toolbox for metagenomics: assessing functional diversity in microbial communities. *BMC Bioinformatics* **23**, 34.
16. Huson, D.H., Fuch, A.F., Qi, J., and Schuster, S.C. (2007) MEGAN analysis of metagenomic data. *Genome Research* **17**, 377–386.
17. Huson, D.H., Richter, D.C., Mitra, S., Auch, A.F., and Schuster, S.C. (2009) Methods for comparative metagenomics. *BMC Bioinformatics* **10** Suppl 1:S12.
18. Mitra, S., Klar, B., and Huson, D. (2009) Visual and statistical comparison of metagenomes. *Bioinformatics* **25**, 1849–1855.
19. Staden, R. (1979) A strategy of DNA sequencing employing computer programs. *Nucleic Acids Res.* **6**, 2601–2610.

Chapter 14

Gene Expression Profiling: Metatranscriptomics

Jack A. Gilbert and Margaret Hughes

Abstract

Metatranscriptomics has been developed to help understand how communities respond to changes in their environment. Metagenomic studies provided a snapshot of the genetic composition of the community at any given time. However, short-timescale studies investigating the response of communities to rapid environmental changes (e.g. pollution events or diurnal light availability) require analysis of changes in the abundance and composition of the active fraction of the community. Metatranscriptomics enables researchers to investigate the actively transcribed ribosomal and messenger RNA from a community. It has been applied to environments as diverse as soil and seawater. This chapter outlines sampling protocols and RNA extraction techniques from these two ecosystems, as well as details a method to enrich mRNA in the extracted nucleic acid. Also, a section is dedicated for outlining a bioinformatic procedure for the analysis of metatranscriptomic datasets.

Key words: Metatranscriptomics, Marine, Soil, Expression

1. Introduction

DNA-based metagenomics has become a standard tool for analysing microbial community structure (1–3) by sequencing random community DNA from environmental samples and subsequent determination of taxonomic and protein-encoding gene diversity. However, understanding how bacterial communities respond to rapid changes in their environment can be better elucidated by analysing community mRNA to explore the expressional profile of functional and taxonomic marker genes (4).

Metatranscriptomic studies have traditionally involved the use of either microarrays (5) or mRNA-derived cDNA clone libraries (expressed sequence tag (EST) libraries) (6). However, more recently, high-throughput sequencing technologies such as pyrosequencing have been applied to metatranscriptomic studies (7–11).

Young Min Kwon and Steven C. Ricke (eds.), *High-Throughput Next Generation Sequencing: Methods and Applications*, Methods in Molecular Biology, vol. 733, DOI 10.1007/978-1-61779-089-8_14, © Springer Science+Business Media, LLC 2011

Two studies of soil communities have sequenced total RNA for the purpose of exploring both community structure, through the analysis of ribosomal RNA (rRNA), and community function, through the study of mRNA (7, 8). Both studies produced extremely valuable information, and the techniques used to extract total RNA from the soil are explored in this chapter. In marine pelagic systems, there has been a strong focus to try and reduce the quantity of rRNA cosequenced with the mRNA (9–11). This mRNA enrichment aims to improve the yield of functional genetic information per sequencing run, so as to better explore specific functional response of a community to specific environmental change. Here, we cover the methodologies of Gilbert and colleagues (10) and Urich and colleagues (8), which represent analyses from seawater and soil. The methodologies of Frias-Lopez et al. (9) and Poretsky et al. (11) are quite similar, and the methodological publication by Poretsky et al. (12) provides an excellent report of this alternative methodology.

2. Materials

2.1. RNA Extraction from Soil

1. DEPC-treated glassware and plasticware.
2. 2-mL RNase-free microcentrifuge tube.
3. Bead beating system (e.g. FastPrep FP120, Bio-101, Vista, Calif.).
4. Hexadecyltrimethylammonium bromide (CTAB) extraction buffer: 10% (wt/vol) CTAB, 0.7 M NaCl, 240 mM potassium phosphate buffer, pH 8.0.
5. Phenol–chloroform–isoamyl alcohol (25:24:1) (pH 8.0).
6. Chloroform–isoamyl alcohol (24:1).
7. 30% (wt/vol) polyethelene glycol 6000 (Fluka BioChemika)–1.6 M NaCl.
8. 70% (vol/vol) ethanol.
9. RNase-free Tris–EDTA buffer, pH 7.4 (Severn Biotech, Kidderminster, UK).

2.2. RNA Extraction from Seawater

1. 140-mm diameter, 1.6-μm pore size GF/A filters (Whatman, USA).
2. Sterivex filter cartridges – 0.22-μm pore size filters (Millipore, UK).
3. Peristaltic pump capable of holding 16-mm Tygon LFL tubing.
4. 142-mm filter rig (Millipore, UK).
5. Liquid nitrogen.

6. SET buffer: 40 mM EDTA, 50 mM Tris–HCl, pH 9, 0.75 M sucrose.
7. 9 mg/ml Lysozyme in 10 mM Tris/HCl, pH 8.
8. 10% SDS.
9. 20 mg/ml Proteinase K in 20 mM Tris/HCl, pH 8.
10. MaxTrack Gel lock tubes (Qiagen, USA).
11. 7.5 M ammonium acetate.
12. Phenol–chloroform–isoamyl alcohol (25:24:1), pH 8 and chloroform–isoamyl alcohol (24:1), pH 8.
13. 100% molecular-grade ethanol.
14. DEPC-treated sterile water.
15. RNA MinElute™ clean-up kit (Qiagen).
16. β-mercaptoethanol.
17. Turbo DNA-free enzyme (Ambion).

2.3. mRNA enrichment

1. Microbe Express Kit (Ambion).
2. TE buffer (10 mM Tris–HCl pH 8.0, 1 mM EDTA).
3. MEGAclear™ kit (Ambion).
4. DEPC-treated sterile water.
5. SuperScript® III enzyme reverse transcriptase kit (Invitrogen).
6. Random hexamer primers (Promega).
7. RiboShredder™ RNase Blend (Epicentre).
8. GenomiPHI™ V2 method (GE Healthcare).
9. S1 nuclease (Invitrogen).
10. 0.5 M EDTA.
11. MinElute column (Qiagen).
12. AMPure beads (Agencourt).

2.4. Pyrosequencing

1. AMPure 60-ml kit (p/n 000130 Agencourt).
2. RNA 6000 Pico Chip kit (p/n 5067-1513 Agilent).
3. DNA 7500 LabChip kit (p/n 5067-1506 Agilent).
4. MinElute PCR purification kit (p/n 28004 Qiagen).
5. RiboGreen RNA Quantitation kit (p/n R-11490 Invitrogen).
6. Quant-iT PicoGreen DNA reagent (p/n P7581 Invitrogen).
7. 3 M sodium acetate buffer.
8. 10 N sodium hydroxide.
9. Isopropanol (reagent grade).
10. Ethanol (reagent and molecular biology grades).
11. GS Titanium General library prep kit (p/n 05233747001 Roche).

12. GS Titanium LV emPCR kit (p/n 05233542001 Roche).
13. GS Titanium SV emPCR kit (p/n 05233615001 Roche).
14. GS Titanium emPCR breaking kit (p/n 05233658001 Roche).
15. GS Titanium emPCR filters (p/n 05233674001 Roche).
16. GS Titanium Sequencing kit (p/n 05233526001 Roche).
17. No Stick 1.5-ml tubes (p/n 2410 Alpha Labs).
18. Nitrogen gas.
19. DynaMag Z (p/n 12321D Invitrogen).
20. Rubber stoppers.

3. Methods

3.1. RNA Extraction from Soil or Sediment (see Notes 1–3)

1. 0.5 g (wet weight) of soil is added to a 2-ml microcentrifuge tube.
2. 0.5 ml of hexadecyltrimethylammonium bromide (CTAB) extraction buffer and 0.5 ml of phenol–chloroform–isoamyl alcohol (25:24:1) (pH 8.0) are added to each extraction.
3. Tubes are shaken in a bead beater system for 30 s at 5.5 m/s.
4. The aqueous phase is separated by centrifugation (16,000 × *g*) for 5 min at 4°C.
5. The aqueous phase was added to an equal volume of chloroform–isoamyl alcohol (24:1).
6. Sample is centrifuged at (16,000 × *g*) for 5 min at 4°C.
7. DNA and RNA are precipitated from the aqueous layer with 2 volumes of 30% (wt/vol) polyethylene glycol 6000–1.6 M NaCl for 2 h at room temperature, followed by centrifugation (18,000 × *g*) at 4°C for 10 min.
8. DNA/RNA pellet is then washed with ice-cold 70% (vol/vol) ethanol by centrifugation at 10,000 × *g* for 20 min.
9. DNA/RNA is then air-dried for 15 min prior to resuspension in 1,000 μl of RNase-free Tris–EDTA buffer.
10. 100 μl of the total RNA should be purified using the RNA MinElute™ clean-up kit (Qiagen) with β-mercaptoethanol added to the RLT buffer.
11. Approximate RNA concentration is determined by nanolitre spectrophotometry and checked for rRNA integrity using an Agilent bioanalyser (RNA nano6000 chip). The integrity of rRNA was demonstrated by highly defined, discrete rRNA peaks, with the 23S rRNA peak being 1.5–2 times higher

than the 16S rRNA peak. Fully intact rRNA is essential for subtractive hybridisation because degraded rRNA molecules will not be fully subtracted from the total RNA pool.

12. DNA contamination was removed from total RNA samples by treating with the Turbo DNA-free enzyme (Ambion).

3.2. RNA Extraction from Seawater (see Notes 1–3)

1. Filter 10–15 L of seawater through a 140-mm diameter, 1.6-μm GF/A filter (Whatman), to reduce eukaryotic cell abundance and maximise the proportion of prokaryotic cells (see Notes 1–3).
2. Apply filtrate directly to a 0.22-μm Sterivex filter (Millipore).
3. Following filtration, each Sterivex was pumped dry and frozen in liquid nitrogen.
4. After thawing on ice, add 1.6 ml of SET lysis buffer directly on top of Sterivex using a 2.5-ml syringe with a 25 G 5/8 in. needle.
5. Add 180 μl of fresh lysozyme and seal the Sterivex (Blu-Tack works well).
6. Incubate at 37°C for 30 min with rotation in a Hybaid oven.
7. Add 200 μl of SDS.
8. Add 55 μl of 20 mg/ml fresh proteinase K.
9. Incubate at 55°C for 2 h with rotation in a Hybaid oven.
10. Withdraw lysate into a 5-ml syringe.
11. Add 1 ml of fresh SET buffer to Sterivex and rotate to rinse.
12. Withdraw rinse buffer into the same 5-ml syringe.
13. Add lysate to 15-ml Maxtract tube (Qiagen) containing 2 ml of phenol–chloroform–isoamyl alcohol (25:24:1), pH 8. Shake gently until mixed and then centrifuge at 1,500 × *g* for 5 min.
14. Add an additional 2 ml of phenol–chloroform–isoamyl alcohol (25:24:1). Shake gently until mixed and centrifuge at 1,500 × *g* for 5 min.
15. Add 2 ml of chloroform–isoamyl alcohol (24:1). Shake gently until mixed and centrifuge at 1,500 × *g* for 5 min.
16. Decant aqueous phase to a sterile and DEPC-treated (if RNA needed) 20-ml centrifuge tube and add 0.5 V of 7.5 M ammonium acetate. Mix briefly and then add 2.5 V of pure ethanol.
17. Mix and leave at –20°C for >1 h (overnight is fine).
18. Centrifuge at 10,000 × *g* for 30 min at 4°C and decant ethanol.
19. Add 2 ml of 80% ethanol and rinse tube, then centrifuge at 10,000 × *g* for 20 min at 4°C and decant ethanol, and repeat.

20. Decant ethanol and leave inverted for 15 min in fume hood (provides air flow).
21. Suspend invisible pellet in 200 μl of DEPC-treated sterile water. Leave on ice for approximately 1 h with frequent finger-tapping to rinse tube walls.
22. Please refer to Subheading 3.1 steps 10–12 for completion of this protocol.

3.3. mRNA Enrichment Techniques

If analysis of total RNA is desired, please follow only steps 4, 5, 8, and 9 to avoid removal of rRNA or other small RNAs, and to produce cDNA ready for pyrosequencing. Alternatively, follow entire protocol to produce mRNA-enriched cDNA ready for pyrosequencing.

1. Total RNA was applied to the subtractive hybridisation method (Microbe Express Kit, Ambion) to remove rRNA from the mRNA. The manufacturer's instructions provide sufficient detail to carry out this procedure.
2. mRNA was eluted in 25 μl of TE buffer
3. Resuspended mRNA was applied to the MEGAclear™ kit (Ambion) to remove small RNAs and small contaminants, as per the manufacturer's instructions. Purified mRNA was eluted in 10 μl of DEPC-treated water.
4. mRNA was then reverse-transcribed to cDNA using the SuperScript® III enzyme (Invitrogen) with random hexamer primers (Promega) following the manufacturer's instructions for random primer transcription.
5. The cDNA was treated with RiboShredder™ RNase Blend (Epicentre) to remove trace RNA contaminants, with incubation at 37°C for 20 min.
6. 1 μl of cDNA was then randomly amplified using the GenomiPHI™ V2 kit (GE Healthcare). Ideally, this reaction is performed 10×, and then these replicates are pooled to remove potential random amplification bias inherent in multiple displacement amplification technology.
7. Amplified samples are treated with S1 nuclease at 2 μ/μg cDNA. The reaction is incubated in supplied buffer at 37°C for 30 min. The reaction is stopped by adding 20 mM final concentration EDTA and then cleaned up through a Qiagen MinElute column. S1 nuclease treatment is required because GenomiPHI produces branched DNA molecules which are recalcitrant to pyrosequencing; the S1 nuclease cuts the branches, leaving unbranched DNA which can then be pyrosequenced.
8. cDNA was nebulised to produce an average size of 500 bp and then cleaned with AMPure beads (Agencourt).
9. cDNA was then pyrosequenced.

3.4. Pyrosequencing

1. The cDNA needs to be accurately measured by fluorescence using a Quant-iT PicoGreen assay. The ideal amount is 3–5 μg for a fragment library but it is possible to use less with success. It also needs to be of a reasonable molecular weight to maximise random shearing.
2. The DNA in a volume of 100 μl is mixed with 500 μl of nebulisation buffer supplied with the library kit. The DNA is sheared by placing it in a nebuliser vessel and connecting the latter to nitrogen gas at 30 psi for 1 min.
3. The DNA is recovered from the nebulisation chamber and cleaned with Qiagen MinElute kit following the instructions supplied by the kit using 2.5 ml of PB buffer and eluting the DNA in 100 μl of elution buffer.
4. At this stage, the small fragments are removed using AMPure beads. A calibrated amount is added to the DNA so that material of less than 300 bp is left in solution. The higher MW DNA bound to the beads is washed with 70% ethanol, dried, and recovered by eluting in 10 mM Tris pH 7.5
5. The DNA is checked (1 μl aliquot) for size distribution on a DNA 7500 LabChip using an Agilent Bioanalyser.
6. Further manipulations are all described in the library kit. The ends of the DNA are polished, the adapters (with or without barcodes) are added, fragments containing adapters are selected on Dynal beads, and the ends are filled. The single-stranded library is recovered by melting from the Dynal beads with NaOH (0.125 N) as per the manufacturer's instructions and cleaned up with a MinElute column.
7. The library is assessed by running on a RNA Pico Chip and the amount determined with RiboGreen assay.
8. A predetermined amount of the library is used to set up an emulsion PCR reaction using either the large or small volume kit from Roche. This involves binding the DNA to capture beads which are specific for one of the adapters.

3.5. Some Thoughts on Bioinformatics

The output from a standard 454-pyrosequencing run on a GS-flx platform includes a sequence quality file (.qual), a binary output file (.sff), and a fasta file (.fasta). Each of these files can be used for specific analysis; however, for the majority of users, the fasta and quality files are the most informative. Many pyrosequencing centres will automatically remove low-quality sequences. The majority of the remaining analysis can be performed using the fasta file. It is important to consider what question you wish to answer with your metatranscriptomic data. In the case of studies in which mRNA has not been enriched, this usually includes two questions: (1) What changes can be observed in the ribosomal RNA between samples? (2) What changes can be observed in the messenger RNA between samples? The first question is aimed at understanding the taxonomy of the active microbial population. The second question is aimed at understanding the function of the active microbial population. In studies with mRNA enrichment, the latter question is the only viable one, and yet taxonomy can still be inferred from nearest-hit protein-encoding transcript annotation. Here, I outline a suggested bioinformatic pipeline for the isolation and analysis of mRNA-derived cDNA from a fasta file.

1. Extraneous sequences resulting from >1 template molecule per picotitre well should be removed from the fasta file – these are identified as having an identical sequence and identical fasta identity tag. Deletion of one of the duplicates is sufficient.
2. Remove sequences with >10% N's. This will remove the sequences of extremely low quality; an N is called if the sequence quality is too low for appropriate base identification. Sequences with <10% N's may still contain valuable information.
3. Remove sequences <75 bp in length. The average sequence length obtained from a pyrosequencing run depends entirely on the quality and integrity of the isolated RNA. Low-quality RNA will reduce the average read length by up to 50%. Even in these samples, it is relevant to remove short fragments as they are difficult to annotate and contained limited information.
4. Remove sequences with >60% of any single base. In some cases, sequencing error or sequencing of highly repetitive transcribed regions can lead to sequences with very limited information content; if a sequence contains more than 60% of a single base, it is likely to be of limited use in any downstream analyses and should be removed.
5. Remove sequences with long repeats. In addition to step 4, sequences with repeating motifs, e.g. ATTATTATTATTA TTATTATT, and no additional genetic information will provide limited information regarding function and hence should be removed. However, these sequences are likely to be microsatellites and hence may be of use in analyses of this sequence group.
6. Remove rRNA sequences. This is an optional step in that if mRNA enrichment was not applied, then the rRNA could contain useful information. However, if mRNA enrichment was applied, then rRNA should be removed, as residual rRNA will be undoubtedly at non-natural abundances. This can be achieved by comparing a database against the NCBI nucleotide database nt and removing all sequences whose most significant alignment is an rRNA sequence. Alternatively, analysing

the metatranscriptome using the MG-RAST pipeline (http://metagenomics.theseed.org) will provide you with a detailed annotation of the rRNA content of the data through comparison against the GreenGenes, RDPII, and SILVA databases. Simply download these sequences as a fasta file, and then remove these fasta sequences from your original database.

7. For more than one sample to be compared so that the regulation of a particular transcript can be determined between samples, each database must have the same sampling effort. Hence, the dataset must be randomly resampled so that each dataset has the same number of sequences. This can be achieved using Daisy_Chopper v1.0 (http://www.genomics.ceh.ac.uk/GeneSwytch/Tools.html).

8. Resampled mRNA fasta files should then undergo open reading frame (ORF) prediction to identify those fragments which contain functional information. ORFs are predicted in all six reading frames, with the rule that a predicted ORF must contain at least 40 amino acids. This approach can result in many non-coding (shadow) ORFs, but this is less likely with short sequence data as the translations from non-coding frames are usually too short (due to random occurrence of stop codons). ORF prediction scripts such as ORF_FINDER are available from the author.

9. To compare between samples, the predicted ORFs can then be grouped into unique clusters using CD-HIT (13) (cd-hit -i yourfilename.fasta -o yourfilename.clstr -c 1.0 -n 5). If all predicted ORF fastas from all samples are compiled into one fasta file prior to clustering, then the relative contribution of each sample to each cluster can be determined by counting the number of sequences from each sample in each cluster. This will provide an abundance matrix for unique transcript sequences among your different samples.

Following this pipeline, it is possible to produce a metatranscriptomic fingerprint of a particular sample. This is extremely important as most transcripts are orphans, i.e. they cannot be annotated against any known sequence or function. Annotation of a metatranscriptome to an existing database such as Pfam (http://PFAM.sanger.ac.uk/) or SEED (http://metagenomics.theseed.org) provides functional interpretation of a small percentage of the available data. Non-parametric statistical comparison of the complete dataset provides more information regarding the differences between two metatranscriptomic profiles with the inclusion of sequences which cannot be currently annotated.

4. Notes

1. Sampling must be done in as sterile a manner as possible; for water sampling, the use of Sterivex cartridges ensures sterility by preventing direct handling of the filter membrane. However, it is important to remember to make sure that all tubing used in the analysis is kept as clean as possible (e.g. using 10% HCl wash). Sampling should also be done as quickly as possible. Unfortunately, relatively large quantities of water, i.e. 1–10 L, are necessary to obtain sufficient quantities of total RNA for these protocols. As the protocol requires concentration of the microbial fraction on a membrane, it is extremely likely that the transcriptional profile of the organisms changes as a result. As transcriptional response can occur in milliseconds, this problem is virtually impossible to overcome in water sampling. This is less of a problem in soil sampling due to the use of RNA stabilisation products such as RNAlater (Ambion), which can be used immediately on an isolated soil fraction.
2. RNA is extremely labile and susceptible to the ubiquitous presence of RNase enzymes. Hence, precautions must be taken when isolating RNA to prevent degradation. The use of RNase inhibitors and cleaning products (e.g. RNaseZAP (Ambion)), constant changing of gloves, and preparation in a sterile, RNase-free fume hood are all important steps. Please refer to appropriate RNA preparation manuals (e.g. Qiagen RNeasy manual) for further advice. In many mRNA enrichment protocols, the quality of the total RNA preparation is vital for ensuring the removal of rRNA. In the protocol outlined in this chapter, the multiple-displacement amplification (MDA) step can also be a significant factor in the selective amplification of mRNA. This is because the Phi29 enzyme responsible for the isothermic amplification during MDA preferentially amplifies DNA with no or limited secondary structure. Intact rRNA maintains significant secondary structure even following reverse transcription, and hence the integrity of the rRNA can significantly influence the level of mRNA enrichment post-MDA amplification.
3. RNA should always be stored at −80°C, the temperature at which it is stable for >1 year.

Acknowledgements

The author would like to thank Margaret Hughes and Neil Hall from the NERC/University of Liverpool Advanced Genomics Facility.

References

1. DeLong EF, Preston CM, Mincer T, Rich V, Hallam SJ, et al. (2006) Community genomics among stratified microbial assemblages in the ocean's interior. *Science* **311**, 496–503.
2. Rusch DB, Halpern AL, Sutton G, Heidelberg KB, Williamson S, et al. (2007) The Sorcerer II Global Ocean Sampling expedition: northwest Atlantic through eastern tropical Pacific. *PLoS Biol.* **5**, 398–431
3. Yooseph S, Sutton G, Rusch DB, Halpern AL, Williamson SJ, et al. (2007) The Sorcerer II Global Ocean Sampling expedition: expanding the universe of protein families. *PLoS Biol.* **5**, 432–466.
4. Handelsman J, Tiedje J, Alvarez-Cohen L, Ashburner M, Cann IKO, et al. (2007) The New Science of metagenomics: revealing the secrets of our microbial planet, Washington, DC: The National Academies Press.
5. Parro V, Moreno-Paz M, Gonzalez-Toril E (2007) Analysis of environmental transcriptomes by DNA microarrays. *Environ. Microbiol.* **9**, 453–464.
6. Poretsky RS, Bano N, Buchan A, LeCleir G, Kleikemper J, et al. (2005) Analysis of microbial gene transcripts in environmental samples. *Appl. Environ. Microbiol.* **71**, 4121–4126.
7. Leininger S, Urich T, Schloter M, Schwark L, Qi J, et al. (2006) Archaea predominate among ammonia-oxidizing prokaryotes in soils. *Nature* **442**, 806–809.
8. Urich T, Lanzén A, Qi J, Huson DH, Schleper C, et al. (2008) Simultaneous Assessment of Soil Microbial Community Structure and Function through Analysis of the Meta-Transcriptome. *PLoS ONE* 3: e2527. doi:10.1371/journal.pone.0002527
9. Frias-Lopez J, Shi Y, Tyson GW, Coleman ML, Schuster SC, et al. (2008) Microbial community gene expression in ocean surface waters. *Proc. Natl. Acad. Sci. USA* **105**, 3805–10.
10. Gilbert JA, Field D, Huang Y, Edwards R, Li W, et al. (2008) Detection of Large Numbers of Novel Sequences in the Metatranscriptomes of Complex Marine Microbial Communities. *PLoS ONE* **3**(8): e3042. doi:10.1371/journal.pone.0003042
11. Poretsky R.S., Hewson I, Sun S, Allen A. E., Zehr J.P. and Moran, M.A. 2009a. Comparative day/night metatranscriptomic analysis of microbial communities in the North Pacific subtropical gyre. *Environ. Microbiol.* **11**, 1358–1375.
12. Poretsky R.S., Gifford S., Rinta-Kanto J., Vila-Costa M., Moran M.A. 2009b. Analyzing Gene Expression from Marine Microbial Communities using Environmental Transcriptomics. JoVE. 24. http://www.jove.com/index/Details.stp?ID=1086, doi: 10.3791/1086
13. Li W & Godzik A (2006) Cd-hit: a fast program for clustering and comparing large sets of protein or nucleotide sequences. *Bioinformatics* **22**, 1658–1659.

Part V

Sequence Profiling for Functional Analysis

Chapter 15

High-Throughput Insertion Tracking by Deep Sequencing for the Analysis of Bacterial Pathogens

Sandy M.S. Wong, Jeffrey D. Gawronski, David Lapointe, and Brian J. Akerley

Abstract

Whole-genome techniques toward identification of microbial genes required for their survival and growth during infection have been useful for studies of bacterial pathogenesis. The advent of massively parallel sequencing platforms has created the opportunity to markedly accelerate such genome-scale analyses and achieve unprecedented sensitivity, resolution, and quantification. This chapter provides an overview of a genome-scale methodology that combines high-density transposon mutagenesis with a *mariner* transposon and deep sequencing to identify genes that are needed for survival in experimental models of pathogenesis. Application of this approach to a model pathogen, *Haemophilus influenzae*, has provided a comprehensive analysis of the relative role of each gene of this human respiratory pathogen in a murine pulmonary model. The method is readily adaptable to nearly any organism amenable to transposon mutagenesis.

Key words: HITS, Himar1, Mariner, Transposon mutagenesis, Bacteria, *Haemophilus influenzae*, Genetic footprinting, Deep sequencing

1. Introduction

Various techniques utilizing "negative selection" strategies have been developed to identify on a genomic scale the bacterial genes essential for growth or survival under a condition of interest in vitro or during infection in a model host. These genes are identified based on the relative decrease in abundance of mutants deficient in specific genes within a large mutant bank exposed to selection. The "signature-tagged mutagenesis" approach utilizes custom-made DNA arrays representing unique hybridization tags that are introduced into each mutant within a library of strains

Young Min Kwon and Steven C. Ricke (eds.), *High-Throughput Next Generation Sequencing: Methods and Applications*, Methods in Molecular Biology, vol. 733, DOI 10.1007/978-1-61779-089-8_15, © Springer Science+Business Media, LLC 2011

to be evaluated (1). The "transposon-site hybridization" and "microarray tracking of transposon mutants" methods use microarrays representing the whole genome of the target organism to track the relative abundance of transposon insertions in each gene under varied selection conditions (2–4). These approaches have identified bacterial genes involved in pathogenesis; however, the generation of large banks of uniquely tagged mutants for many pathogens may be impractical, and in addition whole-genome microarrays may be unavailable for many bacteria. In both methods, hybridization is used to analyze the presence or absence of a given mutation within the pool, and quantification is limited by background hybridization levels and the dynamic range of signal detection. In this chapter, we describe an advancement of the negative selection strategy that generates an output that allows precise noise filtering and a broad dynamic range of detection.

We illustrate a methodology termed HITS (High-throughput *I*nsertion *T*racking by Deep *S*equencing) that utilizes a whole-genome transposon mutant bank in combination with deep sequencing to analyze genes involved in bacterial pathogenesis. We initially demonstrated the use of the HITS procedure to analyze genes required by *Haemophilus influenzae* to resist clearance from the lung (5), a niche colonized during pneumonia and chronic obstructive pulmonary disease (6, 7). The analytical procedures are not specific to this pathogen and can be used with other bacteria with little or no modification to the basic approach. This technique was used in conjunction with genetic footprinting, a rapid method for assessment of the genetic selection procedure prior to more comprehensive analysis. Because massively parallel sequencing is used for detection of transposon insertions in HITS, background signal is easily identified and removed via sequence filters during data analysis, and the dynamic range of detection is limited only by the number of sequencing events, which can be increased as needed.

2. Materials

1. Transposon mutant library constructed with a *Himar1/mariner* derived minitransposon.
2. Oligonucleotide primers.
3. Covaris S2 instrument (Covaris, Inc., Woburn, MA).
4. Covaris 6 mm × 16 mm microTube vessels (Covaris, Inc., Woburn, MA).
5. TE buffer:10 mM Tris–HCl, 1 mM EDTA.

6. 10× T4 DNA ligase buffer [New England Biolabs (NEB) Beverly, MA]: 500 mM Tris–HCL, pH 7.5 at 25°C, 100 mM $MgCl_2$, 100 mM DTT, 10 mM ATP, and 250 μg/ml BSA.
7. T4 DNA polymerase 3,000 U/ml (NEB).
8. *Escherichia coli* polymerase I, Large (Klenow) fragment 5,000 U/ml (NEB).
9. T4 polynucleotide kinase 10,000 U/ml (NEB).
10. 10× EcoPol buffer (NEB): 100 mM Tris–HCl, pH 7.5 at 25°C, 50 mM $MgCl_2$, 75 mM DTT.
11. Deoxyadenosine-5′-triphosphate.
12. Klenow Fragment (3′ to 5′ exo minus) 5,000 U/ml (NEB).
13. LigaFast Rapid DNA Ligation System (Promega).
14. T4 DNA ligase 3 U/μl (Promega).
15. 2× Phusion High-Fidelity PCR Master Mix (Finnzymes).
16. QIAquick PCR Purification Kit (Qiagen).
17. Qiagen MinElute PCR purification.
18. Qiagen QiaQuick Gel Extraction Kit.
19. Thermal cycler.
20. Dynal MyOne Streptavidin C1 beads (Invitrogen).
21. Dynal magnetic particle concentrator (MPC-S; Invitrogen).
22. 2× Binding and Washing buffer: 10 mM Tris–HCl, pH 7.5, 1 mM EDTA, 2 M NaCl.
23. Melt solution: 125 mM NaOH.
24. Neutralization solution (Qiagen PB buffer and acetic acid).
25. Electrophoresis equipment.
26. Rotator.
27. Vortex mixer.
28. Spectrophotometer.
29. Agilent Bioanalyzer 2100 RNA Pico6000 chip (Agilent Technologies, Santa Clara, CA).
30. Illumina Genome Analyzer II™ Sequencing system.

3. Methods

The HITS (*H*igh-throughput *I*nsertion *T*racking by Deep *S*equencing) procedure was originally developed using an experimental animal model of *H. influenzae* infection as the selection condition to define bacterial genes involved in pathogenesis (5). Briefly, a mutant bank consisting of at least 55,935 independent transposon mutants was generated and used to infect mice through the intrapulmonary route. Representation of mutations within the chromosomal DNA of bacteria isolated from infected lungs was then compared to that of the initial mutant bank via the HITS procedure. Generation of the mutant bank exploited highly efficient in vitro transposition by a *mariner*-derived minitransposon and the natural transformability of *H. influenzae* (8). Genetic footprinting by PCR was useful as a tool to validate the comprehensiveness of the transposon mutant bank before and after selection. HITS technology should be widely applicable to any bacterium that is amenable to transposon mutagenesis. Based on our experience with *H. influenzae*, a general procedure applicable to diverse bacteria is described below.

3.1. Overview of HITS

The approach consists of a "negative selection" technique to identify mutants lost from a large mutant pool after exposure to a selective condition of interest. Figure 1 outlines a general scheme for preparing DNA from a transposon mutant library before and after growth or survival under a selective condition of interest, such as a selection for growth and survival of bacteria in an infected host. While in vitro transposition was used for *H. influenzae*, in vivo transposon mutagenesis with *mariner*-derived minitransposons is highly efficient in bacteria, providing a convenient means of generating mutant banks with many pathogens (9). Because *mariner* transposons insert into the TA dinucleotide as their only apparent insertion site specificity (10), these elements produce highly diverse libraries with comprehensive coverage of the target genome (5). Moreover, the TA insertion site specificity is useful for comparing the maximum possible number of insertions per gene to the number of insertions detected. Procedures for in vitro and in vivo transposon mutagenesis of bacteria with *mariner* transposons have been described elsewhere (11). Although the HITS method described below uses a mutant bank made with a *mariner*-derived transposon, other transposons are equally applicable. In fact, any type of insertion mutant library could be used provided that a common primer binding site is present within each mutation (see Note 1).

The overall HITS procedure is described in the following steps. Detailed descriptions of each step are described in Subheadings 3.1.1 through 3.1.9.

1. Isolate and purify genomic DNA from a transposon insertion library before and after exposure to a selective condition (e.g., growth in an infected host).
2. Shear genomic DNA and repair to generate blunt 5′ and 3′ ends.

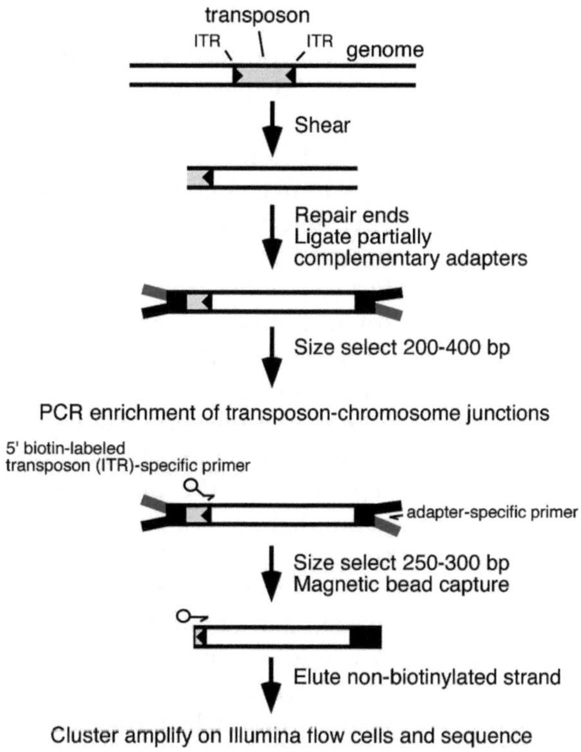

Fig. 1. HITS. Schematic diagram illustrates overview of the preparation of genomic DNA from a transposon mutant library for deep sequencing as performed in *Haemophilus influenzae*. Genomic DNA is isolated from input and output transposon mutant libraries. Purified genomic DNA is sheared and ends are repaired followed by adapter ligation. Size-selected adapter-ligated DNA is used as template in PCR with a primer specific to the inverted terminal repeat (ITR *shown as triangles*) to enrich for transposon–chromosome junctions. Biotinylated templates are captured by magnetic beads. Nonbound single stranded templates are eluted from the beads and used in Illumina flow cell cluster generation for deep sequencing.

3. Add "A" base to the 3′ end of blunt phosphorylated DNA fragments.
4. Ligate partially complementary adapter oligos to end-repaired DNA.
5. Size select DNA and use as template in PCR to enrich for transposon–chromosome junctions with a 5′ biotinylated transposon-specific primer and an adapter-specific primer that anneals to only one oligonucleotide of the partially complementary adapter.
6. Following enrichment, PCR products are size selected again.
7. 5′ biotinylated templates are captured with magnetic beads.
8. Elute nonbound single-stranded templates from beads.

9. Cluster-amplify single-stranded templates in Illumina flow cells for deep sequencing on the Illumina Genome Analyzer II™ Sequencing system.

3.1.1. Exposure of Mutant Libraries to Selection Conditions

The HITS method is designed for comparison of the mutations present in an initially constructed mutant library to those present after selection under a condition of interest. Collecting the bacterial population for analysis before and after selection is the first step of the procedure. Mutants that have sustained a transposon insertion in a gene that is required for optimal growth under a specific growth condition (e.g., in vitro or from an animal host environment) become nonviable under that condition and will be absent after selection. Genes that are essential for growth or survival under standard laboratory conditions in vitro will not be represented in the initial mutant bank. Mutants with insertions in genes that are not required for survival, typically the majority of the population, will remain.

A powerful application of this approach is to identify genes needed during infection of a model host by a bacterial pathogen. A consideration in conducting selections of mutant libraries in experimental models of infection is that some model systems restrict the total number of mutants that can be analyzed. For example, destruction of the bulk of the inoculated bacterial population may occur during early stages of infection, followed by outgrowth of relatively few survivors. Such restriction events, often termed "bottlenecks," would cause the postselection library to represent a random population consisting of far fewer mutants than the original inoculum. Prior to preparing DNA for HITS analysis following genomic DNA isolation, it would be beneficial to first verify retention of library complexity and selection conditions prior to preparation of samples for deep sequencing. Genetic footprinting provides a convenient means and its application is described in Subheading 3.2.

3.1.2. Genomic DNA Isolation

Genomic DNA is purified from mutant libraries before and after selection by standard molecular biology techniques using phenol extraction and ethanol precipitation (12).

3.1.3. Fragment Genomic DNA

1. Shear genomic DNA at a concentration of ~3–6 μg in a total volume of 100 μl of TE buffer with a Covaris S2 instrument using Covaris 6 mm × 16 mm microTube vessels with the following settings: Mode: Power Sweep; Duty Cycle, 5%; Intensity, 5%; Cycles/Burst, 200; Time, 120–135 s.
2. Sheared DNA is concentrated on a Qiagen MinElute PCR purification column and eluted in ~35 μl EB buffer.

3.1.4. Repair DNA Ends

This reaction repairs the 3′ and 5′ overhangs of the sheared DNA resulting in blunt ended fragments with 5′-phosphorylated ends.

1. To a 50 μl reaction containing the ~35 μl sheared DNA (from above step), add 5 μl of 10× T4 DNA ligase buffer, 5 μl of 2.5 mM dNTPs, 2.5 μl of T4 DNA polymerase, 0.5 μl of *E. coli* polymerase I, Large (Klenow) fragment, and 2.5 μl of T4 polynucleotide kinase.
2. Incubate at 20°C for 30 min.
3. Reaction is purified with QIAquick spin columns and eluted in EB buffer.

3.1.5. Addition of "A" Base to 3′ Ends

This reaction adds an "A" base to the 3′ end of the blunt-ended phosphorylated DNA, allowing for subsequent ligation to adapters that carry a "T" base overhang at the 3′ end.

1. To a 50-μl reaction containing the repaired 5′-phosphorylated blunt-ended DNA, add 5 μl of 10× EcoPol buffer, 1 μl of 10 mM dATP, 1.5 μl of Klenow Fragment (3′ to 5′ exo minus). Add water to a final volume of 50 μl.
2. Incubate at 37°C for 30 min.
3. Purify reaction with Qiagen MinElute PCR purification column and elute in ~10 μl of EB buffer.

3.1.6. Adapter Ligation

Ligation of adapters to the ends of the DNA fragments prepares them for attachment to the flow cells used in cluster generation for sequencing. Adapter oligonucleotide sequences:

5′-AGTTCTCCAGGTCTTGCGTTGCTCTTCCGATC*T[a]

(*the phosphorothioate linkage provides greater resistance to excision by 3′–5′ exonucleases. See Note 2)

5′-p-GATCGGAAGAGCTCGTATGCCGTCTTCTGCTTG[b]

[a]Modification of Illumina adapter oligonucleotide sequence.

[b]Oligonucleotide sequences © 2006–2008 Illumina, Inc. All rights reserved. Primer is modified, treated, and purified in a proprietary manner by Illumina, and use of other primers may not guarantee efficient sequencing.

1. To 10 μl of DNA containing "A" base at 3′ end, add 15 μl of 2× LigaFast, 2 μl of T4 DNA ligase, and 3 μl of adapter oligonucleotide mix (50 μM for each oligonucleotide).
2. Incubate at room temperature for 30 min.
3. Purify reaction with QIAquick PCR Purification kit. Elute with 30 μl of EB buffer.
4. Size select DNA fragments to obtain those in the ~200–400 bp size range on 2% agarose TAE gel. Gel-extract DNA with Qiagen QiaQuick Gel Extraction Kit according to the manufacturer's instructions except that the gel is dissolved in QG at room temperature (13). Elute in 50 μl of EB buffer.

3.1.7. Enrichment of Transposon–Chromosomal Junctions with Biotin Tagged Primer

This step uses a short PCR reaction (18 cycles) to enrich for DNA fragments that contain *mariner* transposon/chromosomal junction sequences on one end and adapter molecule on the other end. The transposon-specific primer, PE1MAR, contains a biotin modification attached by a 15 atom tetra-ethyleneglycol spacer (biotin-TEG) to the 5′ end, allowing for isolation of biotinylated templates with transposon insertions. In addition, the 5′ end of PE1MAR consists of sequences required for Illumina/Solexa slide amplification and sequencing, respectively.

Transposon enrichment primer (PE1MAR):

5′-biotinTEG-AATGATACGGCGACCACCGAGATCTACA CTCTTTCCCTACACGACGCTCTTCCGATCTCG GGGACTTATCAGCCAACC[a]

Adapter-specific primer (12):

5′-CAAGCAGAAGACGGCATACGAGCTCTTCCGATC*T[b]

(* phosphorothioate linkage)

[a]Modification of Illumina oligonucleotide PCR PE1.0 (5).

[b]Oligonucleotide sequences © 2006–2008 Illumina, Inc. All rights reserved. Primer is modified, treated, and purified in a proprietary manner by Illumina and use of other primers may not guarantee efficient sequencing.

1. Use portions (or all) of the size selected adapter-ligated DNA in a 50-μl PCR reaction containing 25 μl of 2× Phusion High-Fidelity PCR Master Mix, transposon enrichment and adapter-specific primers at 25–50 pmoles each. Add water to a total volume of 50 μl and amplify under the following conditions: 30 s, 98°C; 18 cycles of 10 s, 98°C, 30 s, 65°C, 30 s, 72°C; 5 min, 72°C; hold at 10°C.
2. Size-select PCR products on a 2% agarose TAE gel at the 250–300 bp range.
3. Extract DNA from the gel using Qiagen MinElute Gel Extraction Kit and elute in 25 μl of Qiagen EB buffer.

3.1.8. Affinity Capture of Biotinylated DNA

Dynal MyOne Streptavidin C1 beads are used to capture the biotinylated amplicons containing transposon/chromosome junction sequences. The nonbiotinylated DNA is dissociated from the biotinylated DNA strand, and the resultant single-stranded templates are ready for sequencing.

1. Wash 50 μl of Dynal MyOne C1 beads with 1× Binding and Washing buffer and the Dynal magnetic particle concentrator (MPC-S) according to the manufacturer's instructions. Add size-selected PCR enriched products to beads, mix by vortexing and place on a rotator for 20 min at room temperature for binding.

2. Wash DNA bound beads three times with 200 μl of 1× Binding and Washing buffer with the Dynal MPC.
3. The nonbiotinylated strand is dissociated from the beads with the addition of 50 μl of Melt Solution and vortexed and placed on a rotator for 10 min at room temperature.
4. Pellet beads with the Dynal MPC and transfer supernatant containing the single stranded DNA to 500 μl of Qiagen PB buffer containing 3.8 μl of 20% acetic acid.
5. Add an additional 50 μl of Melt Solution to beads and repeat step 3. Add supernatant to neutralized PB buffer containing the single-stranded DNA from the previous step. Purify and concentrate the neutralization mixture on a Qiagen MinElute PCR purification column and elute in 20 μl of Qiagen EB buffer.
6. The resulting library of transposon–chromosome junction DNA is quantified on an Agilent Bioanalyzer 2100 RNA Pico6000 chip. Single-stranded templates are cluster-amplified and sequenced on an Illumina GAII as described (14).

3.1.9. Analysis of Illumina Sequencing Data

Enrichment of *mariner* transposon–chromosome junctions entails incorporating *Himar1 mariner* transposon inverted terminal repeat (ITR) sequences (underlined) immediately 3′ of the Illumina primer PCR PE1.0[a] (5′ AATGATACGGCGACCACCGAGATCTACACTCTTTCCCTACACGACGCTCTTCCGATCTCGGGGACTTATCAGCCAACC) ([a] Oligonucleotide sequences © 2006–2008 Illumina, Inc. All rights reserved). Sequencing reads of an amplified transposon–chromosome junction fragment that contains the sequence string ‘cggggacttatcagccaaccTGTta’ are indicative of a transposon insertion where “TGT” is the remainder of the ITR of the *himar1* mariner transposon followed by the chromosomal TA insertion site.

1. The Illumina sequencing reads that contained the exact string of the *Himar1* ITR sequence and the adjacent TA insertion site were identified in the raw fasta files and trimmed of the ITR sequence to leave the TA insertion site at the 5′ end of the sequence.
2. The trimmed sequencing reads are aligned to the reference genome sequence using SOAPv1.11 alignment software using default settings that allow two mismatches per read (15).
3. A custom PERL script was used to parse the TA dinucleotide insertion site coordinates from the SOAP output file to report the number of reads mapped per site and strand orientations of aligned reads (Computer Script below).

```
#!/usr/bin/perl

# Usage from command line: #

# parseSoap.pl <SOAP_output_file | sort –n >parsed_output_ file #


while(<>){

   chomp;

   @flds=split(/\t/);

   $loc=(($flds[6] eq '+')?$flds[8]:$flds[8]+$flds[5]-2);

   $hitloc{$loc}++;

   $strand{$loc}.=$flds[6];

}

printf "Location\tCount\tStrand\n";

foreach $i (sort keys %hitloc){

   printf "%s\t%s\t%s\n",$i,$hitloc{$i},$strand{$i};

}
```

4. Import data into Microsoft Excel and map insertion-site coordinates to positions within protein-coding genes annotated in the protein table in the appropriate RefSeq file (from the National Center for Biotechnology Information: ftp://ftp.ncbi.nih.gov/).
5. Identify the number of insertion sites for each gene and the total number of sequencing reads in the internal 5–80% of the gene by Excel.

3.2. Validation of Mutant Libraries

Before processing genomic DNA for HITS analysis, a quality-control step is incorporated to validate the complexity of mutant libraries and selection conditions by genetic footprinting (Fig. 2). Genetic footprinting is a genome-scale method that allows mapping of transposon mutations within a given gene of interest within a bank of mutants (16) and has been used previously to determine that *Himar1*-derived minitransposons insert efficiently without significant site bias in the genomes of *H. influenzae* and other bacteria, with only the dinucleotide TA as the apparent insertion site specificity (9, 17). This method can be used to interrogate specific genes that are predicted to be required under the selection condition of interest, in addition to specific genes that are known to be nonessential under this condition. For example, in *H. influenzae,* genetic footprinting was applied to the mutant bank prior to inoculation and after in vivo selection to evaluate a

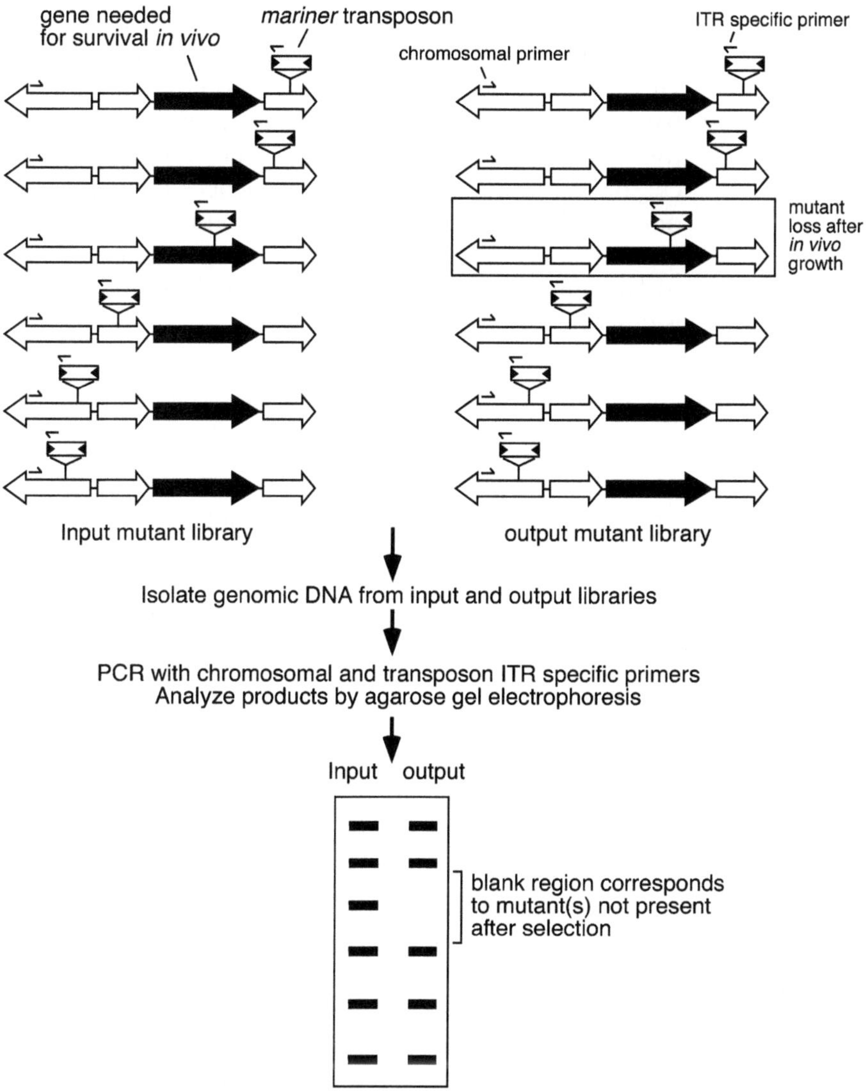

Fig. 2. Genetic footprinting. Genomic DNA is isolated from input and output mutant pools and used as template in PCR with a chromosomal primer and a *mariner* transposon inverted terminal repeat (ITR)-specific primer. The PCR reaction is analyzed by agarose gel electrophoresis to generate a genetic footprint. Mutants that have sustained a transposon insertion in a gene needed for growth or survival in vivo (*shaded in black*) are loss from the output pool and will not be represented by a corresponding PCR product, thus producing a blank region on the gel.

panel of characterized genes required for virulence of *H. influenzae* in an infant rat model of bacteremia (18). For this validation, we chose two genes of LPS biosynthesis *galU* and *orfH* in which mutations resulted in two logs of attenuation relative to wild type in the bacteremia model. Genetic footprinting by PCR detected many transposon insertions within these genes in the input bank, yet the abundance of insertions detected in the output bank was markedly reduced, providing validation of the HITS data which showed pronounced attenuation of *galU* and *orfH* mutants relative to wild type (5). In contrast, a gene of xylose metabolism, *xylA*, known to be nonessential during infection, contained the same approximate number of transposon mutations in the output bank as in the original input library, confirming that passage of the library in the animal model did not result in significant stochastic loss of mutants due to a bottleneck effect. Detailed methodology for application of genetic footprinting to *H. influenzae* and other bacteria has been described previously (11). Another useful approach for validating the complexity of mutant libraries is described in Note 3.

4. Notes

1. Several other variations of the sample preparation method for high-throughput sequencing of transposon–chromosome junctions have also been described and used to address a range of biological questions (19–21). These methods produce similar output and appear to be equally effective.
2. Use of primers containing a phosphorothioate linkage at the 3′ end may improve amplification. This is because some DNA polymerases used in PCR contains 3′–5′ exonuclease activity (22) such as the Phusion DNA polymerase described here. Incorporating a single phosphorothioate bond in the primer decreases degradation of the primer during PCR.
3. Another useful approach to validate the complexity of the transposon mutant banks prior to deep sequencing is to clone a small portion of the size-selected DNA enriched for transposon–chromosome junctions and sequence the cloned inserts. To clone and sequence single-stranded DNA samples, one can perform a linear extension of the template, for example, with a nonbiotinylated transposon-specific primer and DNA polymerase in a thermocycler. This generates the complementary DNA strand, and the resulting double-stranded products can then be cloned. Clone ~ 50 ng of the resultant DNA into pCR-Blunt II-TOPO vector using Zero Blunt TOPO PCR Cloning Kit (Invitrogen) according to the manufacturer's instructions. It is important to use a DNA polymerase in the linear extension reaction that generates blunt-end products (e.g., Phusion DNA polymerase, FinnZymes), as this kit is designed to clone blunt-ended PCR products. Isolate plasmid DNA from a representative set of at least 20 transformants for small-scale sequencing using primers flanking the cloned insert (e.g., M13 Reverse and Forward primers). Analysis of cloned inserts can provide an assessment of successful adaptor ligation and enrichment for transposon –chromosome junction sequences.

Acknowledgments

This work was supported in part by grant from the NIH (AI49437) to B.J.A. Initial development of the method was conducted in collaboration with Georgia Giannoukos and Doyle Ward of the Broad Institute.

References

1. Hensel, M., Shea, J. E., Gleeson, C., Jones, M. D., Dalton, E., and Holden, D. W. (1995) Simultaneous identification of bacterial virulence genes by negative selection, *Science* **269**, 400–403.
2. Sassetti, C. M., Boyd, D. H., and Rubin, E. J. (2001) Comprehensive identification of conditionally essential genes in mycobacteria, *Proc Natl Acad Sci U S A* **98**, 12712–12717.
3. Salama, N. R., Shepherd, B., and Falkow, S. (2004) Global transposon mutagenesis and essential gene analysis of *Helicobacter pylori*, *J Bacteriol* **186**, 7926–7935.
4. Badarinarayana, V., Estep, P. W., 3rd, Shendure, J., Edwards, J., Tavazoie, S., Lam, F., and Church, G. M. (2001) Selection analyses of insertional mutants using subgenic-resolution arrays, *Nat Biotechnol* **19**, 1060–1065.
5. Gawronski, J. D., Wong, S. M., Giannoukos, G., Ward, D. V., and Akerley, B. J. (2009) Tracking insertion mutants within libraries by deep sequencing and a genome-wide screen for *Haemophilus* genes required in the lung, *Proc Natl Acad Sci U S A* **106**, 16422–16427.
6. Vila-Corcoles, A., Ochoa-Gondar, O., Rodriguez-Blanco, T., Raga-Luria, X., and Gomez-Bertomeu, F. (2009) Epidemiology of community-acquired pneumonia in older adults: a population-based study, *Respir Med* **103**, 309–316.
7. Sethi, S., and Murphy, T. F. (2001) Bacterial infection in chronic obstructive pulmonary disease in 2000: a state-of-the-art review, *Clin Microbiol Rev* **14**, 336–363.
8. Akerley, B. J., Rubin, E. J., Novick, V. L., Amaya, K., Judson, N., and Mekalanos, J. J. (2002) A genome-scale analysis for identification of genes required for growth or survival of *Haemophilus influenzae*, *Proc Natl Acad Sci U S A* **99**, 966–971.
9. Rubin, E. J., Akerley, B. J., Novik, V. N., Lampe, D. J., Husson, R. N., and Mekalanos, J. J. (1999) *In vivo* transposition of *mariner*-based elements in enteric bacteria and mycobacteria, *Proc Natl Acad Sci U S A* **96**, 1645–1650.
10. Lampe, D. J., Grant, T. E., and Robertson, H. M. (1998) Factors affecting transposition of the *Himar1 mariner* transposon *in vitro*, *Genetics* **149**, 179–187.
11. Wong, S. M., and Akerley, B. J. (2008) Identification and analysis of essential genes in *Haemophilus influenzae*, *Methods Mol Biol* **416**, 27–44.
12. Ausubel, F. M., Brent, R., Kingston, R. E., Moore, D. D., Seidman, J. G., Smith, J. A., and Struhl, K. (1995), John Wiley and Sons, Inc.
13. Quail, M. A., Kozarewa, I., Smith, F., Scally, A., Stephens, P. J., Durbin, R., Swerdlow, H., and Turner, D. J. (2008) A large genome center's improvements to the Illumina sequencing system, *Nat Methods* **5**, 1005–1010.
14. Bentley, D. R., Balasubramanian, S., Swerdlow, H. P., Smith, G. P., Milton, J., Brown, C. G., et al. (2008) Accurate whole human genome sequencing using reversible terminator chemistry, *Nature* **456**, 53–59.
15. Li, R., Li, Y., Kristiansen, K., and Wang, J. (2008) SOAP: short oligonucleotide alignment program, *Bioinformatics* **24**, 713–714.
16. Singh, I. R., Crowley, R. A., and Brown, P. O. (1997) High-resolution functional mapping of a cloned gene by genetic footprinting, *Proc Natl Acad Sci U S A* **94**, 1304–1309.
17. Akerley, B. J., Rubin, E. J., Camilli, A., Lampe, D. J., Robertson, H. M., and Mekalanos, J. J. (1998) Systematic identification of essential genes by *in vitro mariner* mutagenesis, *Proc Natl Acad Sci U S A* **95**, 8927–8932.
18. Hood, D. W., Deadman, M. E., Allen, T., Masoud, H., Martin, A., Brisson, J. R., Fleischmann, R., Venter, J. C., Richards, J. C., and Moxon, E. R. (1996) Use of the complete genome sequence information of *Haemophilus influenzae* strain Rd to investigate

lipopolysaccharide biosynthesis, *Mol Microbiol* **22**, 951–965.

19. van Opijnen, T., Bodi, K. L., and Camilli, A. (2009) Tn-seq: high-throughput parallel sequencing for fitness and genetic interaction studies in microorganisms, *Nat Methods* **6**, 767–772.

20. Langridge, G. C., Phan, M. D., Turner, D. J., Perkins, T. T., Parts, L., Haase, J., Charles, I., Maskell, D. J., Peters, S. E., Dougan, G., Wain, J., Parkhill, J., and Turner, A. K. (2009) Simultaneous assay of every *Salmonella Typhi* gene using one million transposon mutants, *Genome Res* **19**, 2308–2316.

21. Goodman, A. L., McNulty, N. P., Zhao, Y., Leip, D., Mitra, R. D., Lozupone, C. A., Knight, R., and Gordon, J. I. (2009) Identifying genetic determinants needed to establish a human gut symbiont in its habitat, *Cell Host Microbe* **6**, 279–289.

22. Skerra, A. (1992) Phosphorothioate primers improve the amplification of DNA sequences by DNA polymerases with proofreading activity, *Nucleic Acids Res* **20**, 3551–3554.

Chapter 16

Determining DNA Methylation Profiles Using Sequencing

Suhua Feng, Liudmilla Rubbi, Steven E. Jacobsen, and Matteo Pellegrini

Abstract

Cytosine methylation is an epigenetic mark that has a significant impact on the regulation of transcription and replication of DNA. DNA methylation patterns are highly conserved across cell divisions and are therefore highly heritable. Furthermore, in multicellular organisms, DNA methylation patterning is a key determinant of cellular differentiation and tissue-specific expression patterns. Lastly, DNA demethylases can affect global levels of DNA methylation during specific stages of development. Bisulfite sequencing is considered the gold standard for measuring the methylation state of cytosines. Sodium bisulfite converts unmethylated cytosines to uracils (which after PCR are converted to thymines), while leaving methylated cytosines unconverted. By mapping bisulfite treated DNA back to the original reference genome, it is then possible to determine the methylation state of individual cytosines. With the advent of next generation sequencers during the past few years, it is now possible to determine the methylation state of an entire genome. Here, we describe in detail two protocols for preparing bisulfite treated libraries, which may be sequenced using Illumina GAII sequencers. The first of these uses premethylated adapters, which are not affected by bisulfite treatments, while the second uses a two-stage adapter strategy and does not require premethylation of the adapters. We also describe the specialized protocol for mapping bisulfite converted reads. These approaches allow one to determine the methylation state of each cytosine in the genome.

Key words: DNA methylation, Epigenetics, Next-generation sequencing, Whole-genome methylome

1. Introduction

The methylation state of cytosines in the genome has a profound effect on many biological processes. Most organisms contain maintenance DNA methyltransferases that can preserve the methylation state of CpG dinucleotides during cell division (1). A second class of maintenance methyltransferases are also commonly found, which can often methylate cytosines not in CpG dinucleotides (2, 3). These enzymes are found in a wide variety

Young Min Kwon and Steven C. Ricke (eds.), *High-Throughput Next Generation Sequencing: Methods and Applications*, Methods in Molecular Biology, vol. 733, DOI 10.1007/978-1-61779-089-8_16, © Springer Science+Business Media, LLC 2011

of organisms ranging from plants to animals, and from multicellular organisms to single-celled ones. Although DNA methylation is found very frequently, certain organisms such as *Saccharomyces cerevisiae*, *Drosophila melanogaster*, and *Caenorhabditis elegans* have lost the ability to methylate DNA.

For those organisms that methylate their DNA, this process is usually essential to their survival, and mutants of DNA methyltransferases often lead to nonviable strains (3). The reason for this is that DNA methylation plays a critical role in patterning the transcriptional state of a cell, and DNA methylation mutants typically have aberrant development. By modifying the DNA methylation state of cytosines, it is possible to suppress the transcription of transposable and other repetitive elements, thus constraining unregulated growth of genome sizes (4, 5). Similarly, aberrant methylation of promoters can lead to transcriptional suppression of the downstream genes, as is often seen in diseases such as cancer (5).

Because of the importance of DNA methylation in regulating fundamental biological processes, it is of interest to measure the methylation state of all the cytosines in a genome. Conventional DNA sequencing techniques cannot distinguish between methylated and unmethylated cytosines. It is therefore common to utilize sodium bisulfite to convert unmethylated cytosines to uracils, while leaving the methylated cytosines unconverted. By mapping the converted DNA back to the reference DNA, it is possible to determine the methylation state of each of the cytosines simply by counting the number of cytosines and thymines that align to that position.

While the sequencing of bisulfite-converted DNA is often considered the gold standard for determining the methylation state of DNA, in the past this approach was usually applied to a limited number of loci. However, with the advent of relatively inexpensive high-throughput sequencing, it is now possible to shotgun-sequence an entire bisulfite converted genome, thus enabling genome-wide measurements of cytosine methylation (6–8). Here, we present two protocols for generating libraries of bisulfite-treated DNA, for sequencing on Illumina sequencers. The first of these protocols uses premethylated adapters that are not affected by the bisulfite conversion step (Fig. 1). This protocol has the advantage of requiring fewer overall steps to prepare the library and fewer amplification cycles, thus limiting potential sequence composition biases, and has to date been used more frequently in published papers. The second protocol proceeds in two steps and does not require that the standard library adapters be premethylated (Fig. 2). The advantage of the second approach is that it leads to the sequencing of both the converted DNA strands and its reverse complement, and therefore maintains relatively high frequencies of cytosines in a library even when the

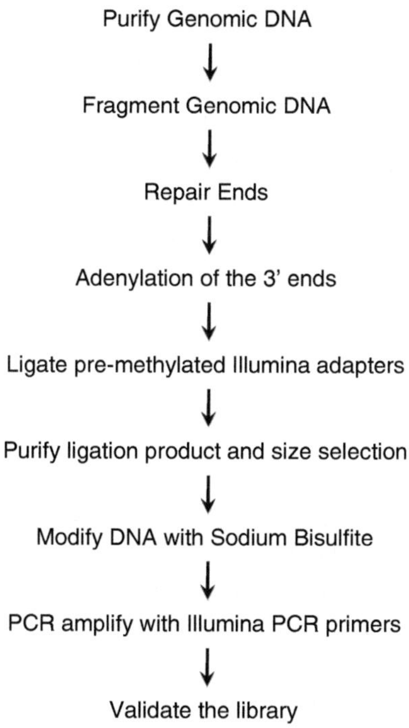

Fig. 1. Protocol I: library generation using premethylated adapters. (Adapted from Illumina protocol.).

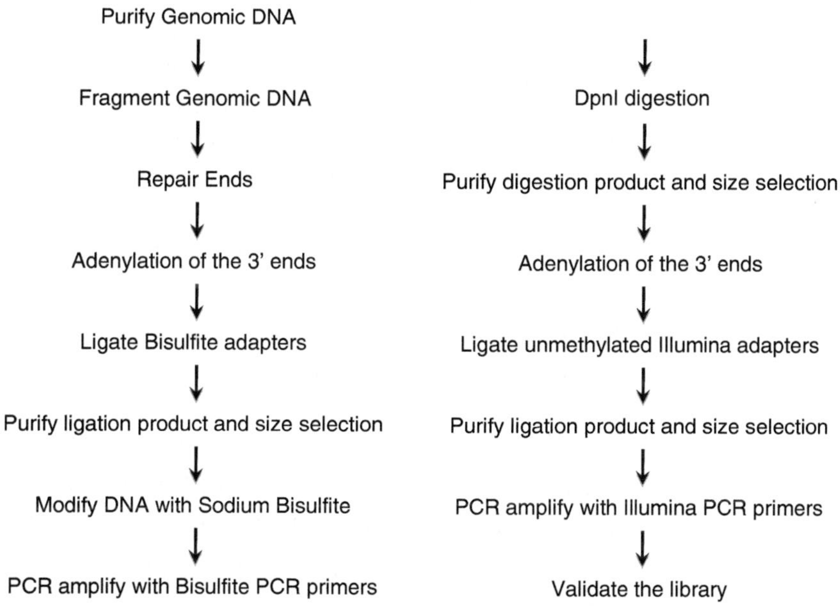

Fig. 2. Protocol II: library generation using unmethylated adapters. (Adapted from Millipore and Illumina protocols.).

genome is mostly unmethylated. This feature potentially improves base calling (which is problematic when very few cytosines are present in a library), but because it requires two rounds of amplification, it can lead to greater sequence composition biases.

Once the library has been prepared and sequenced, the next step requires that the converted reads be aligned to the genome. This results in a nontraditional alignment in which thymines in reads can align to either thymines or cytosines in the genome. To accomplish this, we align a three-letter version of reads to a three-letter genome, and appropriately score correct versus incorrect cytosine to thymine transitions. We briefly describe this approach along with a software we have developed to conduct these alignments.

2. Materials (see Note 1)

2.1. Protocol I: Library Generation Using Premethylated Adapter

2.1.1. DNA Sample Preparation

1. HMB buffer: 25 mM Tris–HCl, pH 7.6, 0.44 M sucrose, 10 mM $MgCl_2$, 0.1% Triton X-100, 10 mM β-mercaptoethanol, 2 mM spermine, 1 mM PMSF, 1 μg/ml pepstatin, 1× EDTA–free protease inhibitors (Roche). Make fresh and keep at 4°C.
2. Miracloth filter (Calbiochem).
3. HBB buffer: 25 mM Tris–HCl, pH 7.6, 0.44 M sucrose, 10 mM $MgCl_2$, 0.1% Triton X-100, 10 mM β-mercaptoethanol. Make fresh and keep at 4°C.
4. 40%/60% percoll gradient: Make 40% and 60% percoll by mixing percoll with HBB. Add 10 ml of 60% percoll into a 30-ml centrifuge tube first and then carefully lay 10 ml of 40% percoll on top of the 60% percoll. Make fresh.
5. Resuspension buffer: 50 mM Tris–HCl (pH 7.5), 20 μg/μl Proteinase K (Roche). Make fresh.
6. Lysis Buffer: 50 mM Tris–HCl (pH 7.5), 15 mM EDTA, 1.5% SDS. Make fresh.
7. QIAquick PCR Purification Kit (Qiagen).
8. Biorupter sonicator (Diagenode).

2.1.2. End Repair

1. T4 DNA polymerase (3 U/μl, New England Biolabs).
2. Klenow DNA polymerase (5 U/μl, New England Biolabs).
3. T4 polynucleotide kinase (10 U/μl, New England Biolabs).
4. dNTP mix (10 mM each, New England Biolabs).
5. 10× T4 DNA ligase buffer with 10 mM ATP (New England Biolabs).

2.1.3. Adenylation

1. Klenow fragment 3′→5′ exo minus (5 U/μl, New England Biolabs).

2. 10× Klenow buffer (Buffer 2, New England Biolabs).
3. 1 mM dATP (New England Biolabs).

2.1.4. Ligation with Premethylated Illumina Adapters

1. Quick T4 DNA ligase (2,000 U/μl) and 2× Quick ligation buffer (New England Biolabs).
2. Premethylated Adapters single-end (Illumina):
   (a) *Premethylated Illumina Adapter 1* (all the cytosines in this oligo are 5-methylated; ask the oligo synthesis company to incorporate 5-methylated cytosines instead of regular unmethylated cytosines during the synthesis of this oligo):

   5'-GATCGGAAGAGCTCGTATGCCGTCTTCTGCTTG-3'
   ↑
   with 5' phosphate

   (b) *Premethylated Illumina Adapter 2* (all the cytosines in this oligo are 5-methylated; ask the oligo synthesis company to incorporate 5-methylated cytosines instead of regular unmethylated cytosines during the synthesis of this oligo):

   5'- ACACTCTTTCCCTACACGACGCTCTTCCGATCT-3'
   ↑
   no 5' phosphate

3. 2% certified low range ultra agarose (Bio-Rad).
4. QIAquick column (Qiagen).

2.1.5. Bisulfite Treatment

1. CpGenome DNA Modification Kit (Millipore).
2. 3 M NaOH (freshly prepared prior to each use). Dissolve 1 g of NaOH pellet in 8.3 ml of water. Use appropriate caution when manipulating this caustic base.
3. Millipore reagent I – Urea solution (freshly prepared prior to each use): use appropriate caution when handling this reagent, as it is irritating to the respiratory system and skin.
4. Millipore reagent II (freshly prepared): excess reagent can be stored in a foil-wrapped container at 2–8°C in the dark for up to 6 weeks.
5. 70% ETOH. Store at –20°C.
6. 20 mM NaOH/90% EtOH (freshly prepared): to prepare 1 ml of this solution, combine 900 μl of 100% EtOH, 93.4 μl of water, and 6.6 μl of 3 M NaOH.
7. 90% EtOH. Store at –20°C.
8. TE buffer. Store at room temperature.

2.1.6. PCR Amplification of the Library

1. Pfu Turbo Cx polymerase and 10× Pfu Turbo Cx buffer (Agilent).
2. dNTPs (2.5 mM each, New England Biolabs).
3. PCR single-end Primers 1.1 and 2.1 (Illumina):
   (a) Illumina PCR primer 1.1:

   5'-
   AATGATACGGCGACCACCGAGATCTACACTCTTT
   CCCTACACGACGCTCTTCCGATCT-3'

   (b) Illumina PCR primer 2.1:

   5'- CAAGCAGAAGACGGCATACGAGCTCTTCCGATCT-3'
4. PCR Purification Kit (Qiagen).

2.1.7. Qualitative and Quantitative Controls of the Library

1. 2% certified low range ultra agarose gel (Bio-Rad).
2. PCR4Blunt-TOPO vector (Invitrogen).
3. Library resuspension Buffer: EB Buffer (Qiagen), 0.1% Tween-20.

2.2. Protocol II: Library Generation Using Unmethylated Adapters

Materials for the following six steps are described in the above sections: *DNA Sample Preparation* (Subheading 2.1.1), *End Repair* (Subheading 2.1.2), *Adenylation* (Subheading 2.1.3), *Bisulfite Treatment* (Subheading 2.1.5), *PCR Amplification of the Library* (Subheading 2.1.6), and *Quantitative and Quantitative Controls of the Library* (Subheading 2.1.7)

2.2.1. Ligation with Bisulfite Adapters

1. Adapters ligation: Quick T4 DNA ligase (2,000 U/μl) and 2× Quick ligation buffer (New England Biolabs)
2. Bisulfite Adapters.
   (a) Bisulfite Adapter 1:

   5'-AGTTATTCCGGACTGTCGAAGCTGAAGTGATC^{m}CGT -3'
   ↑ no 5' phosphate ↑ 5-methylated

   (b) Bisulfite Adapter 2:

   with 5' phosphate ↓
   5'-CGGATC^{m}ACTTCAGCTTCGACAGTCCGGAAT-3'
   ↑ 5-methylated
3. 2% certified low range ultra agarose (Bio-Rad).
4. QIAquick column (Qiagen).

2.2.2. PCR Amplification of the Bisulfite Treated DNA

1. Pfu Turbo Cx polymerase and 10× Pfu Turbo Cx buffer (Agilent).
2. dNTPs (2.5 mM each, New England Biolabs).
3. Bisulfite PCR Primers 1 and 2:
   (a) Bisulfite PCR Primer 1:

   5'-AAACTATCAAAACTAAAATGA^{m}TCCA-3'
   ↑ no 5' phosphate ↑ 6-methylated

   (b) Bisulfite PCR primer 2:

   5'-TTGTTGAAGTTGAAGTGA^{m}TCT-3'
   ↑ no 5' phosphate ↑ 6-methylated

4. 2% certified low range ultra agarose gel (Bio-Rad).

2.2.3. DpnI Digestion of the PCR Products

1. 10× DpnI buffer (Buffer 4, New England Biolabs), BSA (10 mg/ml, New England Biolabs), DpnI (20 U/μl, New England Biolabs).
2. QIAquick column (Qiagen).
3. 2% certified low range ultra agarose gel (Bio-Rad).

2.2.4. Adenylation

1. 10× Klenow buffer (Buffer 2 New England Biolabs) and Klenow fragment 3′→5′ exo minus (5 U/μl, New England Biolabs).
2. dATP (1 mM New England Biolabs).
3. QIAquick MinElute column (Qiagen).

2.2.5. Ligation with Unmethylated Illumina Adapters

1. Quick T4 DNA ligase (2,000 U/μl) and 2× Quick ligation buffer (New England Biolabs).
2. Unmethylated Adapters single-end (Illumina):
   (a) Unmethylated Illumina Adapter 1:

   5'-GATCGGAAGAGCTCGTATGCCGTCTTCTGCTTG-3'
   ↑ with 5' phosphate

   (b) Unmethylated Illumina Adapter 2:

   5'- ACACTCTTTCCCTACACGACGCTCTTCCGATCT-3'
   ↑ no 5' phosphate

3. 2% certified low range ultra agarose gel (Bio-Rad).
4. QIAquick column (Qiagen).

3. Methods (see Note 1)

3.1. Protocol I: Bisulfite Library Generation Using Premethylated Adapters

Genomic DNA can be purified from different sources (e.g., cells, frozen tissue) using various standard DNA purification protocols. It is important to ensure that the DNA sample to be processed is highly pure. In this chapter, we describe the procedure to obtain highly pure DNA from *Arabidopsis thaliana*.

3.1.1. DNA Sample Preparation

1. Ground 1 g of plant tissues into a fine powder in liquid nitrogen and homogenize three times in 10 ml of HBM buffer.
2. Filter through two layers of Miracloth filters (Calbiochem). Centrifuge the homogenate (3,000–3,500 rpm [1,000–1,500 × *g*] for 5 min at 4°C, in an SS-34 rotor), and resuspend the pellet in 5 ml of HBB buffer.
3. To isolate the nuclei load the sample onto 20 ml of 40%/60% percoll gradient HBB and centrifuge for 1,500–2,000 rpm [300–500 × *g*] for 30 min at 4°C without applying the brake. Wash twice with 10 ml of HBB buffer and resuspend in 500 μl of 50 mM Tris–HCl (pH 7.5) containing 20 μl (final μg 20 μg/μl) of Proteinase K (Roche) and incubate at room temperature for 30 min. Add 1 ml of Lysis buffer to lyse the nuclei. Purify DNA by phenol/chloroform extraction and ethanol precipitation. Resuspend DNA in totally 200 μl of Qiagen EB buffer to complete.
4. To fragment the genomic DNA, sonicate 5 μg of DNA (in 100–300 μl solution) (see Note 2), in a 1.5-ml tube, with Biorupter sonicator (Diagenode) for four cycles of 15 min each (30 s "on" and 30 s "off", output H). In between cycles, samples are kept on ice. The DNA fragments are next purified with QIAquick column from QIAquick PCR purification Kit (Qiagen) and eluted in 35 μl of EB buffer (Qiagen) (see Note 3).

3.1.2. End Repair

1. To repair, blunt, and phosphorylate ends ("End Repair reaction"), the DNA fragments are subsequently treated with a mixture of T4 DNA polymerase, *Escherichia coli* DNA polymerase I Klenow fragment, and T4 polynucleotide kinase.

   Set up the End-Repair reaction as follows:

35 μl	DNA from previous step
40 μl	Water
10 μl	10× T4 DNA ligase buffer (with 10 mM ATP)
4 μl	dNTP mix (10 mM each)
5 μl	T4 DNA polymerase (3 U/μl)

1 μl	Klenow DNA polymerase (5 U/μl)
5 μl	T4 PNK (10 U/μl)
100 μl	Total reaction volume
Incubate for 30 min at 20°C.	

2. Purify DNA with QIAquick column (Qiagen) and elute in 32 μl of EB buffer.

3.1.3. Adenylation

1. To add a single "A" base to the 3′ end ("Adenylation reaction"), incubate the repaired DNA fragments from previous step with Klenow exo-fragment (3′→5′ exo^{-}).

   Set up the "Adenylation reaction" as follows:

32 μl	DNA (from previous step)
5 μl	10× Klenow buffer
10 μl	dATP (1 mM)
3 μl	Klenow fragment 3′→5′ exo minus (5 U/μl)
50 μl	Total reaction volume
Incubate for 30 min at 37°C.	

2. Purify DNA with one QIAquick MinElute column (Qiagen) and elute in 14 μl of EB buffer.

3.1.4. Ligation with Premethylated Illumina Adapters

1. To ligate the DNA fragments with the adapters incubate the adenylated fragments from previous step with premethylated, single-end, adapters (Illumina).

   Set up the Adapter Ligation reaction as follows:

14 μl	DNA (from previous step)
30 μl	2× Quick ligation buffer
10 μl	Premethylated Illumina adapters (single-ended) (anneal adapter 1 and adapter 2 at 1:1 molar ratio, adjust the concentration of the adapter duplex so that there is a 10:1 molar ratio of adapter duplex to DNA insert)
6 μl	Quick T4 DNA ligase (2,000 U/μl)
60 μl	Total reaction volume
Incubate for 15 min at room temperature.	

2. Purify DNA with one QIAquick column and elute in 30 μl of Qiagen EB buffer.
3. Run DNA from previous step on 2% certified low range ultra agarose gel (Biorad) for 1 h at 100 V. Excise fragments

ranging from 150 to 300 bp (or according to specific needs, see Note 4) and extract DNA with one QIAquick column. Elute DNA in 30 μL of Qiagen EB buffer.

3.1.5. Bisulfite Treatment

At this step, adapter-ligated DNA is ready for bisulfite treatment. Several commercial bisulfite treatment kits are available. In the following steps, the CpGenome DNA Modification Kit (Millipore) is used. The protocol described here follows the manufacturer's instructions and previously described method (9).

1. Transfer 1 μg of the adapter-ligated DNA to a 1.5–2.0-ml screwcap microcentrifuge tube. Bring the total volume to 100 μl with water (add 2 μl of Millipore Reagent IV if less than 1 μg of DNA is used). Add 7 μl of freshly prepared 3 M NaOH, mix and incubate at 55°C in a heat block for 20 min.
2. Prepare the Millipore Reagent I-Urea solution: for each sample to be modified, dissolve 0.227 g of Millipore Reagent I (warm bottle to room temperature before opening) into 0.464 ml of water. Adjust the pH by adding 20 μl of 3 M NaOH and mix. Check the pH with the pH indicator paper and the pH should be 5.0. Add 0.22 g of urea to 450 μl of this solution. The final volume should be around 650 μl.
3. Add 650 μl of the freshly prepared Millipore Reagent I-Urea solution to the denatured DNA (from step 2), vortex, and incubate at 55°C in a water bath for 24 h. Protect the samples from light exposure.
4. Prepare Millipore Reagent II: add 1 μl of β-mercaptoethanol to 20 ml of deionized water. For each sample to be modified, add 750 μl of this solution to 1.35 g of Millipore Reagent II (warm to room temperature before opening). Mix well until completely dissolved.
5. Resuspend Millipore Reagent III by vortexing vigorously and pipeting up and down for ten times to disperse any remaining clumps. Add 5 μl of well-suspended Millipore Reagent III to the DNA solutions in the tubes. Add 750 μl of freshly prepared Millipore Reagent II, mix briefly, and incubate at room temperature for 15 min.
6. Spin for 30 s at 5,000 × *g* to pellet the Reagent III, discard supernatant (a small white pellet should be visible). Add 1 ml of 70% EtOH, vortex, spin 5,000 × *g* for 30 s, and discard the supernatant. Repeat this step twice. After the last wash has been removed, centrifuge at top speed for 2 min, and remove all the remaining supernatant with a plastic pipette tip.
7. Add 50 μl of the freshly prepared 20 mM NaOH/90% EtOH solution to the samples, vortex briefly, and incubate at room temperature for 10 min.

8. Spin for 30 s at 5,000 × *g*. Add 1 ml of 90% EtOH, vortex to wash the pellet, and again to remove the supernatant. Repeat this step once. Spin at top speed for 3 min, remove all the remaining supernatant with a pipette tip, and let dry for 30 min at room temperature.
9. Add 40 μl of TE, incubate at 55°C for 20 min to elute DNA, centrifuge at top speed for 3 min, and transfer the supernatant to a new tube. Proceed to PCR or store as aliquots at −80°C. Avoid repeated thawing and refreezing.

3.1.6. PCR Amplification of the Library

1. Enrich Adapters-ligated and Bisulfite-modified DNA fragments by PCR reactions.

   Set up four PCR reactions for each sample as follows:

2.5 μl	DNA
5.0 μl	10× PfuTurbo Cx buffer
4.0 μl	dNTPs (2.5 mM each)
1 μl	Illumina PCR primer 1.1 (single-end)
1 μl	Illumina PCR primer 2.1 (single-end)
0.5 μl	PfuTurbo Cx polymerase
36 μl	Water
50 μl	Total reaction volume

   Amplify using the following PCR protocol: 2 min at 98°C. 15 cycles of 10 s at 98°C and 90 s at 60°C, 10 min at 60°C. Hold at 4°C.
2. Follow the instructions in the QIAquick PCR Purification Kit (Qiagen) to purify with one QIAquick column, elute in 30 μl of EB buffer. This is the final library.

3.1.7. Qualitative and Quantitative Controls of the Library

1. Load 10% of the volume of the library (5 μl) on 2% agarose gel to check whether the size range is as expected (see Note 5). It should be slightly larger in size than the size-range excised during the gel purification step (since PCR primers add ~25 bp to the length of the product).
2. Clone 4 μl of the library into a sequencing vector. (e.g., pCR-4Blunt-TOPO, Invitrogen) (see Note 6). Sequence individual clones by conventional Sanger sequencing.
3. Determine the concentration of the library by measuring its absorbance at 260 nm. To determine the molar concentration of the library, examine the gel image (from step 1) and estimate the median size of the library smear. Multiply this size by 650 (the molecular mass of a base pair) to get the molecular weight of the library. Use this number to calculate

the molar concentration. Make 10-nM aliquots of library in Qiagen EB buffer containing 0.1% Tween-20 for Illumina/Solexa sequencing and store at –20°C.

3.1.8. Aligning Bisulfite-Converted Reads

Sodium bisulfite converts unmethylated cytosines to uracils, while leaving methylated cytosines unconverted. After the steps of PCR, uracils are converted to thymines, which is the base associated with unmethylated cytosines in the final reads. When mapping these converted reads back to the reference genome, we cannot therefore use a standard alignment tool, as this would consider the alignment of a thymine in a read to a cytosine in the genome as a mismatch.

To circumvent this limitation, we conduct the alignments in three-letter space, where all cytosines are converted to thymines (or vice versa) in both the reads and the genome. Typically, we restrict our attention to only those reads that map to a unique position in the genome in this three-letter space, although it is possible to include more sophisticated treatments on nonuniquely mapping reads. This approach allows us to correctly align thymines in reads to cytosines in the genome, but also permits incorrect alignments of cytosines in reads to thymines in the genome. Consequently, after the three-letter alignment is complete, we reconvert our sequence to four-letter space, and penalize all alignments that contain read cytosines aligned to genome thymines.

We have implemented a version of this protocol that utilizes a fast short read aligner, Bowtie, to perform the three-letter alignments. Our wrapper around this software then checks for incorrect conversions and penalizes those alignments. The software may be downloaded from http://pellegrini.mcdb.ucla.edu/BS_Seeker/BS_Seeker.html.

3.2. Protocol II: Bisulfite Library Generation Using Unmethylated Adapters

The protocols for the following three steps are described in the above sections: *DNA Sample Preparation* (Subheading 3.1.1), *End Repair* (Subheading 3.1.2), and *Adenylation* (Subheading 3.1.3).

3.2.1. Ligation with Bisulfite Adapters

1. To ligate the DNA fragments with the bisulfite adapters, incubate the adenylated fragments (from previous step) with the Bisulfite Adapters.

   Set up the Adapter Ligation reaction as follows:

14 µl	DNA (from previous step)
30 µl	2× Quick ligation buffer
10 µl	Bisulfite Adapters (anneal adapter 1 and adapter 2 at 1:1 molar ratio, adjust the concentration of the adapter duplex so that there is a 10:1 molar ratio of adapter duplex to DNA insert)

6 μl	Quick T4 DNA ligase (2,000 U/μl)
60 μl	Total reaction volume
Incubate for 15 min at room temperature.	

2. Purify DNA with one QIAquick column and elute in 30 μl of Qiagen EB buffer.
3. Run DNA from previous step on 2% certified low range ultra agarose gel (Biorad) for 1 h at 100 V. Excise fragments ranging from 100 bp and up, purify using three QIAquick columns, and elute DNA from each column with 30 μL of EB buffer.
4. The protocol for *Bisulfite Treatment* is described in Subheading 3.1.5.

3.2.2. PCR Amplification of the Bisulfite Treated DNA

1. Enrich Adapters-ligated and Bisulfite-modified DNA fragments by PCR reactions.

   Set up eight PCR reactions for each sample as follows:

2.5 μl	DNA
5.0 μl	10× PfuTurbo Cx buffer
4.0 μl	dNTPs (2.5 mM each)
1 μl	Bisulfite PCR primer 1
1 μl	Bisulfite PCR primer 2
0.5 μl	Pfu Turbo Cx polymerase
36 μl	Water
50 μl	Total reaction volume

   Use the following PCR conditions: 2 min at 98°C. Eight cycles of 10 s denaturation at 94°C, 30 s annealing at temperatures from 55°C to 52°C (2 cycles at each temperature) and 4 min extension at 60°C. Then, seven cycles of 10 s denaturation at 94°C, 30 s annealing at 51°C, and 4 min extension at 60°C.

   Finally, 10 min at 60°C and hold PCR samples at 4°C.
2. Purify PCR product with two QIAquick columns, elute each column with 30 μl of Qiagen EB buffer and combine.
3. Run 5 μl of purified of the purified PCR product on 2% certified low range ultra agarose gel (Bio-Rad) to check the DNA size.

3.2.3. DpnI Digestion of the PCR Products

1. The DpnI enzyme digests GATC site with 6-methylated adenosine, which is formed immediately flanking the library DNA insert after PCR amplification with bisulfite PCR

primers (previous step). Thus, DpnI digestion will cleave the bisulfite adapters off PCR products, which are ligated to unmethylated Illumina adapter in subsequent steps.

Set up the DpnI digestion as follows:

55 µl	(DNA from previous step)
12 µl	10× DpnI Buffer
1.2 µl	BSA (10 mg/ml)
12 µl	DpnI (20 U/µl)
39.8 µl	Water
120 µl	Total reaction volume
Incubate overnight at 37°C.	

2. Purify the digestion products with one QIAquick column and elute in 30 µl of Qiagen EB buffer.
3. Run DNA from previous step on 2% certified low range ultra agarose gel (Bio-Rad) for 1 h at 100 V. Excise DNA fragments from 40 bp and up, purify DNA with four QIAquick MinElute columns, and elute each column with 16 µl of Qiagen EB buffer and combine.

3.2.4. Adenylation

1. Addition of an "A" to 3′ end of DNA fragments, "adenylation reaction":

   Set up the "Adenylation reaction" as follows:

64 µl	DNA from previous step
20 µl	10× Klenow buffer
10 µl	dATP (1 mM)
6 µl	Klenow fragment 3′→5′ exo minus (5 U/µl)
100 µl	Total reaction volume
Incubate at 37°C for 30 min.	

2. Purify DNA with one QIAquick MinElute column and elute in 14 µl of Qiagen EB buffer.

3.2.5. Ligation with Unmethylated Illumina Adapters

1. Ligation with unmethylated single-end adapters (Illumina).

   Set up the Adapter Ligation reaction as follows:

14 µl	DNA from previous step
30 µl	2× Quick ligation buffer
10 µl	Unmethylated Illumina adapters (single-end) (anneal adapter 1 and adapter 2 at 1:1 molar ratio, adjust the concentration of the adapter duplex so that there is a 10:1 molar ratio of adapter duplex to DNA insert)

6 μl	Quick T4 DNA ligase (2,000 U/μl)
60 μl	Total reaction volume
Incubate for 15 min at room temperature.	

2. Purify DNA with one QIAquick column and elute in 30 μl of Qiagen EB buffer.
3. Run DNA from previous step on 2% certified low range ultra agarose gel (Biorad) for 1 h at 100 V. Excise fragments ranging from 150 to 300 bp (or sizes suitable for specific need) and extract DNA with one QIAquick column. Elute DNA in 30 μL of Qiagen EB buffer.
4. The protocols for *PCR Amplification of the Library* and *Qualitative and Quantitative Controls of the Library* are described in Subheadings 3.1.6 and 3.1.7, respectively.

3.2.6. Aligning Bisulfite-Converted Reads

When constructing libraries using this second protocol, reads may be observed in either of four forms. This second protocol generates a forward read (+FW) from the Watson strand, the reverse complement (+RC) of +FW, a forward read (−FW) from the Crick strand, and the reverse complement (−RC) of −FW. The FW reads all start with the tag "TCTGT" (the remnant of the DPNI-digested first tag of the protocol), while the RC reads contain the tag "TCCAT". The tags are used to determine whether the read is of type FW or RC, after which the tag is removed from the read for alignment to the genome. The alignment program BS Seeker first converts all Cs to Ts on FW reads and both strands of the reference genome so that the subsequent mapping is performed using only three letters, A, T, G. Similarly, G/A conversion is performed on RC reads and both strands of the reverse complement of the reference genome. Then, it uses Bowtie to map the C/T converted FW reads to the C/T converted Watson and Crick strands, and the G/A converted RC reads to the two G/A converted reverse complements of the Watson and Crick strands. Reads that do not have a tag are treated as if they could be both FW and RC reads. During each of the four runs of Bowtie, the mapped positions for each read are recorded. After all the runs of Bowtie are complete, only unique alignments are retained; such alignments as those that have no other hits with the same or fewer mismatches in the three-letter alignment (between the converted read and the converted genomic sequence). Finally, we calculate the number of mismatches. For this calculation, we consider a read T that aligns to a genomic C as a match, while a read C that aligns to a genomic C is considered a mismatch. Similarly, when aligning RC reads, a read A that aligns to a genomic G is considered a match, while a read G that aligns to a genomic A is considered a mismatch. Low-quality alignments with the number of mismatches larger than the user-defined value are discarded.

4. Notes

1. Wear gloves at all times. Use clean equipment. Try to separate equipment and reagents for library generation from the rest of the lab.
2. Follow the yield of each step by Nanodrop.
3. When eluting QIAquick columns, make sure to add Qiagen buffer EB onto the center of the membrane of the columns.
4. During size selection, leave at least one blank well between samples to avoid cross-contamination. Make sure to change razor blades between samples.
5. If primer dimers are seen after PCR, it is very likely that the library is not good. In this case, it is not recommended to just purify away the primer dimers; instead, it is better to repeat the library construction procedures.
6. The 5′ ends of the library molecules are not phosphorylated and therefore require a phosphorylated vector for cloning.

References

1. Goll, M.G. and T.H. Bestor (2005) Eukaryotic cytosine methyltransferases. *Annu Rev Biochem.* **74**, 481–514.
2. Chan, S.W., Zilberman, D., Xie, Z., Johansen, L. K., Carrington, J. C., and Jacobsen, S. E. (2004) RNA silencing genes control *de novo* DNA methylation. *Science* **303**, 1336.
3. Jones, P.A. and Takai, D. (2001) The role of DNA methylation in mammalian epigenetics. *Science* **293**, 1068–1070.
4. Martienssen, R.A. and V. Colot (2001) DNA methylation and epigenetic inheritance in plants and filamentous fungi. *Science* **293**, 1070–1074.
5. McCabe, M. T., Brandes, J. C., and Vertino, P. M. (2009) Cancer DNA methylation: molecular mechanisms and clinical implications. *Clin. Cancer Res.* **15**, 3927–3937.
6. Cokus, S.J., Feng, S., Zhang, X., Chen, Z., Merriman, B., Haudenschild, C. D., et al., (2008) Shotgun bisulphite sequencing of the Arabidopsis genome reveals DNA methylation patterning. *Nature* **452**, 215–219.
7. Lister, R., O'Malley, R. C., Tonti-Filippini, J., Gregory, B. D., Berry, C. C., Millar, A. H., and Ecker, J. R. (2008) Highly integrated single-base resolution maps of the epigenome in Arabidopsis. *Cell* **133**, 523–536.
8. Lister, R., Pelizzola, M., Dowen, R. H., Hawkins, R. D., Hon, G., Tonti-Filippini, J., et al. (2009) Human DNA methylomes at base resolution show widespread epigenomic differences. *Nature* **462**, 315–322.
9. Meissner, A., Gnirke, A., Bell, G. W., Ramsahoye, B., Lander, E. S., and Jaenisch, R. (2005) Reduced representation bisulfite sequencing for comparative high-resolution DNA methylation analysis. *Nucleic Acids Res.* **33**, 5868–5877.

Part VI

Sequencing Library Preparation

Chapter 17

Preparation of Next-Generation Sequencing Libraries Using Nextera™ Technology: Simultaneous DNA Fragmentation and Adaptor Tagging by In Vitro Transposition

Nicholas Caruccio

Abstract

DNA library preparation is a common entry point and bottleneck for next-generation sequencing. Current methods generally consist of distinct steps that often involve significant sample loss and hands-on time: DNA fragmentation, end-polishing, and adaptor-ligation. In vitro transposition with Nextera™ Transposomes simultaneously fragments and covalently tags the target DNA, thereby combining these three distinct steps into a single reaction.

Platform-specific sequencing adaptors can be added, and the sample can be enriched and bar-coded using limited-cycle PCR to prepare di-tagged DNA fragment libraries. Nextera technology offers a streamlined, efficient, and high-throughput method for generating bar-coded libraries compatible with multiple next-generation sequencing platforms.

Key words: Next-generation sequencing, DNA library preparation, Roche/454, Illumina/Solexa, Nextera

1. Introduction

Next-generation sequencing (variously known as deep sequencing, second-generation sequencing, or high-throughput sequencing) has led to genomic data acquisition on a scale never seen before, enabling "hypothesis-free" studies and changing the approaches used to answer the fundamental scientific questions ((1) and references therein). Currently available sequencing platforms, as well as platforms anticipated in the near future, utilize three basic chemistries to sequence DNA, including pyrosequencing, sequencing by synthesis (both cyclic and real time), and sequencing by ligation (recently reviewed in (2–4)).

Young Min Kwon and Steven C. Ricke (eds.), *High-Throughput Next Generation Sequencing: Methods and Applications*, Methods in Molecular Biology, vol. 733, DOI 10.1007/978-1-61779-089-8_17, © Springer Science+Business Media, LLC 2011

Despite the diversity in sequencing chemistries used for sequencing polony arrays or single molecules, upstream sample preparation is generally very similar and follows a common three-step workflow. In the first step, large DNAs are sheared to appropriately sized fragments. Next, the DNA fragments are "repaired" to have blunt ends or A tails. In the final step, platform-specific adaptors are ligated onto the repaired fragments in order to attach the library to a solid surface in a spatially separated array via a complementary sequence (e.g., a tagged glass slide or bead). Each of these steps consists of multiple and distinct handling, incubation, and clean-up procedures, which result in a complex, multi-step protocol. Thus, current sample preparation methods require microgram amounts of genomic DNA, significant hands-on time, and suffer from limited sample throughput (5). In vitro transposition using Nextera technology can be exploited to perform all of these steps in a single, 5-min reaction.

In a classical transposition reaction, a hyperactive transposase enzyme is used under conditions that catalyze near-random insertion of excised transposons into DNA targets with high efficiency in vitro. When Transposome™ complexes are assembled with free transposon-end DNA instead, the target DNA is simultaneously fragmented and tagged with the transposon-end sequence, thereby generating a tagged DNA fragment library in a single reaction (Fig. 1).

Sequencing libraries are prepared using a simple, two-step process. First, genomic DNA is fragmented and tagged using in vitro transposition. Second, platform-specific adaptors and

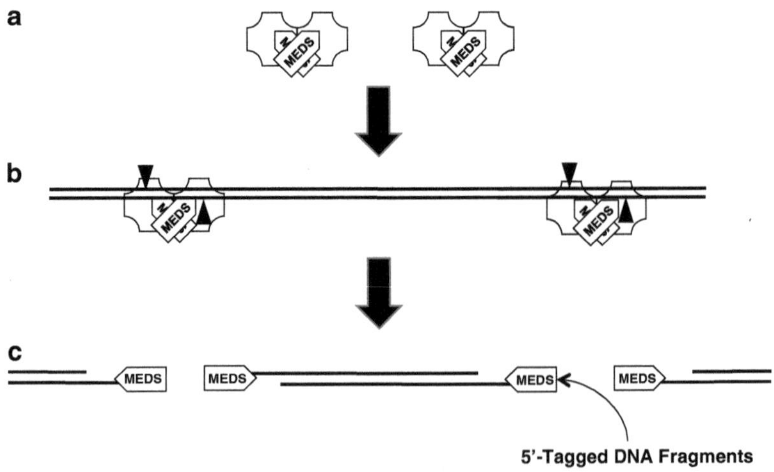

Fig. 1. Simultaneous fragmenting and tagging of DNA by in vitro transposition. Transposomes assembled with free, double-stranded transposon end DNA (**a**) are sufficient to bind and integrate into target DNA (**b**). The resulting tagged fragments (**c**) serve as the foundation for subsequent adaptor addition.

optional bar codes are added and the library is enriched using limited-cycle PCR.

Methods are described to prepare DNA fragment libraries compatible with the Roche/454 GS FLX and FLX-Titanium platforms and the Illumina/Solexa GAII platform, from 50 ng of genomic DNA in less than 2 h. These methods can be readily adapted to produce DNA fragment libraries that are compatible with other sequencing platforms.

2. Materials

2.1. Roche/454-Compatible Library Production

1. Target DNA – in water or TE, free of contaminating proteins and salts (see Note 1).
2. 12.5 μM Nextera™ Enzyme Mix, Roche FLX or Ti-compatible (EZ-Tn5™ Transposase bound to double-stranded transposon-end DNA, EPICENTRE Biotechnologies, Madison, WI).

   `"METS" 5'-AGATGTGTATAAGAGACAG-3'`

   `"pMENTS" 3'-TCTACACATATTCTCTGTC-PO4-5'`
3. 5× Tagmentation reaction buffer [50 mM Tris-OAc pH 8.0, 25 mM $Mg(OAc)_2$].
4. ZYMO DNA Clean & Concentrator™-5 (ZYMO Research, Cat. No. D4013 or equivalent).
5. ZYMO DNA-Binding Buffer (ZYMO Research Cat. No. D4003-1-L or equivalent).
6. ZYMO Wash Buffer (ZYMO Research Cat. No. D4003-2-6 or equivalent).
7. Nuclease-free water.
8. 2× Nextera™ PCR Buffer (EPICENTRE Biotechnologies, Madison, WI).
9. PCR adaptors and primers.

   Standard FLX-Compatible Adaptors and Primers

   - Adaptor 1 (FLX-Compatible Library) – 0.5 μM in water

     `5'-GCCTCCCTCGCGCCATCAGAGATGTGTATAAGA GACAG-3'`
   - Adaptor 2 (FLX-Compatible Library) – 0.5 μM in water

     `5'-GCCTTGCCAGCCCGCTCAGAGATGTGTATAA GAGACAG-3'`
   - Primer 1 (FLX-Compatible Library) – 10 μM in water

     `5'-GCCTCCCTCGCGCCATCAG-3'`

- Primer 2 (FLX-Compatible Library) – 10 μM in water
  5'-GCCTTGCCAGCCCGCTCAG-3'

FLX Titanium-Compatible Adaptors and Primers

- Adaptor 1 (Ti-Compatible Library) – 0.5 μM in water
  5'-CCATCTCATCCCTGCGTGTCTCCGACTCAGAGATG TGTATAAGAGACAG-3'
- Adaptor 2 (Ti-Compatible Library) – 0.5 μM in water
  5'-CCTATCCCCTGTGTGCCTTGGCAGTCTCAGAGA TGTGTATAAGAGACAG-3'
- Primer 1 (Ti-Compatible Library) – 10 μM in water
  5'-CCATCTCATCCCTGCGTGTCTCCGAC-3'
- Primer 2 (Ti-Compatible Library) – 10 μM in water
  5'-CCTATCCCCTGTGTGCCTTGGCAGTC-3'

10. Nextera™ PCR Enzyme (EPICENTRE Biotechnologies, Madison, WI).

2.2. Illumina/Solexa-Compatible Library Production

1. Target DNA – in water or TE, free of contaminating proteins and salts (see Note 1).
2. 12.5 μM Nextera™ Enzyme Mix, Illumina GAII-compatible (EZ-Tn5™ Transposase bound to appended, double-stranded transposon-end DNA, EPICENTRE Biotechnologies, Madison, WI). Equal ratios of:

   "A-METS" 5'-GCCTCCCTCGCGCCATCAGAGATGTGTATAAG AGACAG-3'

   "pMENTS" 3'-TCTACACATATTCTCTGTC-PO4-5'

   and

   "B-METS" 5'-GCCTTGCCAGCCCGCTCAGAGATGTGTATAAG AGACAG-3'

   "pMENTS" 3'-TCTACACATATTCTCTGTC-PO4-5'
3. 5× Tagmentation reaction buffer [50 mM Tris-OAc pH 8.0, 25 mM $Mg(OAc)_2$].
4. ZYMO DNA Clean & Concentrator™-5 (ZYMO Research Cat. No. D4013 or equivalent).
5. ZYMO DNA-Binding Buffer (ZYMO Research Cat. No. D4003-1-L or equivalent).
6. ZYMO Wash Buffer (ZYMO Research Cat. No. D4003-2-6 or equivalent).
7. Nuclease-free water.
8. 2× Nextera™ PCR Buffer (EPICENTRE Biotechnologies, Madison, WI).

9. PCR adaptors and primers.

   Illumina GAII-Compatible Adaptors and Primers[1]

   - Adaptor 1 (GAII-Compatible Library) – 0.5 μM in water

     5'-AATGATACGGCGACCACCGAGATCTACACGCCTCC CTCGCGCCATCAG-3'
   - Adaptor 2 (GAII-Compatible Library) – 0.5 μM in water

     5'-CAAGCAGAAGACGGCATACGAGATCGGTCTGCC TTGCCAGCCCGCTCAG-3'
   - Primer 1 (GAII-Compatible Library) – 10 μM in water

     5'-AATGATACGGCGACCACCGA-3'
   - Primer 2 (GAII-Compatible Library) – 10 μM in water

     5'-CAAGCAGAAGACGGCATACGA-3'
   - Nextera Read 1 Primer – 100 μM in water

     5'-GCCTCCCTCGCGCCATCAGAGATGTGTATAAGAGA CAG-3'
   - Nextera Read 2 Primer – 100 μM in water

     5'-GCCTTGCCAGCCCGCTCAGAGATGTGTATAAGAG ACAG-3'
   - Nextera Index Read Primer – 100 μM in water

     5'-CTGTCTCTTATACACATCTCTGAGCGGGCTGGCAA GGCAGACCG-3'
10. Nextera™ PCR Enzyme (EPICENTRE Biotechnologies, Madison, WI).
11. Illumina paired-end sequencing mixes (HP1 and HP2).
12. Illumina GAII Hybridization Buffer (Illumina Cat. No. GA0084204-HT1).
13. 5× SSC, 0.05% Tween®-20.
14. Illumina single-read sequencing mix (HP4).

3. Methods

Using in vitro transposition to simultaneously fragment and tag DNA in a single-tube reaction is a significant improvement upon current procedures, which generally consist of distinct DNA fragmentation, end-polishing, and adaptor-ligation steps. The following library preparation procedure combines these steps into one,

[1]The sequences of Primer 1 and Primer 2 and portions of Adaptor 1 and Adaptor 2 correspond to Illumina bPCR sequences and are copyrighted by Illumina, Inc.

uses only 50 ng of starting DNA, and allows incorporation of platform-specific adaptors and optional bar codes.

This chapter details methods to prepare libraries for the Roche/454 (GS FLX and FLX-Titanium) and Illumina/Solexa (GAII) sequencing platforms. The basic design can be applied to other adaptor-tagged DNA libraries.

3.1. Roche/454-Compatible Library Preparation

Target DNA is fragmented and tagged with Transposomes containing free transposon ends. Limited-cycle PCR with a four-primer reaction adds Roche/454-compatible adaptor sequences (GS FLX or FLX-Titanium). Optional bar coding is added between the upstream emulsion PCR (emPCR) adaptor and the transposon end (Fig. 2).

The Tagmentation reaction is the same for standard Roche GS FLX or for FLX-Titanium-compatible sequencing libraries. Specific adaptors for GS FLX or FLX-Titanium-compatible libraries are used during the limited-cycle PCR.

Roche/454 Tagmentation Reaction

Tagmentation with Roche/454-compatible Transposomes should predominantly yield fragments approximately 500–2,000 bp

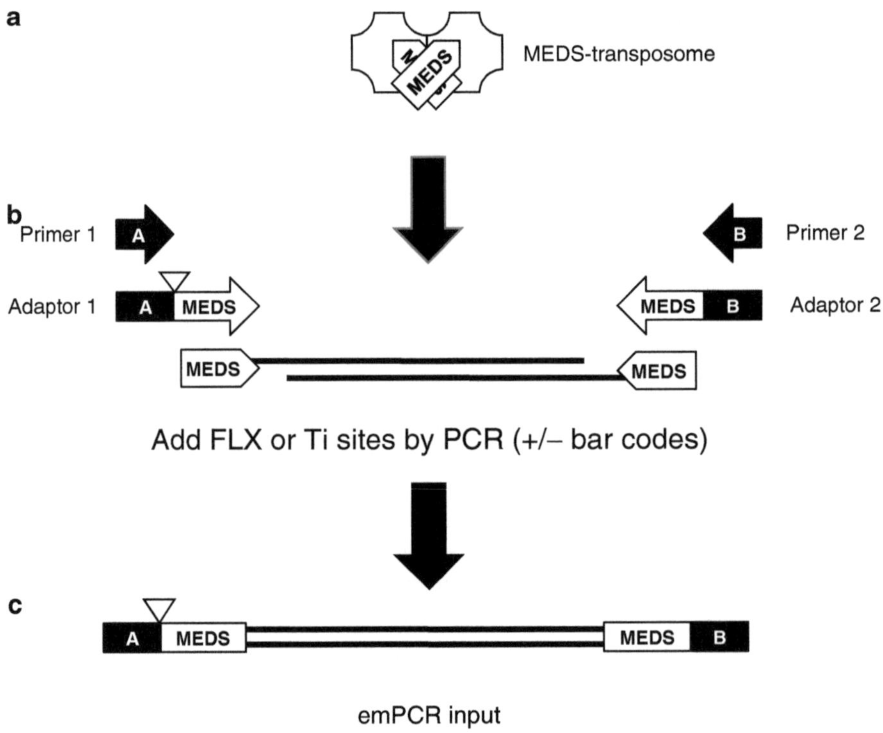

Fig. 2. Schematic representation of Roche/454-compatible sequencing library preparation. Transposomes assembled with free ends (**a**) are used in a tagmentation reaction to produce a 5′-tagged DNA fragment library. Limited-cycle PCR with a four-primer reaction adds Roche/454-compatible adaptor sequences (**b**). Optional bar coding (*triangle*) is incorporated into Adaptor 1 and added between the upstream emPCR adaptor and the transposon end (**c**).

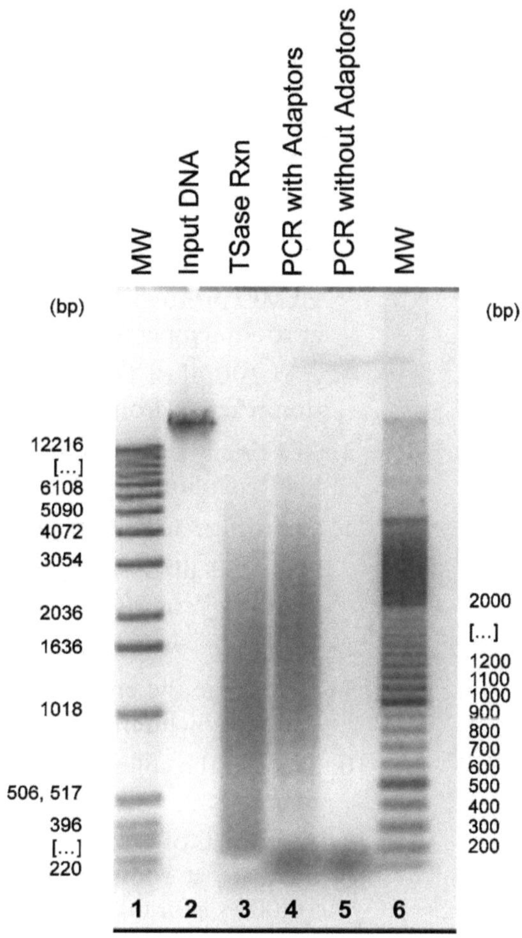

Fig. 3. Preparation of a Roche/454-compatible sequencing library. HeLa DNA (*lane 2*) was tagmented with Transposomes assembled with free ends (*lane 3*). Limited-cycle PCR was used to add Roche/454 GS FLX-compatible adaptors (*lane 4*). The recovered DNA was used directly in emPCR prior to pyrosequencing. Control PCR without Adaptor 1 and Adaptor 2 (including Primer 1 and Primer 2) confirmed reaction specificity (*lane 5*).

(Fig. 3). The actual fragment size distribution will depend on a number of factors, including the quantity and quality of starting DNA.

1. Assemble the following reaction components on ice, in the order listed (see Note 2)

x μl	Nuclease-free water
50 ng	Target DNA (see Note 1)
4 μl	5× Tagmentation reaction buffer
1 μl	Nextera Transposomes (Roche FLX or Ti-Compatible)
20 μl	Total reaction volume

2. Mix briefly by vortexing and incubate at 55°C for 5 min. To prevent evaporation, the reaction should be carried out in a thermocycler with a heated lid or the reaction should be overlaid with mineral oil. Adding DNA-Binding Buffer in step 3 stops the reaction.
3. Purify the tagmented DNA using a Zymo DNA Clean & Concentrator-5 Kit (or equivalent) according to the manufacturer's instructions (see Note 3). A synopsis of the ZYMO protocol is shown in steps 4–13. Perform these steps at room temperature.
4. Add 100 μl of DNA-binding buffer to the 20 μl tagmentation reaction from step 2.
5. Mix briefly by vortexing and transfer the mixture to a Zymo-Spin™ Column in a Collection Tube.
6. Centrifuge at 10,000 × *g* for 60 s. Discard the flow-through.
7. Add 250 μl of wash buffer to the column. Centrifuge at 10,000 × *g* for 60 s. Discard the flow-through.
8. Repeat the wash step once for a total of two washes.
9. Centrifuge the empty column again at 10,000 × *g* for 60 s to dry and to eliminate any residual wash buffer.
10. Transfer the column to a clean and sterile 1.5-ml microcentrifuge tube.
11. Add 11 μl of nuclease-free water directly to the column and incubate at room temperature for 1–2 min. Centrifuge at 10,000 × *g* for 60 s to elute the DNA into the clean microcentrifuge tube.
12. The final eluted volume should be ~10 μl. Use 5 μl of recovered DNA as PCR template in step 14.
13. The recovered DNA may be stored at −20°C.

Addition of emPCR-Compatible Adaptors and Library Enrichment

14. Assemble the following reaction components at room temperature:

15 μl	Nuclease-free water
5 μl	Recovered DNA fragment library (from step 12)
25 μl	2× Nextera PCR buffer
1 μl	0.5 μM Adaptor 1 (FLX or Ti-compatible)
For a bar-coded Roche-compatible library, replace Adaptor 1 with a bar-coded Adaptor 1 with the sequence composition, see Note 4. The other three primers and the cycling conditions remain unchanged.	
1 μl	0.5 μM Adaptor 2 (FLX or Ti-compatible)

1 µl	10 µM Primer 1 (FLX or Ti-compatible)
1 µl	10 µM Primer 2 (FLX or Ti-compatible)
1 µl	Nextera PCR enzyme (2.5 U/µl)
50 µl	Total reaction volume

15. Incubate the sample in a thermocycler under the following conditions (see Note 5):

    72°C for 3 min, 95°C for 30 s, followed by 15 cycles of (95°C for 10 s, 55°C for 30 s, 72°C for 3 min), Hold at 4°C.

16. Purify the DNA sequencing library using a Zymo DNA Clean & Concentrator-5 kit, or equivalent. The anticipated yield is ~300–500 ng of PCR-amplified DNA (Fig. 3).
17. The recovered DNA is ready to be quantified and used as input for Roche/454 emPCR and sequencing. No changes to the standard Roche/454 protocol are needed (see Note 6).

3.2. Illumina/Solexa-Compatible Library Preparation

Target DNA is fragmented and tagged with Transposomes containing transposon ends appended with sequencing tags. Limited-cycle PCR with a four-primer reaction adds Illumina/Solexa bridge PCR (bPCR)-compatible adaptor sequences. Optional bar coding is added between the upstream bPCR adaptor and the sequencing tag on Adaptor 2 (see Fig. 4).

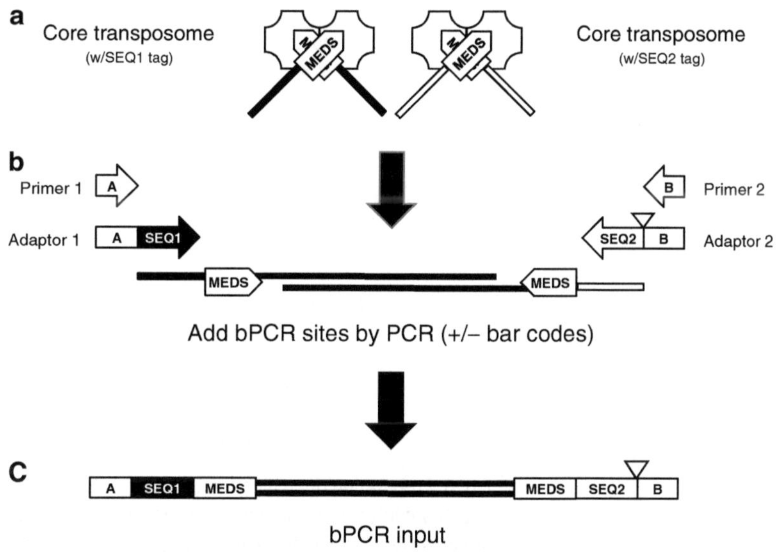

Fig. 4. Schematic representation of Illumina/Solexa-compatible sequencing library preparation. Transposomes assembled with free ends appended with sequencing tags (**a**) are used in a tagmentation reaction to produce a 5′-tagged DNA fragment library. Limited-cycle PCR with a four-primer reaction adds Illumina/Solexa bPCR-compatible adaptor sequences (**b**). Optional bar coding (*triangle*) is incorporated into Adaptor 2 and added between the upstream bPCR adaptor and sequencing tag 2 (**c**).

Illumina–Solexa Tagmentation Reaction

Tagmentation with Illumina GAII-compatible Transposomes should predominantly yield fragments approximately 250–1,000 bp (Fig. 5). The actual fragment size distribution will depend on a number of factors, including the quantity and quality of starting DNA.

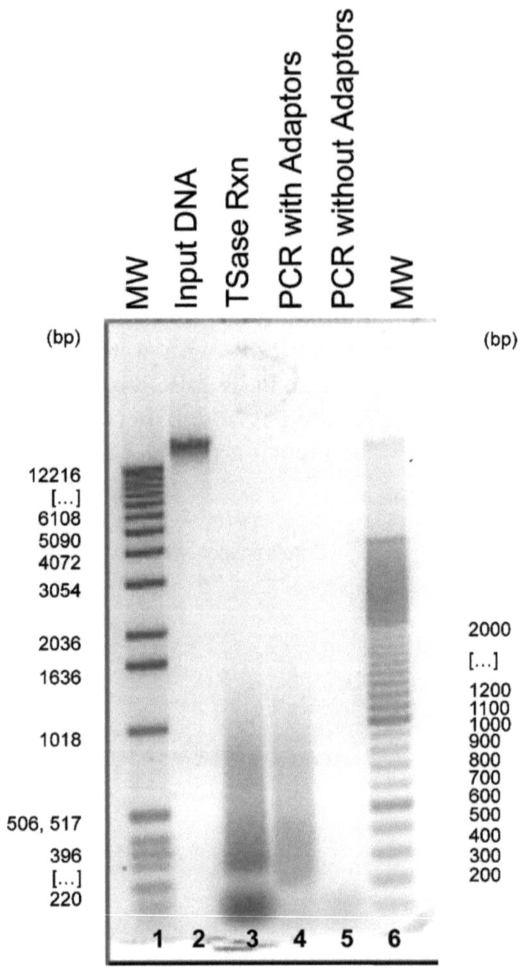

Fig. 5. Preparation of an Illumina GAII-compatible sequencing library. HeLa DNA (*lane 2*) was tagmented with Transposomes assembled with free ends appended with sequencing tags (*lane 3*). Limited-cycle PCR was used to add Illumina/Solexa bPCR-compatible adaptors (*lane 4*). The recovered DNA was used directly in bPCR prior to sequencing on the Illumina GAII instrument. Control PCR without Adaptor 1 and Adaptor 2 (including Primer 1 and Primer 2) confirmed reaction specificity (*lane 5*).

1. Assemble the following reaction components on ice, in the order listed (see Note 2)

x µl	Nuclease-free water
50 ng	Target DNA (see Note 1)
4 µl	5× Tagmentation reaction buffer
1 µl	Nextera Transposomes (Illumina GAII-Compatible)
20 µl	Total reaction volume

2. Mix briefly by vortexing and incubate at 55°C for 5 min. To prevent evaporation, the reaction should be carried out in a thermocycler with a heated lid or the reaction should be overlaid with mineral oil. Adding DNA-Binding Buffer in step 3 stops the reaction.
3. Purify the DNA sequencing library using a Zymo DNA Clean & Concentrator-5 Kit (or equivalent) according to the manufacturer's instructions (see Note 3). A synopsis of the ZYMO protocol is shown in steps 4–13. Perform these steps at room temperature.
4. Add 100 µl of DNA-binding buffer to the 20 µl tagmentation reaction from step 2.
5. Mix briefly by vortexing and transfer the mixture to a Zymo-Spin™ Column in a Collection Tube.
6. Centrifuge at 10,000 × *g* for 60 s. Discard the flow-through.
7. Add 250 µl of wash buffer to the column. Centrifuge at 10,000 × *g* for 60 s. Discard the flow-through.
8. Repeat the wash step once for a total of two washes.
9. Centrifuge the empty column again at 10,000 × *g* for 60 s to dry and to eliminate any residual wash buffer.
10. Transfer the column to a clean and sterile 1.5-ml microcentrifuge tube.
11. Add 11 µl of nuclease-free water directly to the column and incubate at room temperature for 1–2 min. Centrifuge at 10,000 × *g* for 60 s to elute the DNA into the clean microcentrifuge tube.
12. The final eluted volume should be ~10 µl. Use 5 µl of recovered DNA as PCR template in step 14.
13. The recovered DNA may be stored at −20°C.

Addition of bPCR-Compatible Adaptors and Library Enrichment

14. Assemble the following reaction components at room temperature:

15 µl	Nuclease-free water
5 µl	Recovered DNA Fragment Library (from step 3)
25 µl	2× Nextera PCR Buffer
1 µl	0.5 µM Adaptor 1 (Illumina GAII-Compatible)
1 µl	0.5 µM Adaptor 2 (Illumina GAII-Compatible, see Note 7)
For a bar-coded Illumina-compatible library, replace Adaptor 2 with a bar-coded Adaptor 2 with the sequence composition, see Note 7. The other three primers and the cycling conditions remain unchanged.	
1 µl	10 µM Primer 1 (Illumina GAII-Compatible)
1 µl	10 µM Primer 2 (Illumina GAII-Compatible)
1 µl	Nextera PCR Enzyme (2.5 U/µl)
50 µl	Total reaction volume

15. Incubate the samples in a thermocycler under the following conditions (see Note 5): 72°C for 3 min, 95°C for 30 s, followed by 12 cycles of (95°C for 10 s, 62°C for 30 s, 72°C for 3 min), hold at 4°C.

16. Purify the tagged DNA fragments using a Zymo DNA Clean & Concentrator-5 kit, or equivalent. The anticipated yield is ~300–500 ng of PCR-amplified DNA (Fig. 5).

17. The recovered DNA is ready to be used as input for Illumina GAII bPCR and cluster formation. No changes to the standard Illumina protocol are needed for cluster formation. However, alternate sequencing primers must be included during the sequencing reactions as described below (see Note 8).

    For Paired-End Reads: Illumina paired-end sequencing mixes (HP1 and HP2) contain the Illumina GAII sequencing primers. The Nextera sequencing primers are compatible with the Illumina GAII sequencing primers and can be used together in the same flow cell channel.

    - Paired-End Read 1: Dilute 100 µM Nextera Read 1 Primer 1:200 into Illumina GAII Sequencing Mix HP1.
    - Index Read: Dilute 100 µM Nextera Index Read Primer 1:200 into Illumina GAII Hybridization Buffer.
    - Paired-End Read 2: Dilute 100 µM Nextera Read 2 Primer 1:200 into Sequencing Mix HP2.
    - Alternatively, if it is required to perform sequencing in the presence of only the Nextera primers, the 100 µM Nextera sequencing primers can be diluted 1:200 into Illumina GAII Hybridization Buffer or 5× SSC, 0.05% Tween®-20.

For Single Reads: Illumina single-read sequencing mix (HP4) does not contain the Illumina sequencing primer.

- Single-Read Sequencing: Dilute Nextera Read 1 Primer 1:200 into Sequencing Mix HP4.
- Index Read: Dilute Nextera Index Read Primer 1:200 into Illumina GAII Hybridization Buffer or 5× SSC, 0.05% Tween®-20.

4. Notes

1. The quality of the starting DNA is critical. Contaminants such as protein and RNA may interfere with accurate quantification and/or inhibit the transposome fragmentation and tagging reaction if present in the DNA preparation. If DNA purity is in question, the DNA should be cleaned using Zymo Genomic DNA Clean & Concentrator or equivalent (ZYMO Research, Cat. No. D4010) prior to the Tagmentation reaction.
2. The tagmentation reaction does occur, although very slowly, at room temperature. Therefore, it is recommended that the components be assembled on ice and immediately incubated at 55°C.
3. The tagmentation reaction results in the simultaneous tagging and fragmentation of the target DNA. However, the transposase remains tightly bound to the ends of the resulting fragment. It is necessary to remove the transposase with chaotropic salts prior to tagging the 3′ ends and adding platform-specific adaptors by PCR. Equivalent kits can also be used; however, care must be taken not to denature the dsDNA fragments.
4. For a bar-coded Roche-compatible library, replace Adaptor 1 with a bar-coded Adaptor 1 with the sequence composition shown below. The other three primers and the cycling conditions remain unchanged.
   - Standard GS FLX-compatible Adaptor 1 with bar code.

     `5'-GCCTCCCTCGCGCCATCAG-[BAR CODE]-AGATGTGTATAAGAGACAG-3'`
   - Standard FLX Titanium-compatible Adaptor 1 with bar code.

     `5'-CCATCTCATCCCTGCGTGTCTCCGACTCAG-[BAR CODE]-AGATGTGTATAAGAGACAG-3'`
5. It is critical to perform the 72°C extension step to tag the 3′ ends of the DNA fragments *before* denaturing the PCR

templates. The 3′ ends are tagged by polymerase extension from the 3′ end of the genomic fragment and displacement of the MENTS oligonucleotide.

6. Nextera-generated and "standard" libraries can be sequenced in the same region of a plate. Perform the sequencing reactions with the standard Roche/454 sequencing primers. Sequencing reads will contain any added bar code sequence and the 19-base transposon-end sequence at the 5′ end of all reads. The bar code sequence can be used to bin the sequence reads. The 19-base transposon-end sequence should be filtered out prior to assembly and analysis (19-base transposon-end sequence: 5′-AGATGTGTATAAGAGACAG-3′).
7. For a bar-coded Illumina-compatible library, replace Adaptor 2 with a bar-coded Adaptor 2 with the sequence composition shown below. The other three primers and the cycling conditions remain unchanged.
   - Illumina GAII-compatible Adaptor 2 with bar code.

     5'-CAAGCAGAAGACGGCATACGAGA-[BAR CODE]-TC GGTCTGCCTTGCCAGCCCGCTCAG-3'
8. It is *critical* to use the Nextera Read 1, Nextera Read 2, and Index Read primers during cluster sequencing. While the resulting libraries are compatible with Illumina GA II bPCR and cluster formation, the transposon-derived sequencing primer sites are *not compatible with Illumina GAII sequencing primers.* The 19-bp transposon DNA sequence is present at the 5′ end of all Illumina-compatible libraries. However, the 19-bp transposon DNA sequence is NOT sequenced on the Illumina platform. The Nextera Read 1 and Read 2 primers anneal to this sequence so that the first nucleotide sequenced is target DNA.

Acknowledgments

The author would like to thank Jay Shendure and Hilary Morrison for their work characterizing the Illumina-compatible and Roche-compatible libraries and Haiying Grunenwald for technical advice and assistance during the development of this method.

Illumina and *Solexa* are registered trademarks of Illumina Inc., San Diego, CA. *Tween* is a registered trademark of ICI Americas Inc., Wilmington, DE. *454* and *GS FLX* are trademarks of Roche, Nutley, NJ. *DNA Clean & Concentrator* and *Zymo-Spin* are trademarks of Zymo Research., Orange, CA. *Nextera* and *Tagmentation* are trademarks of EPICENTRE, Madison, WI. Patent applications assigned to EPICENTRE and by US

Patent Nos. 5,965,443, and 6,437,109; European Patent No. 0927258, and related patents and patent applications, exclusively licensed to EPICENTRE, cover Nextera™ Products.

References

1. Kahvehian, V., Quackenbush, J., and Thompson, J. F. (2008) What would you do if you could sequence everything? *Nature Biotechnology.* **26**, 1125–1133.
2. Metzker, M. L. (2010) Sequencing Technologies – the next generation. *Nature Reviews – Genetics.* **11**, 31–46.
3. Shendure, J. and Henlee, J. (2008) Next-generation DNA Sequencing. *Nature Biotechnology.* **26**, 1135–1145.
4. Mardis, E. (2008) Next-Generation DNA Sequencing Methods. (2008) *Annual Review of Genomics and Human Genetics.* **9**, 387–402.
5. Fuller, C. W., Meddendorf, L. R., Benner, S. A., Church, G. M., Harris, T., Huang, X., Jovanovich, S. B., Nelson, J. R., Schloss, J. A., Schwartz, D. C., and Vezenov, D. V. (2009) The challenges of sequencing by synthesis. *Nature Biotechnology.* **27**, 1013–1023.

Chapter 18

Amplification-Free Library Preparation for Paired-End Illumina Sequencing

Iwanka Kozarewa and Daniel J. Turner

Abstract

The library preparation step is of critical importance for the quality of next-generation sequencing data. The use of the polymerase chain reaction (PCR) as a part of the standard Illumina library preparation protocol causes an appreciable proportion of the obtained sequences to be duplicates, making the sequencing run less efficient. Also, amplification introduces biases, particularly for genomes with high or low GC content, which reduces the complexity of the resulting library. To overcome these difficulties, we developed an amplification-free library preparation. By the use of custom adapters, unamplified, ligated samples can hybridize directly to the oligonucleotides on the flowcell surface.

Key words: Next-generation sequencing, Amplification-free library preparation, Malaria, (A + T)-rich genome

1. Introduction

Next-generation (NGS) sequencing technologies are revolutionizing biological sciences and medicine by providing unprecedented amounts of inexpensive and accurate sequencing data. Like other short-read NGS sequencing technologies, the Illumina Genome Analyzer is capable of the simultaneous sequencing of millions of immobilized clonally amplified template strands (1) and can generate more than 1×10^9 bases from a single lane of a flowcell in a 36-base-paired run.

As a part of the Illumina library preparation protocol, the genomic DNA is fragmented, end-repaired, and A-tailed (1). Following this, partially double-stranded adapters are ligated via a T-overhang. Finally, the ligated fragments are amplified by

Young Min Kwon and Steven C. Ricke (eds.), *High-Throughput Next Generation Sequencing: Methods and Applications*, Methods in Molecular Biology, vol. 733, DOI 10.1007/978-1-61779-089-8_18, © Springer Science+Business Media, LLC 2011

10–12 cycles of polymerase chain reaction (PCR). In addition to enriching successfully ligated fragments within the library, the PCR step uses tailed primers which introduce the flowcell annealing sequences to all amplified fragments.

Though PCR has been widely accepted as the standard technique to amplify specific sections of DNA exponentially, amplification tends to be biased toward some DNA sections vs. others: PCR can preferentially amplify some wild-type vs. mutant sequences as well as one allele vs. another (2, 3). This bias has hindered PCR-based diagnostics and is not easily remedied by alterations to the PCR protocols.

To prevent amplification-related bias and to reduce the formation of primer dimers, the Illumina library preparation protocol employs universal PCR primers, which allow all fragments within the library to be amplified simultaneously (1). Regardless of this, the use of PCR can lead to poor library quality, especially for (G+C)-biased templates. When prepared according to the standard, PCR-based library preparation protocol, we observed that for *Plasmodium falciparum* (mean exonic (A+T) content >75% (4)) the quality of read mapping was so low, and with such a high proportion of duplicate reads and uneven coverage that de novo assembly was not possible (5).

To address this, we developed a genomic DNA sample preparation method that omits the PCR step entirely (termed no-PCR method). Using this method, we not only improved our single-nucleotide polymorphism (SNP) detection but also obtained a *P. falciparum* assembly of 20.8 Mb with N50 = 1.28 kb (5). We also tested our method on (G+C)-neutral and (G+C)-rich microbial genomes, where we obtained de novo assemblies with similar contig sizes to the ones obtained from standard Illumina libraries. For human DNA, the use of the no-PCR method gave 20–100× more even coverage than the standard library preparation method as well as a very low percentage of duplicates.

Our no-PCR library preparation method uses custom adapters, which are longer than the standard Illumina ones (Fig. 1) and which contain the complete sequence required for hybridization of the templates to the flowcell surface, and for annealing of sequencing primers. The structure of the custom adapters is the same as the one of the standard ones (i.e., partially double stranded), ensuring that after ligation, each strand receives a unique adapter sequence at either end. As with the standard ligation, the no-PCR method can generate products that are partially or non-ligated, but whereas in the standard library prep, PCR is used to enrich for fully ligated templates, in the no-PCR approach, this enrichment step is performed on the flowcell surface: incompletely ligated fragments are inert in this step and will not form clusters.

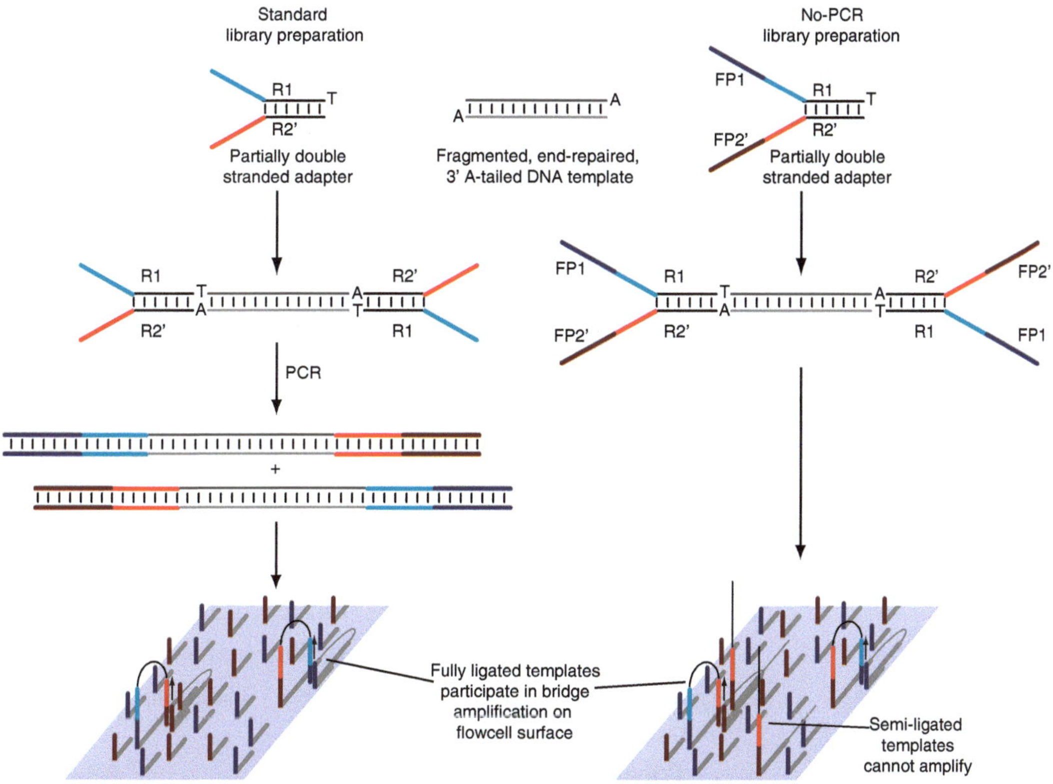

Fig. 1. Comparison of standard adapters and standard library preparation method (*on the left*) vs. the no-PCR adapters and method (*on the right*). In both cases, partially complementary adapters are ligated onto the template via a 3′ thymine (T) overhang. Whereas standard adapters contain only the sections to which the sequencing primers hybridize (R1 and R2′), the no-PCR adapters include also the sequences necessary for hybridization to the oligonucleotides on the flowcell surface (FP1 and FP2′). Thus, the latter sequences do not have to be added by PCR to the no-PCR adapters. Reprinted with permission from Macmillan Publishers Ltd: Kozarewa, I. *et al.* Nature Methods 6(4), 291-5(2009).

Since the PCR step is not retained in the no-PCR library preparation method, the libraries obtained are often less concentrated than the standard ones and may contain more incompletely ligated fragments. As a result, it is crucial to quantify only those templates that will actually amplify on the flowcell, rather than the entire library. For this, we use a TaqMan-based quantitative PCR assay, in which the unknown library is quantified against a standard library with the same insert size and base composition, that has already been sequenced and for which both the concentration and the cluster number are known. The qPCR primers are designed to target these adapter sequences that are required for hybridization of the template to the flowcell oligonucleotides (Fig. 2). As a result, only fully ligated fragments are quantified, which allows accurate estimation of the cluster density without the need for titration runs.

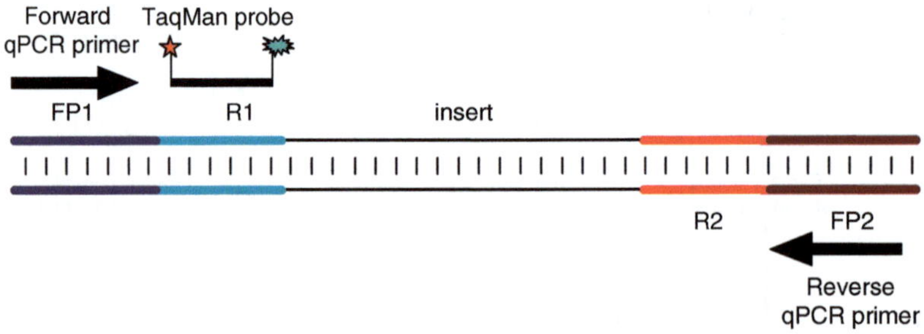

Fig. 2. Annealing sites of the qPCR primers and the TaqMan probe on the ligated template. Both primers anneal to the required for hybridization sequences, which allows only the fully ligated templates to be amplified and quantified.

2. Materials

2.1. Library Preparation

1. QIAquick PCR purification kit (Qiagen, Valencia, CA).
2. MinElute PCR purification kit (Qiagen, Valencia, CA).
3. Paired-end DNA sample prep kit (Illumina, San Diego, CA).
4. Agilent DNA 1000 chips (Agilent Technologies, Santa Clara, CA).
5. T4 Polynucleotide kinase.
6. T4 DNA ligase buffer with 10 mM ATP.
7. HPLC-purified No-PCR adapters

   A_adapter_t

   5′AATGATACGGCGACCACCGAGATCTACACTCTTTC CCTACACGACGCTCTTCCGATC*T

   *indicates phosphorothioate

   A_adapter_b

   5′GATCGGAAGAGCGGTTCAGCAGGAATGCCG AGACCGATCTCGTATGCCGTCTTCTGCTTG
8. Agarose.
9. 10× TBE.
10. SafeView (NBS Biologicals, Huntington, UK).
11. 5× Qiagen GelPilot loading dye (Qiagen, Valencia, CA).
12. Low molecular weight ladder (NEB, Ipswich, MA).
13. MinElute gel extraction kit (Qiagen, Valencia, CA).
14. Isopropanol.
15. Tween 80.

2.2. Quantification

1. Platinum *Taq* polymerase (Invitrogen, Carlsbad, CA).
2. dNTP set.

3. qPCR primers and TaqMan probe

c_qPCR_v2.1	5′ AATGATACGGCGACCACCGAGATC
PE_qPCR_v2.2	5′ CAAGCAGAAGACGGCATACGAGATC
TaqMan probe	5′ (6-FAM) CCCTACACGACGCTCTTCCGAT CT (TAMRA)

PCR primers c_qPCR_v2.1 and PE_qPCR_v2.2 are desalted, whereas the TaqMan probe is HLPC purified.

4. ROX dye (Invitrogen, Carlsbad, CA).

2.3. Equipment

1. Vacuum dryer.
2. Covaris S2 (Kbiosciences, Herts, UK).
3. Microtubes, AFA fiber with snap-cap (Kbiosciences, Herts, UK).
4. Crimp seals (Kbiosciences, Herts, UK).
5. Set of pipettes.
6. Thermal cycler with 96-well blocks.
7. Agilent 2100 Bioanalyzer (Agilent Technologies, Santa Clara, CA).
8. Vortex mixer.
9. Microcentrifuge.
10. Plate centrifuge.
11. 50-ml Falcon tubes.
12. 15-ml Falcon tubes.
13. Thin-walled 200-μl PCR tubes.
14. Horizontal mini electrophoresis gel tank.
15. Electrophoresis power supply.
16. UV transilluminator.
17. Fridge.
18. Microwave.
19. Real-time PCR system.
20. Real-time PCR-compatible 96-well reaction plates and plate seals.

3. Methods

3.1. Sample Fragmentation

1. The DNA (300 ng–5 μg) is supplied suspended in water, Elution buffer (EB), or TE buffer (10 mM Tris–HCl, pH 7.5/1 mM EDTA).If the sample volume is bigger than 75 μl, reduce it using a SpeedVac.

2. If necessary, add ultra-pure water to make the final volume 75 μl.
3. Mix thoroughly and transfer to a 100-μl Covaris microtube.
4. Shear the sample for 180 s with the following program (see Note 1):

   Duty cycle: 20%

   Intensity: 5

   Cycle burst: 200

   Power: 37 W

   Temperature: 7°C

   Mode: freq sweeping

   PAUSE POINT: reactions can be stored at −20°C for 4 weeks.

3.2. End-Repair and A-Tailing

1. Clean the fragmented sample using QIAquick PCR purification kit, following the recommended protocol. Elute in 30 μl elution buffer (EB).

   PAUSE POINT: the cleaned reactions can be stored at −20°C for several weeks.

2. For each fragmented sample prepare a master mix (MM) containing the reagents below. Add the reagents in the specified order.

10× T4 DNA ligase buffer with 10 mM ATP	10 μl
10 mM dNTPs mix	4 μl
3 U/μl T4 DNA polymerase	5 μl
10 U/μl T4 PNK	5 μl
5 U/μl Klenow DNA polymerase	1 μl
Water	40 μl

   Mix briefly and spin down (see Note 2). Use 200-μl thin-walled PCR tubes.

3. Incubate for 30 min at 20°C in a thermal cycler.
4. Clean the sample using QIAquick PCR purification kit, following the recommended protocol. Elute in 32 μl EB.
5. For each purified, end-repaired DNA sample obtained in the preceding step prepare MM containing the following reagents, added in the following order:

1 mM dATP	10 μl
10× Klenow buffer	5 μl
5 U/μl Klenow exo (3′–5′ exo minus)	3 μl

Mix briefly and spin down. Use 200-μl thin-walled PCR tubes.

6. Incubate for 30 min at 37°C in a thermal cycler (use heated lid).
7. Clean the A-tailed sample using MinElute PCR purification kit and elute in 12 μl EB.

3.3. Ligation

1. While doing the end-repair and A-tailing, prepare the custom adapters. Combine the following in one 200-μl PCR tube:

100 μM oligo A_adapter_t	20 μl
100 μM oligo A_adapter_b	20 μl
10× T4 ligase buffer with 10 mM ATP	5 μl
10 U/μl T4 PNK	5 μl

Vortex, spin and place the tube in the thermal cycler.

2. Phosphorylate and anneal the adapters using the following program:

30 min at 37°C

Ramp PCR machine at 0.5°C/s to 97.5°C

Hold at 97.5°C for 150 s then 97.5°C for 5 s and temp drop of (−)0.1°C per cycle for 775 cycles (i.e., decrease temperature from 97.5°C by 0.1°C every 5 s).

4°C indefinite hold

Store the phosphorylated oligos in 8 μl aliquots at −20°C until use.

3. Run 1 μl of the DNA sample on Agilent chip and use the "integrated peak" function to measure the concentration of the sample following manufacturer's recommended protocol.
4. To optimize the ligation, a molar ratio of 20:1 of adapters: A-tailed sample is used. The moles of A-tailed sample are calculated according to the formula:

$$\text{Moles} = \text{DNA mass in g}/(\text{average fragment size} \times 650)$$

5. Add 25 μl 2× Illumina DNA ligase buffer, adapter and water to A-tailed DNA (in 1.5-ml Eppendorf tube). Vortex and spin down (see Note 3).
6. Add 5 μl 2,000 U/μl Illumina DNA ligase, vortex and spin down.
7. Incubate for 15 min at room temperature (start timing after ligase has been added to the last sample).

PAUSE POINT: reactions can be stored at −20°C for up to 1 week.

8. Clean reactions up using QIAquick PCR purification kit, eluting in 30 μl EB.

   PAUSE POINT: reactions can be stored at –20°C for up to 6 months.

3.4. Size Selection Step

1. Prepare 150 ml of a 2% agarose gel in 1× TBE (3 g agarose; see Note 4). This is enough for one mini gel. Use one gel per library unless doing several identical libraries. In the latter case, use one gel for up to three libraries. Use the midi-comb (eight sample slots).
2. Once the agarose has dissolved completely, immediately add 7.5 μl (5 μl per 100 ml) SafeView. Pour the gel and allow to solidify at room temperature. Once the gel has solidified, allow to cool for 10 min at 4°C in the fridge.
3. Mix by pipetting 10 μl of 5× Qiagen GelPilot loading dye with the 30 μl sample and load in one well of the gel.
4. Mix well by pipetting 8 μl of NEB low molecular weight ladder with 3 μl of the above-mentioned loading dye and load in one well of the gel. Load the ladder in the first on the left well of the gel. Leave one or two wells empty between the ladder and the samples (see Note 5).
5. Run the gel at 60 V for 2 h using chilled (kept at 4°C) 1× TBE buffer. After the first hour, replace the buffer with fresh, chilled TBE. Stop electrophoresis once the orange marker reaches the bottom of the gel (see Note 6).
6. For an *X*bp insert library, cut gel slice between *X*+100 and *X*+150 bp (e.g., for a 200 bp insert library, cut the gel slice between 300 and 350 bp and for a 500 bp insert library, cut the gel slice between 600 and 750 bp).

   PAUSE POINT: the gel slices can be stored at –20°C for up to a week.
7. Weigh the gel slice and transfer to a 15-ml falcon tube. If the gel slice weighs >400 mg, use two MinElute columns for purification and do final elution in 15 μl EB + 0.1% Tween for each column. If the gel slice is <400 mg, use only one column and elute in 30 μl (see Note 7).
8. Add 3× volume of QG buffer. Clean reaction up using MinElute gel extraction kit, incubating gel slice and QG buffer at room temperature.

   PAUSE POINT: the ready libraries can be stored at –20°C for up to 6 months.

3.5. Quantification by qPCR

1. For each library, prepare 1:80 and 1:200 dilutions in EB + 0.1% Tween.
2. For each PCR reaction, mix the following reagents (final concentrations are given in parentheses):

10× Platinum *Taq* buffer	2.5 µl (→ 1×)
50 mM $MgCl_2$	0.75 µl (→ 1.5 mM)
Template DNA	1 µl
10 µM TaqMan probe	0.625 µl (→ 250 nM)
50× Rox	0.5 µl (→ 1×)
10 µM c_qPCR_v2.1	75 µl (→ 300 nM)
10 µM PE_qPCR_v2.2	0.75 µl (→ 300 nM)
mM dNTPs	2 µl (→ 200 µM)
5 U/µl Platinum Taq	0.1 µl (→ 0.02 U/µl)
H_2O	16.025 µl

3. Conduct the qPCR using the following cycling conditions:

   94°C for 2 min

   94°C for 15 s

   62°C for 15 s ×40

   72°C for 32 s
4. Establish the concentration of the libraries using standards. The latter are libraries with identical insert size range to that of the unknown libraries that have already been sequenced and for which the cluster number at given loading concentration is known.
5. The standards used have concentrations: 1, 10, and 100 pM, based on the concentrations measured on the Agilent Bioanalyzer 2100. Both the standards and the libraries with unknown concentration are run in triplicate.
6. Prior to sequencing, dilute the libraries to 5-8 pM concentration following the current Illumina cluster generation protocol.

3.6. Timing

1. Sample fragmentation: 5 min per sample
2. End-repair and A-tailing: 1.5 h
3. Ligation: 1.5 h
4. Size selection step

   4.1. Gel casting: 1 h

   4.2. Gel set up and running: 3 h

   4.3. Gel slices purification: 1 h per library
5. qPCR quantification

   5.1. Reaction set up: 0.5 h

   5.2. qPCR: 1.5 h

   5.3. Analysis: 0.5 h

4. Notes

1. This program produces libraries with mean fragment size of 200 bp. To obtain libraries with mean insert size of ~350 bp reduce the shearing time to 90 s.
2. When preparing a master mix, add additional 20% of the reagents.
3. If the volume of the adapters required is >10 μl, prepare the ligation in 70 μl volume instead of 50 μl. Add the following reagents in the following order:

2× Illumina DNA ligase buffer	35 μl
A-tailed DNA	10 μl
40 μM Adapter mix	up to 19 μl
Water	(19 μl – volume of adapter mix)
2,000 U/μl Illumina DNA ligase	6 μl

4. The gels can be prepared up to 24 h before use and stored at 4°C.
5. For larger insert size libraries (≥400 bp), load also 10 μl of Bioline Hyperladder IV in the second start from the left. Leave one or two lanes empty between the second ladder and the samples.
6. For larger insert size libraries (≥400 bp), allow the orange marker to run out completely from the gel.
7. If the starting amount of DNA is low (<1 μg), elute each column with 10 μl EB + 0.1% Tween. If the starting amount of DNA is >1 μg but <5 μg, follow the standard protocol (Subheading 3.4, step 7). If the starting amount of DNA is >5 μg, elute each column with 20 μl EB + 0.1% Tween.

References

1. Bentley D. R., Balasubramanian S., Swerdlow H. P., Smith G. P., Milton J., Brown C. G. et al. (2008) Accurate whole human genome sequencing using reversible terminator chemistry. *Nature* **456**, 53–59.
2. Barnard R., Futo V., Pecheniuk N., Slattery M. and Walsh T. (1998) PCR bias toward the wild-type k-ras and p53 sequences: implications for PCR detection of mutations and cancer diagnosis. *BioTechniques* **25**, 684–691.
3. Hahn S., Garvin A. M., Di Naro E. and Holzgreve W. (1998) Allele drop-out can occur in alleles differing by a single nucleotide and is not alleviated by preamplification or minor template increments. *Genetic Testing* **2**, 351–355.
4. Gardner M. J., Hall N., Fung E., White O., Berriman M., Hyman R. W. et al. (2002) Genome sequence of the human malaria parasite Plasmodium falciparum. *Nature* **419**, 498–511.
5. Kozarewa I., Ning Z., Quail M. A., Sanders M. J., Berriman M. and Turner D. J. (2009) Amplification-free Illumina sequencing-library preparation facilitates improved mapping and assembly of (G+C)-biased genomes. *Nature Methods* **6**, 291–295.

Chapter 19

Target-Enrichment Through Amplification of Hairpin-Ligated Universal Targets for Next-Generation Sequencing Analysis

Pallavi Singh, Rajesh Nayak, and Young Min Kwon

Abstract

With rapid development of next-generation sequencing (NGS) technologies, it is becoming increasingly feasible to sequence entire genomes of various organisms from virus to human. However, in many occasions, it is still more practical to sequence and analyze only small regions of the entire genome that are informative for the purpose of the experiment. Although many target-enrichment or target capture methods exist, each method has its own strength and weakness in terms of the number of enriched targets, specificity, drop-off rate, and uniformity in capturing target DNA sequences. Many applications require a consistently low drop-off rate and high uniformity of enriched targets for routine collection of meaningful data. Here, we describe a simple and robust PCR-based protocol that can allow simultaneous amplification of numerous target regions. This method employs target-specific hairpin selectors to create DNA templates that contain target regions flanked by common universal priming sequences. We demonstrated the utility of this method by applying it for simultaneous amplification of 21 targets in the range of 191–604 bp from 41 different *Salmonella* strains using bar-coded universal primers. Analysis of 454 FLX pyrosequencing data demonstrated the promising performance of this method in terms of specificity and uniformity. This method, with great potential for robust amplification of hundreds of targets, should find broad applications for efficient analysis of multiple genomic targets for various experimental goals.

Key words: Target-enrichment, Multiple targets, Next-generation sequencing, Hairpin selector, PCR amplification

1. Introduction

Recent development of next-generation sequencing (NGS) in various platforms has enabled processing and sequencing of an unprecedented amount of DNA sequences. These NGS technologies have been used to decipher whole genome sequences of various organisms of different sizes from virus to humans, making it feasible to compare multiple individual organisms for the entire genome sequences (1). In many situations, however, only small

Young Min Kwon and Steven C. Ricke (eds.), *High-Throughput Next Generation Sequencing: Methods and Applications*, Methods in Molecular Biology, vol. 733, DOI 10.1007/978-1-61779-089-8_19, © Springer Science+Business Media, LLC 2011

portions of the genomes carry important meaningful information for the given analysis. For example, during genotypic analysis of single nucleotide polymorphisms (SNPs) associated with a phenotype of interest, sequence data are desired only from the SNP regions known to be associated with the phenotype, whereas other sequences may not provide meaningful information or insight. Therefore, it is often convenient and practical to enrich, capture, or amplify only predefined targets from individual genomic DNA samples, which can then be analyzed using NGS. Current methods for target-enrichment (or target capturing) are mainly based on two different underlying principles: DNA hybridization and PCR amplification (2). Each method has its own advantage and disadvantage. In general, hybridization-based methods can capture a much larger set of targets, while many targets are missed during capturing. On the contrary, with amplification-based methods, the portion of missing targets is relatively much lower. However, the numbers of the targets that can be captured are much smaller in amplification-based methods as compared to those in hybridization-based methods (3, 4). Therefore, the method of choice for target-enrichment should be chosen carefully, depending on the purpose of the given application. When it comes to the application where a limited number of targets (less than hundreds) are analyzed and the same set of the target sequences are necessary for routine and meaningful data analysis and interpretation, PCR amplification-based methods are more suitable.

In this chapter, we describe a robust and reproducible method using target-specific hairpin selectors to create targets to which universal primer-binding sites are attached on both sides. The multiple target sequences then can be amplified using a universal primer pair. In principle, generation of universal targets is not limited by size of targets and the limitations posed are only due to general PCR mechanism that would determine the size limit and uniformity of target amplification. Although the upper limit of multiplexing capacity and target length were not rigorously evaluated, we found that 21 targets, up to ~600 bp, can be amplified from a *Salmonella* genome using this method in a highly reproducible manner. When necessary, barcodes can be easily incorporated into universal primer design to allow pooling of multiple samples for NGS analysis.

2. Materials

2.1. Genomic DNA Isolation and Restriction Enzyme Digestion

1. QIAamp DNA mini Kit for genomic DNA isolation (Qiagen, Valencia, CA).
2. MspI or other restriction enzymes (New England BioLabs, Ipswich, MA). Store at −20°C.

3. Double-distilled water autoclaved at 121°C for 20 min. Store at 4°C.

2.2. Phosphorylation of Hairpin Selectors and Ligation Reaction

1. TE buffer: 10 mM Tris–HCl, pH 8.0, 1 mM EDTA.
2. Oligonucleotides for two hairpin selectors are designed and synthesized for each target DNA region (Integrated DNA Technologies, Coralville, IA). Resuspend the oligonucleotides in 1× TE buffer at the final concentration of 1 mM. Store at −20°C. Oligonucleotides for two hairpin selectors to amplify *fimA* gene are shown here as an example (Fig. 2c).
   (a) Hairpin selector I for *fimA* (91 bp)
   5′-AGTACCGCCACTCACGCTCACCGCATGAATTC
   C-ATCCGAATTCAGGCCGGCCGGGCGCGCCC
   GG-CCGGCCTGAATTCGGATGGAATTCATG-3′
   (b) Hairpin selector II for *fimA* (89 bp)
   5′-CATTCTAGACATCCGAATTCAGCCGGGCCCC
   GGCCGGGGCCCGGCTGAATTCGGATGTCTA
   GAATGGTTGCGGTAGTGCTATTGTCCGC-3′
3. 10 mM ATP (Invitrogen, Carlsbad, CA). Store at −20°C.
4. T4 Polynucleotide Kinase (PNK) and 10× reaction buffer (New England BioLabs, Ipswich, MA). Store at −20°C.
5. Ampligase Thermostable DNA ligase (5 U/μl) and 10× reaction buffer (Epicentre Biotechnologies, Madison, WI). Store at −20°C.
6. Double-distilled water autoclaved at 121°C for 20 min. Store at 4°C.
7. Thermal cycler.

2.3. PCR Amplification, Gel Electrophoresis, and Gel-Purification

1. Cloned *pfu* DNA polymerase and 10× reaction buffer (Stratagene, La Jolla, CA). Store at −20°C.
2. 2.5 mM dNTPs mixture. Store at −20°C.
3. Two universal primers specific to hairpin selectors for amplification of the hairpin selector-ligated targets:
   (a) Universal primer I (23 bp): 5′-CCTGAATTCGGATGG AATTCATG-3′
   (b) Universal primer II (22 bp): 5′-CTGAATTCGGAT GTCTAG-AATG-3′
4. Double-distilled water autoclaved at 121°C for 20 min. Store at 4°C.
5. 1× TBE electrophoresis buffer: 89 mM Tris base, 89 mM boric acid, 2 mM EDTA, pH 8.0. Store at room temperature.
6. Novex 6% TBE gels (Invitrogen, Carlsbad, CA). Store at 4°C.

7. DNA size marker. Store at 4°C.
8. Gel loading dye (5× or 6×). Store at 4°C.
9. SyberGreen.
10. QIAquick PCR purification kit (Qiagen, Valencia, CA).

3. Methods

3.1. Overview

The overall schematic diagram of this multiplex amplification method is shown in Fig. 1. The first step is to digest the genomic DNA with an appropriate restriction enzyme that defines the boundary of the target region to be amplified. Two hairpin selectors specifically designed to anneal to each restriction enzyme-digested target fragment are ligated to the plus strand of the target fragment. Once ligation happens, the target fragments carry common hairpin sequences attached to both the ends. These ligated fragments are then digested with the same restriction enzyme, initially used to digest the genomic DNA, to remove hairpin structures on both ends. The next step is to amplify multiple target sequences using a pair of universal primers. For NGS analysis, appropriate platform-specific sequences are attached to

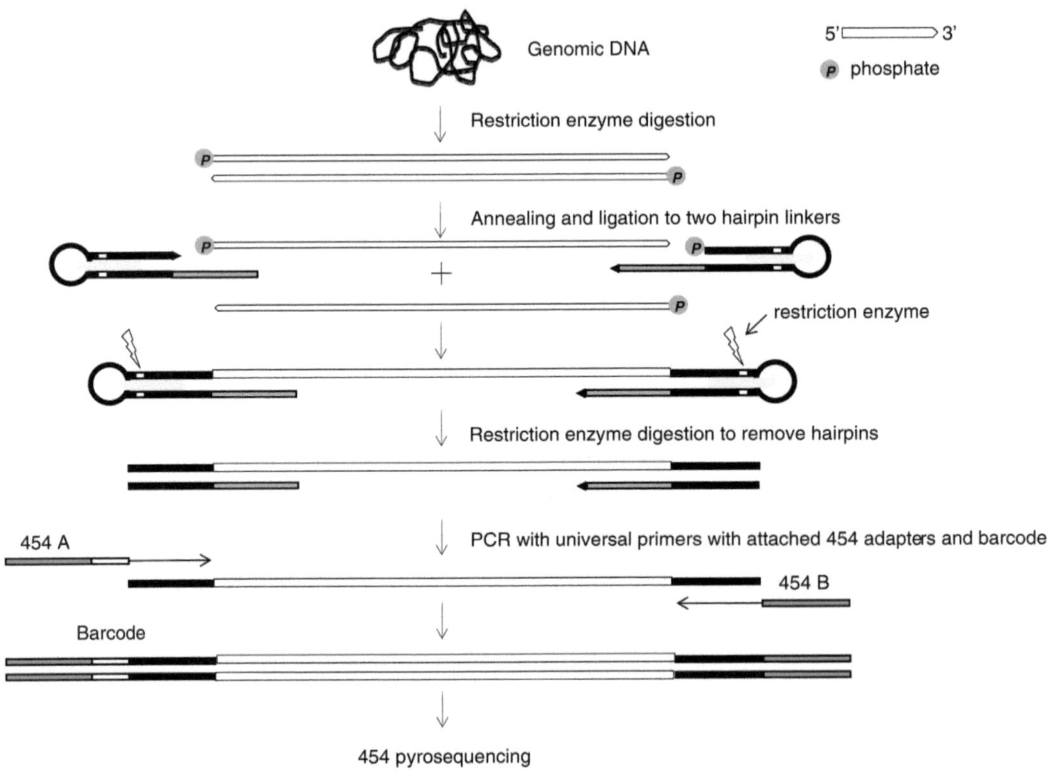

Fig. 1. Schematic diagram of the multiple target amplification protocol using hairpin selectors.

the 5′ end of the universal primers. For analysis of multiple samples, barcodes can be incorporated into one or both of the universal primers between the NGS platform-specific sequence and universal primer sequence. In our example shown here, we used 454 FLX Titanium pyrosequencing to cover the entire target lengths in the range of 191–604 bp by bidirectional sequencing.

3.2. Hairpin Selector Design

1. The first step is to select the target regions to be amplified from genome sequence of the organism of interest based on the purpose of the analysis. In our experiment shown here for illustration, we chose 21 target genes previously used for multi-locus sequence typing analysis of *Salmonella* strains (5–7).
2. Once the targets are selected, *in silico* restriction enzyme digestion analysis is carried out to search for the suitable restriction enzyme using the NEBcutter software V2.0 (http://tools.neb.com/NEBcutter2/). Restriction enzymes are searched and analyzed for which the recognition sequence is present at multiple sites in each and every target. In our example, we chose the restriction enzyme MspI (5′-C↓CGG-3′/3′-GGC↑C-5′) (Fig. 2a) (see Note 1).

a

>*fimA* (*Salmonella* Typhimurium LT2)
agggaaatccATGAAACATAAATTAATGACCTCTACTATTGCGAGTCTGATGTTTGTCGCTGGCGCAGCGGTTGCGGCTGATCCTACT **CCGG**TG
AGCGTGAGTGGCGGTACTATTCATTTCGAAGGTAAACTGGTTAATGCAGCCTGTGCCGTCAGCACTAAATCCGCCGATCAAACGGTGACGCTGG
GTCAATACCGTACCGCCAGCTTTACGGCGATTGGTAATACGACTGCGCAGGTGCCTTTCTCCATCGTCCTGAATGACTGCGATCCGAAAGTGGC
GGCCAACGCTGCCGTGGCTTTCTCTGGTCAGGCAGATAACACCAACCCTAATTTGCTGGCTGTCTCCTCTGCGGACAATAGCACTACCGCAA**CC**
GGCGTCGGGATTGAGATTCTTGATAATACCTCTTCACCGTTGAAG **CCGG**ACGGCGCGACCTTCTCGGCGAAGCAGTCGCTGGTTGAAGGCACCA
ATACGCTGCGTTTTACCGCACGCTATAAGGCAACCGCCGCCGCCACGACGCCAGGCCAGGCTAATGCCGACGCCACCTTTATCATGAAATACGA
ATAAtcccgtcagg

b

CGGTGAGCGTGAGTGGCGGTACTATTCATTTCGAAGGTAAACTGGTTAATGCAGCCTGTGCCGTCAGCACTAAATCCGCCGATCAAACGGTGAC
GCTGGGTCAATACCGTACCGCCAGCTTTACGGCGATTGGTAATACGACTGCGCAGGTGCCTTTCTCCATCGTCCTGAATGACTGCGATCCGAAA
GTGGCGGCCAACGCTGCCGTGGCTTTCTCTGGTCAGGCAGATAACACCAACCCTAATTTGCTGGCTGTCTCCTCTGCGGACAATAGCACTACCG
CAA**C**

c

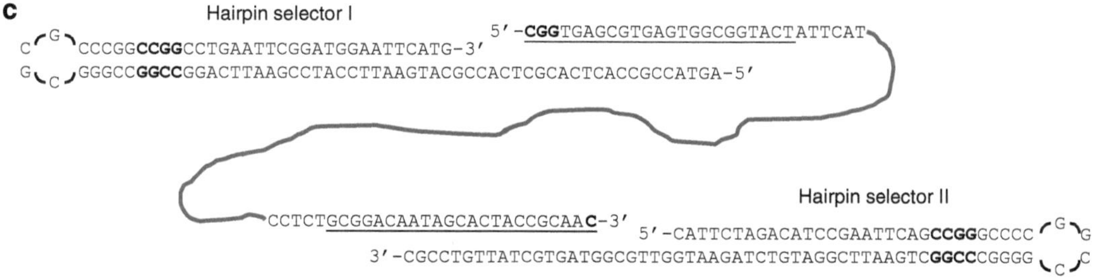

Titanium A-universal primer I
5′-CGTATCGCCTCCCTCGCGCCATCAG-8bp barcode-CCTGAATTCGGATGGAATTCATG-3′
Titanium B-universal primer II
5′-CTATGCGCCTTGCCAGCCCGCTCAG-8bp barcode-CTGAATTCGGATGTCTAGAATG-3′

Fig. 2. Hairpin selector design. (**a**) Nucleotide sequence of *fimA* gene in *Salmonella* Typhimurium LT2 strain. The coding region is shown in capital letters and the 10-bp upstream and downstream sequences are shown in small letters. Three MspI sites are shown in bold type. (**b**) The sequence of targeting fragment. The priming sequences are underlined. (**c**) The sequence of the *fimA* hairpin selectors (I and II) and universal primers I and II to which the 454 FLX Titanium-specific sequence and 8 bp barcodes are attached at 5′ ends.

3. Among multiple fragments produced by restriction enzyme digestion for each target, one target fragment should be chosen based on the information content and the length (Fig. 2b). The lengths from all targets should be within a certain range to minimize the bias in amplification step due to the length variation. In our experiment, MspI digestion-generated fragments are in the range of 191–604 bp in all 21 target genes.
4. Once the target fragments are defined (one fragment/target), two hairpin selectors are designed for each target fragment as shown in Fig. 2c. A hairpin selector consists of the following four components: (1) target-binding region of minimum length of 20 bp complementary to the sequence at the end of the target fragment, (2) a stem-loop region that forms a hairpin structure, (3) universal primer binding site, which is a part of the stem region, and (4) a restriction enzyme site that will be used to remove the hairpin structure outside the universal primer-binding site after the hairpin selector is ligated to the target. In our design of hairpin selectors, MspI recognition site was inserted in the hairpin selectors as shown in Fig. 2 (see Note 2). The length of target-binding regions should be adjusted to have relatively uniform melting temperatures, so that the ligation reaction can be carried out at the same high stringent temperature for all targets.

3.3. Restriction Enzyme Digestion

1. Isolate genomic DNA from the cells using QIAamp DNA mini Kit following the manufacture's instruction. Other methods for genomic DNA isolation can be used.
2. Determine the concentration and purity of isolated genomic DNA (e.g. NanoDrop).
3. Set up the following 50 μl reaction for digestion of genomic DNA (see Note 3):

Double-distilled water	13 μl
10× NEBuffer 4	5 μl
Genomic DNA (100–300 ng/μl)	30 μl
MspI (20 U/μl)	2 μl
Total volume	50 μl

4. Incubate the reaction tube at 37°C for 2–3 h.
5. Heat inactivate MspI by incubating at 80°C for 20 min.
6. Digested DNA can be stored at −20°C.

3.4. Phosphorylation of the Hairpin Selector II

1. For each hairpin selector, prepare a stock solution in 1×TE buffer at the final concentration of 100 mM. Also, prepare a working solution in 1×TE at the final concentration of 1 mM.
2. Set up the following reaction for 5′-phosphorylation of hairpin selector II to allow ligation between target fragments and hairpin selector II molecules (see Note 4):

10 mM dATP	5 μl
PNK buffer	5 μl
PNK	2 μl
Hairpin selector II (1 mM)	38 μl
Total volume	50 μl

3. Incubate the reaction mixture at 37°C for 30 min.
4. Inactivate PNK by incubation at 65°C for 20 min.

3.5. Ligation Reaction and Digestion with a Restriction Enzyme

1. To carry out ligation reaction, set up the following reaction (see Note 5):

Water	19.5 μl
10× Ampligase reaction buffer	2.5 μl
Restriction digested genomic DNA (from Subheading 3.3)	1 μl
Hairpin primer I (1 mM)	0.5 μl
5′-Phosphorylated harpin primer II (from Subheading 3.4)	0.5 μl
Ampligase	1 μl
Total volume	25 μl

2. Ligation reaction is carried out using thermal cycler through 40 cycles of 95°C for 1 min and 70°C for 2 min (see Notes 6 and 7).
3. To the ligation reaction, add 3 μl of NEBuffer 4 and 1 μl of MspI enzyme to make a 30-μl digestion reaction. This digestion step after ligation is necessary to remove hairpin structures from the ligation products. This step facilitates amplification of the ligated products by the universal primers, because the hairpin structure would interfere with annealing of the universal primers.

4. This reaction mix is incubated at 37°C for 1 h, followed by enzyme inactivation at 80°C for 20 min.
5. Stored the reaction tube at −20°C.

3.6. Amplification of Multiple Targets Using Universal Primers

1. Multiplex amplification is carried out using the universal primers to amplify the all targets simultaneously.
2. A 50-μl reaction is set up as following (see Note 8):

Double-distilled water	37 μl
10× Cloned *pfu* DNA polymerase buffer	5 μl
2.5 mM dNTPs	4 μl
Universal primer I (350 ng/ μl)	1 μl
Universal primer II (350 ng/μl)	1 μl
MspI digested-ligated template	1 μl
1 μl Cloned *pfu* DNA polymerase	1 μl
Total volume	50 μl

3. Amplification is carried out through 30 cycles of 94°C for 30 s, 60°C for 30 s, 72°C for 60 s followed by extension at 72°C for 10 min.
4. Novex TBE gel is prepared and used to analyze the amplified products.
5. The amplified products are stained with SyBrGreen for 15 min and visualized on UV light (Fig. 3a).

3.7. Sample Preparation for 454 FLX Pyrosequencing

1. If multiple strains or samples are being analyzed, then 454 pyrosequencing is a cost-effective method for analyzing numerous strains or samples. For this, we used universal primers tagged with 454 FLX Titanium-specific sequences and a 10-bp barcode to distinguish between sequence reads from different strains or samples (Fig. 2c). Amplification can be done as above (Subheading 3.6) using universal barcoded fusion primers (Fig. 3b).
2. The amplified products tagged with different barcodes are visually inspected for the intensity of the bands on TBE gels, and then appropriate amounts of amplicons from different strains or samples are combined to make a pooled amplicons after normalization.

3. Purify the combined PCR amplicons using QIAguick PCR purification kit and elute in 30 μl of EB buffer.
4. Check the final combined amplicons using Agilent Bioanalyzer before 454 FLX sequencing run (Fig. 3c).

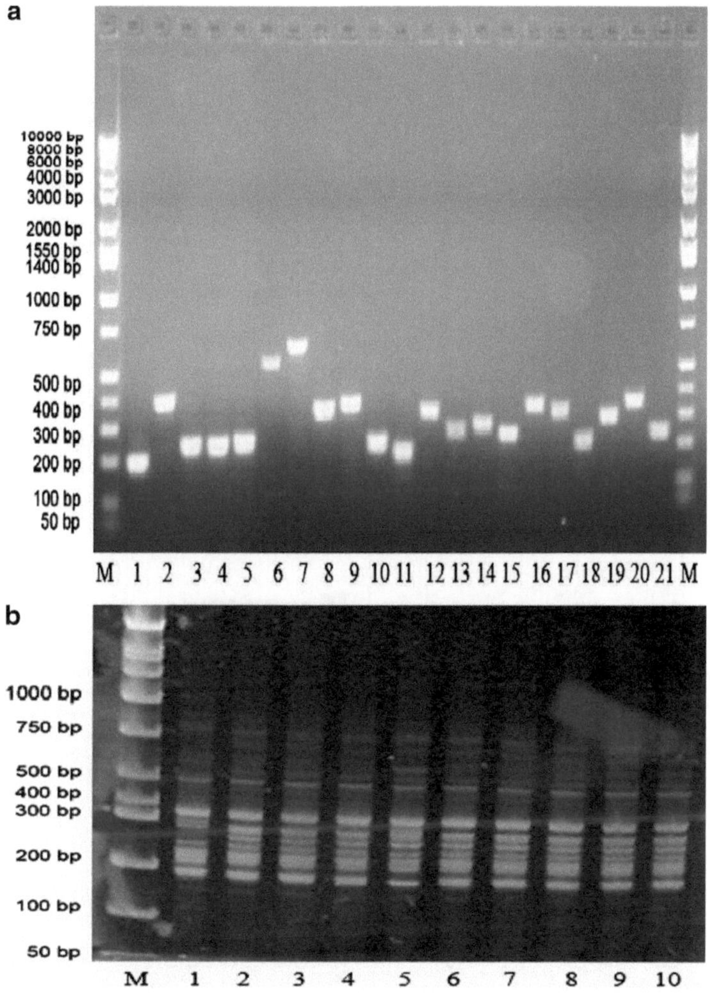

Fig. 3. Gel electrophoresis of multiplex amplicons. (**a**) Individual PCR amplification products of 21 targets from *S.* Typhimurium ATCC 14028 genome were separated on 2% agarose gel. M. Hi-Lo marker, 1. *thrA* (191 bp), 2. *aroC* (378), 3. *hisD* (229), 4. *purE* (229), 5. *sucA* (229), 6. *dnaN* (522), 7. *hemD* (604), 8. *fimA* (331), 9. *mdh* (349), 10. *manB* (223), 11. *panB* (198), 12. *fliC* (324), 13. *gyrB* (265), 14. *atpD* (283), 15. *aceK* (250), 16. *icdA* (351), 17. *fljB* (330), 18. *pduF* (223), 19. *glnA* (301), 20. *glpF* (354), 21. *pgm* (240). (**b**) Amplification of all 21 target genes from ten *Salmonella* strains. The amplified products were separated on 6% TBE gel and stained with SyberGreen.

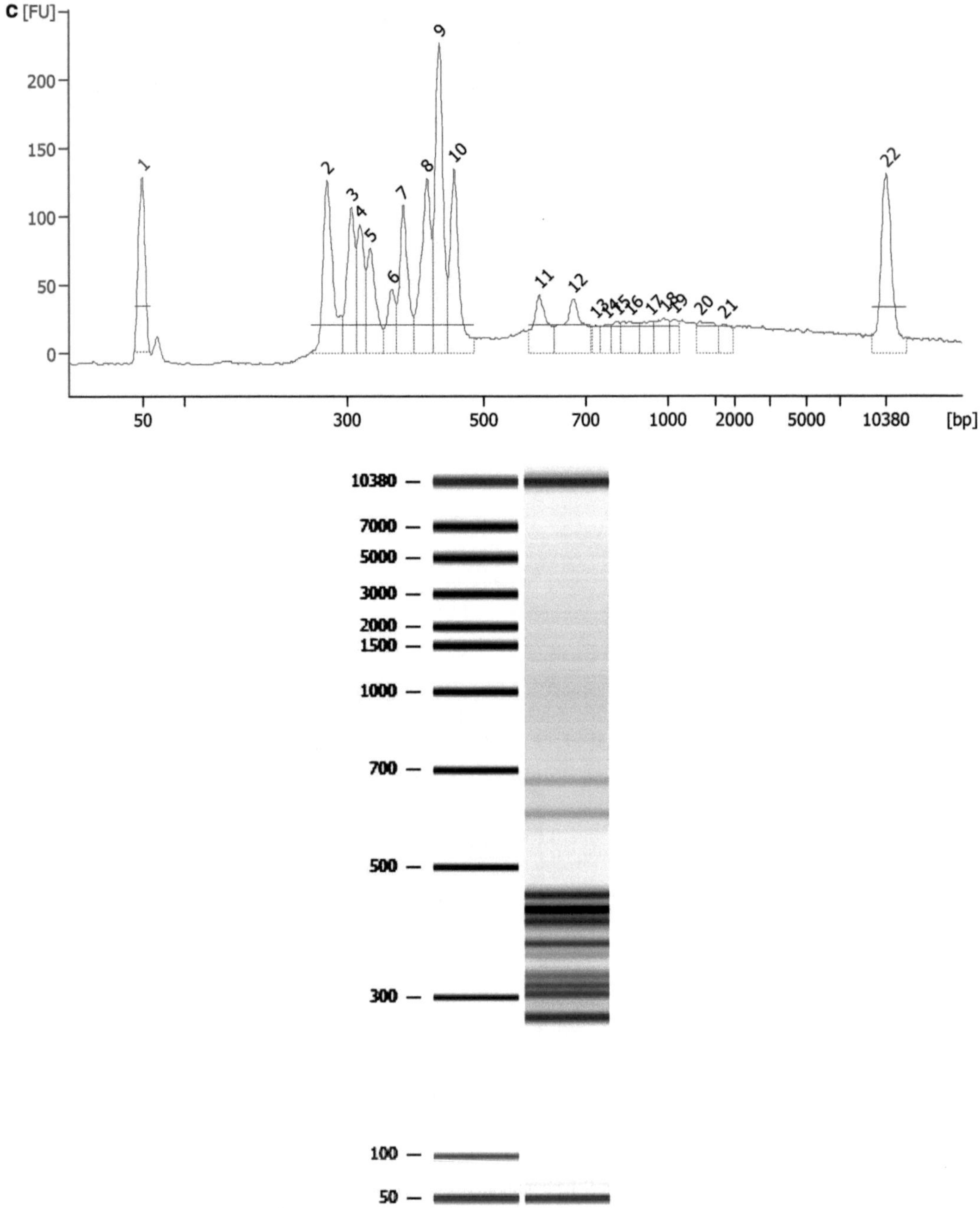

Fig. 3. (continued) (**c**) The electropherogram and simulated gel view obtained from analysis of the combined amplicons of 41 *Salmonella* strains using Agilent Bioanalyzer.

4. Notes

1. Only one restriction enzyme, MspI, was used in our experimental protocol. Although more than one restriction enzyme can be used, we limited to one enzyme for simplicity of hairpin selector design and sample processing.
2. We used the recognition site for the same enzyme, MspI, initially used for genomic DNA digestion by inserting this enzyme site into the design of all hairpin selectors. This can ensure that there is no internal recognition site in all target fragments.
3. Even though only 1 μl of the digestion reaction is used for ligation in the next step, a 50-μl aliquot is set up for use in multiple reactions. 30 μl of genomic DNA is digested to achieve high DNA concentration.
4. Phosphorylation of 5′ end of hairpin selector II can be done during oligonucluotide synthesis. However, by performing 5′-phosphorylation, we can reduce the time and cost required for synthesis substantially, especially when there are many targets for which a different hairpin selector II should be prepared.
5. When the ligation reaction is performed for multiple targets, a large number of hairpin selectors (two selectors/target) should be added in the reaction. When it was necessary to include all hairpin selectors without increasing the total reaction volume, we prepared and used the working solution for each hairpin selector with higher final concentration than was normally used. For convenience, we prepared a fresh mixture of the hairpin selectors for all targets and used it to set up the ligation reactions.
6. We initially tried the ligation reaction by incubating overnight at 70°C. We found that cyclic ligation significantly increased the ligation efficiency.
7. We initially used phenol:chloroform extraction to inactivate Ampligase enzyme before final restriction digestion, because this enzyme cannot be heat-inactivated. But this step was later on eliminated, as we found that the enzyme activity does not interfere with the later steps.
8. For the target amplification, other thermostable DNA polymerases can be used. But on comparison with *Taq* DNA polymerase from NEB, Cloned *pfu* DNA polymerase performed better in specific amplification of multiple targets.

Acknowledgments

This research was supported by the USDA Food Safety Consortium grant.

Disclaimer

The views expressed in this manuscript do not necessarily reflect those of the US Food and Drug Administration.

References

1. Metzker, M. L. (2010) Sequencing technologies – the next generation. *Nat. Rev. Genet.* **11**, 31–46.
2. Mamanova, L., Coffey, A. J., Scott, C. E., Kozarewa, I., Turner, E. H., Kumar, A., Howard, E., Shendure, J., and Turner, D. J. (2010) Target-enrichment strategies for next-generation sequencing. *Nat. Methods* **7**, 111–118.
3. Dahl, F., Stenberg, J., Fredriksson, S., Welch, K., Zhang, M., Nilsson, M., Bicknell, D., Bodmer, W. F., Davis, R. W., and Ji, H. (2007) Multigene amplification and massively parallel sequencing for cancer mutation discovery. *Proc. Natl. Acad. Sci. USA.* **104**, 9387–9392.
4. Varley, K. E., and Mitra, R. D. (2008) Nested Patch PCR enables highly multiplexed mutation discovery in candidate genes. *Genome Res.* **18**, 1844–1850.
5. Kotetishvili, M., Stine, O. C., Kreger, A., Morris, J. G. Jr, and Sulakvelidze, A.(2002) Multilocus sequence typing for characterization of clinical and environmental salmonella strains. *J Clin Microbiol.* **40**, 1626–1635.
6. Sukhnanand, S., Alcaine, S., Warnick, L. D., Su, W. L., Hof, J., Craver, M. P., McDonough, P., Boor, K. J., and Wiedmann, M. (2005) DNA sequence-based subtyping and evolutionary analysis of selected *Salmonella* enterica serotypes. *J Clin Microbiol.* **43**, 3688–3698.
7. Tankouo-Sandjong, B., Sessitsch, A., Liebana, E., Kornschober, C., Allerberger, F., Hächler, H., and Bodrossy, L. (2007) MLST-v, multilocus sequence typing based on virulence genes, for molecular typing of *Salmonella enterica* subsp. enterica serovars. *J Microbiol Methods.* **69**, 23–36.

Chapter 20

96-Plex Molecular Barcoding for the Illumina Genome Analyzer

Iwanka Kozarewa and Daniel J. Turner

Abstract

Next-generation sequencing technologies have a massive throughput, which dramatically reduces the cost of sequencing per gigabase, compared to standard Sanger sequencing. To make the most efficient use of this throughput when sequencing small regions or genomes, we developed a barcoding method, which allows multiplexing of 96 or more samples per lane. The method employs 8 bp tags, incorporated into each sequencing library during the library preparation enrichment polymerase chain reaction (PCR), pooling bar-coded libraries in equimolar ratios based on quantitative PCR, and sequencing using the three-read Illumina method.

Key words: Next-generation sequencing, Library preparation, Multiplexing, Indexing, Barcoding

1. Introduction

The emergence of next-generation sequencing (NGS) technologies has drastically altered our perception of what sequencing technology is and what it is capable of. The decreasing sequencing cost coupled with increasing data output has enabled a wide range of NGS applications, including de novo sequencing of small (e.g. bacteria, viral) genomes. Currently (March 2010), one lane of an Illumina Genome Analyzer II can yield in excess of two gigabases (Gb, 1×10^9 bases) in a 36-bp-run, which represents >377-fold coverage of the 5.3 Mb genome of the bacterium *Escherichia coli* strain 042.

Since the yield per lane from Illumina sequencing has been steadily increasing, it is becoming time and cost inefficient to use a single lane of an Illumina flowcell to sequence single libraries, when prepared from small genomes, or from sequence capture experiments that target less than 10 Mb. To remediate this, several approaches to sample barcoding have been developed, which allow

Young Min Kwon and Steven C. Ricke (eds.), *High-Throughput Next Generation Sequencing: Methods and Applications*, Methods in Molecular Biology, vol. 733, DOI 10.1007/978-1-61779-089-8_20, © Springer Science+Business Media, LLC 2011

sequencing libraries from multiple samples to be mixed and sequenced simultaneously. The first NGS barcoding technique was used with 454/Roche sequencing, with barcodes being part either of the adapters (1, 2) or the amplification primers (3, 4). In 2008, two groups reported the use of molecular barcoding for Illumina GAI sequencing (5, 6). In both cases, the barcodes were part of adapters that were ligated onto A-tailed templates as part of the library preparation procedure. Both studies observed good uniformity across the targeted region for a given individual, but varying representation of each barcode. Also, in 2008, Illumina released a commercial barcoding kit that allows multiplexing of 12 samples per lane. Identical ligation adapters are used for all sequencing libraries, and the barcodes are introduced during the enrichment polymerase chain reaction (PCR) (http://www.illumina.com/technology/multiplexing_sequencing_assay.ilmn), which results in a relatively uniform representation of each barcode in the pool.

The yield obtained from a single sequencing lane is likely to continue to increase, with the result that 12-plex sample pooling may become inadequate for efficient bacterial whole-genome sequencing in the near future, but additionally, for smaller regions, such as long PCR products, it is already desirable to increase the degree of multiplexing to 96 or even 384 samples per lane. With this aim in mind, we developed a 96-well library prep and molecular barcoding method for Illumina library preparation. In a similar way to the Illumina protocol, we use barcodes which are part of the amplification primers. These barcodes have been designed using Hamming codes (7), which allow single nucleotide sequencing errors to be corrected, and double errors and single indels to be detected without ambiguity. We employ several optimizations in our indexing protocol:

- Our tags are 8 bp long, which allows us to design larger numbers of barcodes with error-correcting capability.
- We introduce our bar codes in a regular PCR, not with the 3-primer approach used by Illumina. This simplifies the PCR step and allows us to use as few as six cycles.
- Before pooling, we measure the relative concentration of each sample library by quantitative PCR (qPCR). This allows us to pool libraries together accurately, which improves the uniformity of their representation.

For a proof-of-principle experiment, we used our molecular barcoding strategy to analyze global variation around *ACTN3*, a gene linked with human muscle performance. To characterize the genetic diversity of *ACTN3*, we used a long PCR to amplify 25 kb of genomic DNA sequence encompassing *ACTN3* and its cis-sequences, from 1,457 individuals from 57 worldwide populations. We created pools of 96 samples and sequenced each pool on two flowcell lanes of Illumina GAII Analyzer using 50-bp paired end runs. The average sample coverage ranged from 550 to 790×, with

50–80% of the samples falling into the twofold range of the median, and we were able to detect rare, population-specific variants.

This bar coding method is equally applicable to bacterial and viral whole-genome and targeted sequencing. The average size of a bacterial genome (5 Mb) makes the -fold coverage from 1 lane of Illumina GAII much greater than is required for reliable single nucleotide polymorphism detection. In addition, many studies aiming to investigate microbial and viral diversity in different environmental and pathogenic communities (metagenomic studies) only use 16S ribosomal RNA sequences for phylogenetic classifications (8). These sequences are usually obtained via PCR and thus are highly amenable to multiplexing.

2. Materials

2.1. Library Preparation

1. T4 Polynucleotide kinase.
2. T4 DNA polymerase.
3. Klenow DNA polymerase.
4. T4 DNA ligase buffer with 10 mM ATP.
5. dNTP set.
6. Klenow fragment (3′–5′ minus).
7. Agencourt AMPure SPRI beads (Beckman Coulter, Brea, CA).
8. QIAquick PCR purification kit (Qiagen, Valencia, CA).
9. Quick ligation kit (NEB, Ipswich, MA).
10. HPLC-purified indexing adapters:

PE_t_adapter	5′-ACACTCTTTCCCTACACGACGCTCTTCC GATC*T
Ind_adapter_b	5′[Phosphate]GATCGGAAGAGCGGTTCAGCA GGAATGCCGAGACCGATCTC

(*indicates phosphorothioate)

11. Agilent DNA 1000 chips (Agilent Technologies, Santa Clara, CA).
12. HPLC-purified enrich_PCR_common_primer:

    5′AATGATACGGCGACCACCGAGATCTACACTCTTTC CCTACACGACGCTCTTCCGATC*T

    (*indicates phosphorothioate)
13. PAGE-purified enrich_PCR_indexing_primers:

    5′CAAGCAGAAGACGGCATACGAGAT[index] GAGATCGGTCTCGGCATTC

    For a list of index tags and primers see Table 1.
14. Platinum Pfx Taq polymerase (Invitrogen, Carlsbad, CA).

Table 1
Error-correcting barcode tags and PCR primers

	Single correcting, double and shift detecting octamers	PCR primers	Sequence obtained
1	ACAAGCTA	CAAGCAGAAGACGGCATACGAGATACAAGCTAGAGATCGGTCTCGGCATTC	TAGCTTGT
2	AAACATCG	CAAGCAGAAGACGGCATACGAGATAAACATCGGAGATCGGTCTCGGCATTC	CGATGTTT
3	ACATTGGC	CAAGCAGAAGACGGCATACGAGATACATTGGCGAGATCGGTCTCGGCATTC	GCCAATGT
4	ACCACTGT	CAAGCAGAAGACGGCATACGAGATACCACTGTGAGATCGGTCTCGGCATTC	ACAGTGGT
5	AACGTGAT	CAAGCAGAAGACGGCATACGAGATAACGTGATGAGATCGGTCTCGGCATTC	ATCACGTT
6	CGCTGATC	CAAGCAGAAGACGGCATACGAGATCGCTGATCGAGATCGGTCTCGGCATTC	GATCAGCG
7	CAGATCTG	CAAGCAGAAGACGGCATACGAGATCAGATCTGGAGATCGGTCTCGGCATTC	CAGATCTG
8	ATGCCTAA	CAAGCAGAAGACGGCATACGAGATATGCCTAAGAGATCGGTCTCGGCATTC	TTAGGCAT
9	CTGTAGCC	CAAGCAGAAGACGGCATACGAGATCTGTAGCCGAGATCGGTCTCGGCATTC	GGCTACAG
10	AGTACAAG	CAAGCAGAAGACGGCATACGAGATAGTACAAGGAGATCGGTCTCGGCATTC	CTTGTACT
11	CATCAAGT	CAAGCAGAAGACGGCATACGAGATCATCAAGTGAGATCGGTCTCGGCATTC	ACTTGATG
12	AGTGGTCA	CAAGCAGAAGACGGCATACGAGATAGTGGTCAGAGATCGGTCTCGGCATTC	TGACCACT
13	AACAACCA	CAAGCAGAAGACGGCATACGAGATAACAACCAGAGATCGGTCTCGGCATTC	TGGTTGTT
14	AACCGAGA	CAAGCAGAAGACGGCATACGAGATAACCGAGAGAGATCGGTCTCGGCATTC	TCTCGGTT
15	AACGCTTA	CAAGCAGAAGACGGCATACGAGATAACGCTTAGAGATCGGTCTCGGCATTC	TAAGCGTT
16	AAGACGGA	CAAGCAGAAGACGGCATACGAGATAAGACGGAGAGATCGGTCTCGGCATTC	TCCGTCTT
17	AAGGTACA	CAAGCAGAAGACGGCATACGAGATAAGGTACAGAGATCGGTCTCGGCATTC	TGTACCTT
18	ACACAGAA	CAAGCAGAAGACGGCATACGAGATACACAGAAGAGATCGGTCTCGGCATTC	TTCTGTGT
19	ACAGCAGA	CAAGCAGAAGACGGCATACGAGATACAGCAGAGAGATCGGTCTCGGCATTC	TCTGCTGT

20	ACCTCCAA	CAAGCAGAAGACGGCATACGAGATACCTCCAAGAGATCGGTCTCGGCATTC	TTGGAGGT
21	ACGCTCGA	CAAGCAGAAGACGGCATACGAGATACGCTCGAGAGATCGGTCTCGGCATTC	TCGAGCGT
22	ACGTATCA	CAAGCAGAAGACGGCATACGAGATACGTATCAGAGATCGGTCTCGGCATTC	TGATACGT
23	ACTATGCA	CAAGCAGAAGACGGCATACGAGATACTATGCAGAGATCGGTCTCGGCATTC	TGCATAGT
24	AGAGTCAA	CAAGCAGAAGACGGCATACGAGATAGAGTCAAGAGATCGGTCTCGGCATTC	TTGACTCT
25	AGATCGCA	CAAGCAGAAGACGGCATACGAGATAGATCGCAGAGATCGGTCTCGGCATTC	TGCGATCT
26	AGCAGGAA	CAAGCAGAAGACGGCATACGAGATAGCAGGAAGAGATCGGTCTCGGCATTC	TTCCTGCT
27	AGTCACTA	CAAGCAGAAGACGGCATACGAGATAGTCACTAGAGATCGGTCTCGGCATTC	TAGTGACT
28	ATCCTGTA	CAAGCAGAAGACGGCATACGAGATATCCTGTAGAGATCGGTCTCGGCATTC	TACAGGAT
29	ATTGAGGA	CAAGCAGAAGACGGCATACGAGATATTGAGGAGAGATCGGTCTCGGCATTC	TCCTCAAT
30	CAACCACA	CAAGCAGAAGACGGCATACGAGATCAACCACAGAGATCGGTCTCGGCATTC	TGTGGTTG
31	CAAGACTA	CAAGCAGAAGACGGCATACGAGATCAAGACTAGAGATCGGTCTCGGCATTC	TAGTCTTG
32	CAATGGAA	CAAGCAGAAGACGGCATACGAGATCAATGGAAGAGATCGGTCTCGGCATTC	TTCCATTG
33	CACTTCGA	CAAGCAGAAGACGGCATACGAGATCACTTCGAGAGATCGGTCTCGGCATTC	TCGAAGTG
34	CAGCGTTA	CAAGCAGAAGACGGCATACGAGATCAGCGTTAGAGATCGGTCTCGGCATTC	TAACGCTG
35	CATACCAA	CAAGCAGAAGACGGCATACGAGATCATACCAAGAGATCGGTCTCGGCATTC	TTGGTATG
36	CCAGTTCA	CAAGCAGAAGACGGCATACGAGATCCAGTTCAGAGATCGGTCTCGGCATTC	TGAACTGG
37	CCGAAGTA	CAAGCAGAAGACGGCATACGAGATCCGAAGTAGAGATCGGTCTCGGCATTC	TACTTCGG
38	CCGTGAGA	CAAGCAGAAGACGGCATACGAGATCCGTGAGAGAGATCGGTCTCGGCATTC	TCTCACGG
39	CCTCCTGA	CAAGCAGAAGACGGCATACGAGATCCTCCTGAGAGATCGGTCTCGGCATTC	TCAGGAGG
40	CGAACTTA	CAAGCAGAAGACGGCATACGAGATCGAACTTAGAGATCGGTCTCGGCATTC	TAAGTTCG

(continued)

Table 1 (continued)

	Single correcting, double and shift detecting octamers	PCR primers	Sequence obtained
41	CGACTGGA	CAAGCAGAAGACGGCATACGAGATCGACTGGAGAGATCGGTCTCGGCATTC	TCCAGTCG
42	CGCATACA	CAAGCAGAAGACGGCATACGAGATCGCATACAGAGATCGGTCTCGGCATTC	TGTATGCG
43	CTCAATGA	CAAGCAGAAGACGGCATACGAGATCTCAATGAGAGATCGGTCTCGGCATTC	TCATTGAG
44	CTGAGCCA	CAAGCAGAAGACGGCATACGAGATCTGAGCCAGAGATCGGTCTCGGCATTC	TGGCTCAG
45	CTGGCATA	CAAGCAGAAGACGGCATACGAGATCTGGCATAGAGATCGGTCTCGGCATTC	TATGCCAG
46	GAATCTGA	CAAGCAGAAGACGGCATACGAGATGAATCTGAGAGATCGGTCTCGGCATTC	TCAGATTC
47	GACTAGTA	CAAGCAGAAGACGGCATACGAGATGACTAGTAGAGATCGGTCTCGGCATTC	TACTAGTC
48	GAGCTGAA	CAAGCAGAAGACGGCATACGAGATGAGCTGAAGAGATCGGTCTCGGCATTC	TTCAGCTC
49	GATAGACA	CAAGCAGAAGACGGCATACGAGATGATAGACAGAGATCGGTCTCGGCATTC	TGTCTATC
50	GCCACATA	CAAGCAGAAGACGGCATACGAGATGCCACATAGAGATCGGTCTCGGCATTC	TATGTGGC
51	GCGAGTAA	CAAGCAGAAGACGGCATACGAGATGCGAGTAAGAGATCGGTCTCGGCATTC	TTACTCGC
52	GCTAACGA	CAAGCAGAAGACGGCATACGAGATGCTAACGAGAGATCGGTCTCGGCATTC	TCGTTAGC
53	GCTCGGTA	CAAGCAGAAGACGGCATACGAGATGCTCGGTAGAGATCGGTCTCGGCATTC	TACCGAGC
54	GGAGAACA	CAAGCAGAAGACGGCATACGAGATGGAGAACAGAGATCGGTCTCGGCATTC	TGTTCTCC
55	GGTGCGAA	CAAGCAGAAGACGGCATACGAGATGGTGCGAAGAGATCGGTCTCGGCATTC	TTCGCACC
56	GTACGCAA	CAAGCAGAAGACGGCATACGAGATGTACGCAAGAGATCGGTCTCGGCATTC	TTGCGTAC
57	GTCGTAGA	CAAGCAGAAGACGGCATACGAGATGTCGTAGAGAGATCGGTCTCGGCATTC	TCTACGAC
58	GTCTGTCA	CAAGCAGAAGACGGCATACGAGATGTCTGTCAGAGATCGGTCTCGGCATTC	TGACAGAC

59	GTGTTCTA	CAAGCAGAAGACGGCATACGAGATGTGTTCTAGAGATCGGTCTCGGCATTC	TAGAACAC
60	TAGGATGA	CAAGCAGAAGACGGCATACGAGATTAGGATGAGAGATCGGTCTCGGCATTC	TCATCCTA
61	TATCAGCA	CAAGCAGAAGACGGCATACGAGATTATCAGCAGAGATCGGTCTCGGCATTC	TGCTGATA
62	TCCGTCTA	CAAGCAGAAGACGGCATACGAGATTCCGTCTAGAGATCGGTCTCGGCATTC	TAGACGGA
63	TCTTCACA	CAAGCAGAAGACGGCATACGAGATTCTTCACAGAGATCGGTCTCGGCATTC	TGTGAAGA
64	TGAAGAGA	CAAGCAGAAGACGGCATACGAGATTGAAGAGAGAGATCGGTCTCGGCATTC	TCTCTTCA
65	TGGAACAA	CAAGCAGAAGACGGCATACGAGATTGGAACAAGAGATCGGTCTCGGCATTC	TTGTTCCA
66	TGGCTTCA	CAAGCAGAAGACGGCATACGAGATTGGCTTCAGAGATCGGTCTCGGCATTC	TGAAGCCA
67	TGGTGGTA	CAAGCAGAAGACGGCATACGAGATTGGTGGTAGAGATCGGTCTCGGCATTC	TACCACCA
68	TTCACGCA	CAAGCAGAAGACGGCATACGAGATTTCACGCAGAGATCGGTCTCGGCATTC	TGCGTGAA
69	AACTCACC	CAAGCAGAAGACGGCATACGAGATAACTCACCGAGATCGGTCTCGGCATTC	GGTGAGTT
70	AAGAGATC	CAAGCAGAAGACGGCATACGAGATAAGAGATCGAGATCGGTCTCGGCATTC	GATCTCTT
71	AAGGACAC	CAAGCAGAAGACGGCATACGAGATAAGGACACGAGATCGGTCTCGGCATTC	GTGTCCTT
72	AATCCGTC	CAAGCAGAAGACGGCATACGAGATAATCCGTCGAGATCGGTCTCGGCATTC	GACGGATT
73	AATGTTGC	CAAGCAGAAGACGGCATACGAGATAATGTTGCGAGATCGGTCTCGGCATTC	GCAACATT
74	ACACGACC	CAAGCAGAAGACGGCATACGAGATACACGACCGAGATCGGTCTCGGCATTC	GGTCGTGT
75	ACAGATTC	CAAGCAGAAGACGGCATACGAGATACAGATTCGAGATCGGTCTCGGCATTC	GAATCTGT
76	AGATGTAC	CAAGCAGAAGACGGCATACGAGATAGATGTACGAGATCGGTCTCGGCATTC	GTACATCT
77	AGCACCTC	CAAGCAGAAGACGGCATACGAGATAGCACCTCGAGATCGGTCTCGGCATTC	GAGGTGCT
78	AGCCATGC	CAAGCAGAAGACGGCATACGAGATAGCCATGCGAGATCGGTCTCGGCATTC	GCATGGCT
79	AGGCTAAC	CAAGCAGAAGACGGCATACGAGATAGGCTAACGAGATCGGTCTCGGCATTC	GTTAGCCT

(continued)

Table 1 (continued)

	Single correcting, double and shift detecting octamers	PCR primers	Sequence obtained
80	ATAGCGAC	CAAGCAGAAGACGGCATACGAGATATAGCGACGAGATCGGTCTCGGCATTC	GTCGCTAT
81	ATCATTCC	CAAGCAGAAGACGGCATACGAGATATCATTCCGAGATCGGTCTCGGCATTC	GGAATGAT
82	ATTGGCTC	CAAGCAGAAGACGGCATACGAGATATTGGCTCGAGATCGGTCTCGGCATTC	GAGCCAAT
83	CAAGGAGC	CAAGCAGAAGACGGCATACGAGATCAAGGAGCGAGATCGGTCTCGGCATTC	GCTCCTTG
84	CACCTTAC	CAAGCAGAAGACGGCATACGAGATCACCTTACGAGATCGGTCTCGGCATTC	GTAAGGTG
85	CCATCCTC	CAAGCAGAAGACGGCATACGAGATCCATCCTCGAGATCGGTCTCGGCATTC	GAGGATGG
86	CCGACAAC	CAAGCAGAAGACGGCATACGAGATCCGACAACGAGATCGGTCTCGGCATTC	GTTGTCGG
87	CCTAATCC	CAAGCAGAAGACGGCATACGAGATCCTAATCCGAGATCGGTCTCGGCATTC	GGATTAGG
88	CCTCTATC	CAAGCAGAAGACGGCATACGAGATCCTCTATCGAGATCGGTCTCGGCATTC	GATAGAGG
89	CGACACAC	CAAGCAGAAGACGGCATACGAGATCGACACACGAGATCGGTCTCGGCATTC	GTGTGTCG
90	CGGATTGC	CAAGCAGAAGACGGCATACGAGATCGGATTGCGAGATCGGTCTCGGCATTC	GCAATCCG
91	CTAAGGTC	CAAGCAGAAGACGGCATACGAGATCTAAGGTCGAGATCGGTCTCGGCATTC	GACCTTAG
92	GAACAGGC	CAAGCAGAAGACGGCATACGAGATGAACAGGCGAGATCGGTCTCGGCATTC	GCCTGTTC
93	GACAGTGC	CAAGCAGAAGACGGCATACGAGATGACAGTGCGAGATCGGTCTCGGCATTC	GCACTGTC
94	GAGTTAGC	CAAGCAGAAGACGGCATACGAGATGAGTTAGCGAGATCGGTCTCGGCATTC	GCTAACTC
95	GATGAATC	CAAGCAGAAGACGGCATACGAGATGATGAATCGAGATCGGTCTCGGCATTC	GATTCATC
96	GCCAAGAC	CAAGCAGAAGACGGCATACGAGATGCCAAGACGAGATCGGTCTCGGCATTC	GTCTTGGC

The second column contains octamer barcodes, and the third column contains PCR primers that contain these bar codes. The fourth column contains the reverse complement of the octamer barcodes. This is the sequence that is obtained from the barcodes in the sequencing reaction, and which are robust to sequencing errors

2.2. Normalization

1. Desalted normalization qPCR primers:

Forw_norm primer: e.g. Act_10_f	5′TGCATCCGCCGTATGGTGCC
Rev_norm primer: e.g. Act_10_r	5′CCCTTGCTCCGAGGGACGGA

2. 2× SYBRgreen master mix.
3. Tween 80.

2.3. Size-Selection Step

1. Agarose.
2. 10× TBE.
3. SafeView (NBS Biologicals, Huntington, UK).
4. 5× Qiagen GelPilot loading dye (Qiagen, Valencia, CA).
5. Low-molecular weight ladder (NEB, Ipswich, MA).
6. MinElute gel extraction kit (Qiagen, Valencia, CA).
7. Isopropanol.

2.4. Absolute Quantification

1. Platinum Taq polymerase (Invitrogen, Carlsbad, CA).
2. qPCR primers and TaqMan probe:

c_qPCR_v2.1	5′AATGATACGGCGACCACCGAGATC
PE_qPCR_v2.2	5′CAAGCAGAAGACGGCATACGAGATC
TaqMan probe	5′[6FAM]CCCTACACGACGCTCTTCCGATCT[TAMRA]

PCR primers c_qPCR_v2.1 and PE_qPCR_v2.2 are desalted, whereas TaqMan is HLPC purified.

3. ROX dye (Invitrogen, Carlsbad, CA).
4. Index-sequencing primer:

Ind_seq	5′-GTTCAGCAGGAATGCCGAGACCGATCTC-3′ (HPLC)

2.5. Equipment

1. Covaris S2 (Kbiosciences, Herts, UK).
2. Microtubes, AFA fiber with snap-cap (Kbiosciences, Herts, UK).
3. Crimp seals (Kbiosciences, Herts, UK).
4. Real-time PCR system.
5. MicroAmp optical 96-well reaction plate (Applied Biosystems, Foster City, CA).

6. Real-time PCR-compatible plate seals.
7. Microseal "F" film (Bio-Rad, Hercules, CA).
8. 0.2-ml Thin-walled strip tubes and caps.
9. Thermal cycler with 96-well blocks.
10. Agilent 2100 Bioanalyzer (Agilent Technologies, Santa Clara, CA).
11. Vortex mixer.
12. Plate centrifuge.
13. 50-ml Falcon tubes.
14. 15-ml Falcon tubes.
15. Thin-walled 200 μl PCR tubes.
16. Electronic multichannel pipettes.
17. Manual multichannel pipettes.
18. Racked tips, ART XLP.
19. RT-L200XF extended length LTS filter tips (Rainin, Columbus, OH).
20. RT-L10F LTS filter tips (Rainin, Columbus, OH).
21. SPRIPlate 96R-ring magnet plate (Beckman Coultier, Brea, CA).
22. Horizontal mini electrophoresis gel tank.
23. Electrophoresis power supply.
24. UV transilluminator.
25. Fridge.
26. Microwave.

3. Methods

3.1. Experimental Design

3.1.1. Ligation

A molar ratio of 20:1 of adapters: A-tailed sample is used. Since the protocol is for 96-well plates, we calculate an average molar concentration for all the samples by selecting six at random from the plate after A-tailing, and running 1 μl of each on an Agilent Bioanalyzer 2100.

In the standard Illumina library preparation protocol, the ligation step is preceded by an A-tailing reaction. The products from this step are cleaned up using a spin column, and are eluted in 10 μl of a 10 mM Tris–HCl buffer. It is possible to clean reactions up in this way during the indexing protocol, but for larger numbers of samples (e.g. 96), it becomes unfeasible to use spin columns. 96-well spin plates are available, but tend to be expensive and can be difficult to work with, so we prefer to clean reactions up with SPRI beads. The drawback with this approach is that successful elution requires at least 17 μl of elution buffer (EB) to be used, which limits the volume of adapters that can be added to the ligation reaction to 3 μl or less. In cases where a greater volume of adapters is required, the reaction volume can be increased to accommodate this, so long as the correct buffer and enzyme concentrations are maintained.

3.1.2. Normalization

To pool 96 samples in equimolar concentrations, we measure their relative quantities using qPCR with SYBRgreen. It is not advisable to quantify at this stage using the Illumina adapter sequences, because adapter dimers will have formed during the ligation step. These will be removed before sequencing, and it is important to calculate pooling volumes based upon the insert sequences. We design locus- or genome-specific qPCR primers using mPrimer3 online software (http://bioinfo.ebc.ee/mprimer3), ideally 20 bp in length and with 60–62°C melting temperatures. We then perform relative quantification using a standard dilution series with three dilutions of a concentration standard. The sample chosen as a standard is the one with the highest concentration according to the Agilent DNA 1000 traces from the end of step 5, Subheading 3.5. At least two technical replicates are required per sample in the qPCR, so for >45 samples, it is necessary to use more than one plate in the qPCR. The concentration standard should be run on each plate.

After the samples are quantified against the standard, the sample with the highest concentration (highest quantity, lowest Ct) is determined. The relative quantities of all the other samples with respect to the one with the highest concentration are calculated using the formula:

$$\text{Ratio} = \frac{\text{Quantity of the sample with the highest concentration}}{\text{Quantity of any given sample}}$$

For pooling, e.g. 1.5 μl of the sample with highest concentration is taken. The required volumes for all the other samples are calculated by multiplying the ratio obtained above by the volume of the highest concentration sample (1.5 μl in this example).

If more than one qPCR plate has been run, relative abundances for pooling from all plates are calculated against a single sample. If two sets of primers are used for normalization, the volume used for pooling is the average of the two volumes calculated with each set of primers.

3.2. Sample Fragmentation

1. 300 ng–5 μg DNA in EB or TE buffer (10 mM Tris, pH 7.5/1 mM EDTA).
2. If necessary, add EB buffer to make the final volume of 75 μl.

3. Transfer the samples to 100 μl Covaris microtubes (see Note 1).
4. Shear the samples for 150 s with the following program (see Note 2):

Duty cycle	20%
Intensity	5
Cycle burst	200
Power	37 W
Temperature	7°C
Mode	Freq sweeping

5. Transfer the samples to a MicroAmp optical 96-well plate (see Note 3).
6. Choose six samples at random and run 1 μl of each on an Agilent DNA 1000 chip to check the quality of fragmentation. A 150–250 bp smear should be detected with 70–90% of the initial DNA amount recovered (Fig. 1).

   PAUSE POINT: Fragmented samples can be stored at –20°C for up to 4 weeks.

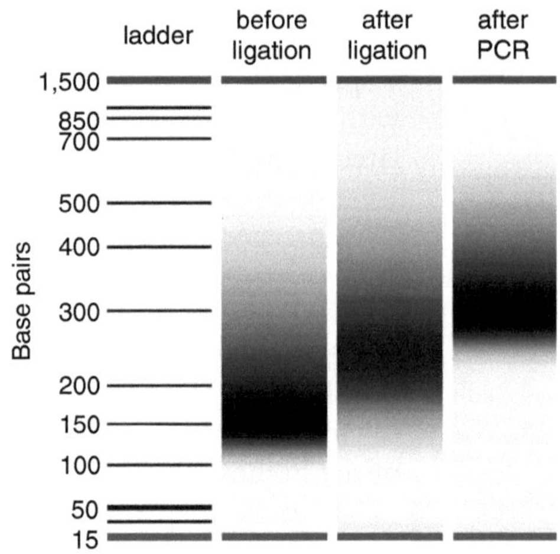

Fig. 1. Agilent traces of a randomly chosen sample. Initial DNA quantity before shearing: ~3 μg. Lane 1 (*first lane* on the *left*) shows a DNA ladder. Lane 2 shows a trace of a sample after fragmentation for 150 s with 20/5/200 program. A smear with well-defined peak around 150–250 bp is visible. Lane 3 is a trace of the same sample after adapter ligation. Clear shift of the peak with 50 bp up is detected, indicative of successful ligation. Lane 4 is after the indexing enrichment PCR. Another shift of the peak up indicative of successful PCR is observed.

3.3. End-Repair, A-Tailing and Ligation

1. For each sample, prepare an end-repair master mix (end-repair MM) containing the reagents below. For a full 96-well plate (104 samples), mix in a reservoir:

	1×	104×
10× T4 DNA ligase buffer with 10 mM ATP	10 μl	1,040 μl
10 mM dNTPs mix	4 μl	416 μl
3 U/μl T4 DNA polymerase	5 μl	520 μl
5 U/μl Klenow DNA polymerase	1 μl	104 μl
10 U/μl T4 PNK	5 μl	520 μl

2. Add 25 μl end-repair MM to each well. Cover the plate with a transparent cover. Vortex briefly and spin down. Incubate for 30 min at 20°C in a thermal cycler.
3. Clean the samples using SPRI beads (see Subheading 3.4) and elute each sample in 32 μl EB.
4. For each sample, prepare A-tailing master mix (A-tailing MM) containing the reagents below. For a full 96-well plate (104 samples), mix in a reservoir:

	1×	104×
10× NEB buffer2	5 μl	520 μl
1 mM dATP	10 μl	1,040 μl
5 U/μl Klenow fragment (3′–5′ exo minus)	3 μl	312 μl

5. Add 18 μl A-tailing MM to each well. Cover the plate with a transparent cover. Vortex briefly and spin down. Incubate for 30 min at 37°C in a thermal cycler.
6. Clean the samples using SPRI beads and elute each sample in 10 μl EB + 8.5 μl water (see Note 4).
7. Run 1 μl of the same six samples as in step 1 on Agilent DNA 1000 chip and use the "integrated peak" function to measure the concentration of the samples following the manufacturer's recommended protocol. Calculate the average for all six samples' concentration.
8. Use the formula specified in the experimental design to calculate the average number moles of the samples and, following this, the required adapter and water quantities.

9. While doing the end-repair and A-tailing, prepare the custom adapters. Combine in one 200-μl PCR tube the following:

100 μM PE_t_adapter	20 μl
100 μM Ind_adapter_b	20 μl
50 mM Tris/50 mM NaCl pH 7.0	10 μl

Vortex, spin, and place the tube in the thermal cycler.

10. Anneal the adapters using the following program:
Ramp PCR machine at 0.5°C/s to 97.5°C.

Hold at 97.5°C for 150 s then 97.5°C for 5 s and temperature drop of (−)0.1°C per cycle for 775 cycles (i.e. decrease temperature from 97.5°C by 0.1°C every 5 s).
4°C Indefinite hold.
Store the oligos in 8 μl aliquots at −20°C until use.

11. For each sample, prepare ligation master mix (ligation MM) containing the reagents below. For a full 96-well plate (104 samples), mix in a reservoir:

	1×	104×
2× Quick DNA ligase buffer (2×)	25 μl	2,600 μl
40 μM Adapter oligo mix	1.5–3 μl	156–312 μl (see Note 5)

12. Add the required volume of ligation MM to each well (dependent on the volume of the adapters used). Cover the plate with a transparent cover. Vortex for 30 s and spin down.

13. Pipette 80 μl of 2,000 U/μl Quick ligase to each tube in a strip of eight PCR tubes. Use manual 8-channel pipette to dispense 5 μl into each sample. Pipette several times to mix. Incubate at 20–22°C (room temperature) for 15 min.

PAUSE POINT: The ligations can be stored at −20°C for up to 1 week.

14. Clean the samples using SPRI beads and elute each sample in 20 μl EB (see Note 6).

15. Run 1 μl of the same six samples as in step 1 on Agilent DNA 1000 chip to check the success of the ligation. If the ligation is successful, the average molecular size of the smears detected will be 50–200 bp bigger than before ligation (Fig. 1).

PAUSE POINT: The cleaned ligations can be stored at −20°C for up to 6 months.

3.4. SPRI-Bead Cleanup

1. Dispense 1.8× the sample volume of SPRI beads into each well (e.g. 180 μl following end repair and 90 μl following A-tailing and ligation). Cover the plate with a transparent cover. Vortex for 30 s.

2. Place the reaction plate on the magnetic SPRIPlate for 10 min to separate beads from solution. Proceed to the next step only when the solution is completely clear, even if you need to wait >10 min.
3. Aspirate out the cleared solution from the reaction plate and discard.
4. Dispense 200 μl of 70% ethanol to each well and incubate for 30 s at room temperature. Aspirate out the ethanol and discard. Repeat for a total of two washes.
5. Dry the reaction plate for 15 min at room temperature in a thermocycler. Leave the plate uncovered to allow ethanol to evaporate. Do not over-dry.
6. Elute with the specified volume at each step volume of EB (e.g. 32 μl in step 3 and 18.5 μl in step 6 of Subheading 3.3). Use the manual multichannel pipette. Mix by pipetting 6–8 times, cover the plate and vortex for 30 s. Incubate the plate on the bench top for 3 min. Place the reaction plate on magnet SPRIPlate for 3 min to separate beads from solution.
7. Transfer the cleared solution to a new plate.

3.5. Indexing Enrichment PCR

1. For each sample, prepare MM containing the reagents below (final concentrations are given in parentheses). For a full 96-well plate (104 samples), mix in a reservoir (see Note 7):

	1×	104×
10× Pfx buffer	10 μl (→1×)	1,040 μl
2.5 mM dNTPs	20 μl (→500 μM)	2,080 μl
50 mM MgSO4	4 μl (→2 nM)	416 μl
10 μM enrich_PCR_common Primer	8 μl (→800 nM)	832 μl
2.5 U/μl Platinum Pfx Taq polymerase	1 μl (→0.025 U/μl)	104 μl
Water	49 μl	5 096 μl

2. Dispense 92 μl MM into each 200 μl PCR tube. Add to each tube 8 μl of the respective enrich_PCR_indexing_primer. Add 1.2 μl of the cleaned ligations (see Note 8). Mix well by vortexing and spin down.
3. Transfer tubes to a thermal cycler and perform PCR with the following cycling conditions:

12 cycles of	94°C for 2 min 94°C for 15 s 58°C for 45 s
Then	4°C indefinitely

4. Clean the samples using SPRI beads and elute each sample in 18 μl EB (see Note 9).
5. Run 1 μl of the same six samples as in step 1 on Agilent DNA 1000 chip to check the success of the indexing enrichment PCR (Fig. 1).

PAUSE POINT: The PCRs can be stored at –20°C for up to 6 months.

3.6. Normalization and Pooling

1. Prepare the standard dilutions by diluting the sample with the highest concentration according to the Bioanalyzer assay in step 5, Subheading 3.5. Dilute the standard 1:5 with EB + 0.1% Tween and do three more consecutive 1:5 dilutions. Dilute all other samples 1:80 with EB + 0.1% Tween (see Note 10).
2. For each sample, prepare qPCR MM containing the reagents below:

	1×
2× SYBRgreen master mix	12.5 μl (→1×)
10 μM Forw_norm primer	1 μl (→00 nM)
10 μM Rev_norm primer	1 μl (→400 nM)

3. Dispense 24 μl of the qPCR MM to each well of a MicroAmp fast optical 96-well plate. Add 1 μl diluted sample. Mix by pipetting and spin down.
4. Transfer the plate to a real-time qPCR machine and perform qPCR with the following cycling conditions:

40 cycles of	95°C for 10 min
	95°C for 15 s
	60–62°C for 1 min

5. Calculate the relative quantities of the samples as outlined in the experimental design.
6. Pool the calculated volumes of the samples into one 1.5-ml tube.

PAUSE POINT: The pooled volumes can be stored at –20°C for up to 2 weeks.

3.7. Size-Selection Step

1. Measure the volume of the pooled samples. Separate the combined samples into 120 μl aliquots. Add 40 μl of 5× Qiagen GelPilot Loading Dye to each aliquot. Mix by pipetting. Separate the samples again into 160 μl aliquots.
2. Prepare the required number of 2% agarose gels in 1× TBE (one gel is sufficient for each 160 μl aliquot). Use the midi-comb (eight sample slots).

3. Once the agarose has dissolved completely, immediately add 5 µl per 100 ml SafeView. Pour the gel and allow to solidify at room temperature. Once the gel has solidified, allow to cool for 10 min at 4°C in the fridge.
4. Mix by pipetting 8 µl of NEB low-molecular weight ladder with 3 µl of the above-mentioned loading dye and load in one well of the gel. Load the ladder in the first well on the left of the gel. Load the samples, 40 µl in each well, leaving two lanes empty between the first sample and the ladder.
5. Run the gel at 60 V for 2 h using chilled (kept at 4°C) 1× TBE buffer. After the first 1 h, replace the buffer with fresh, chilled TBE. Stop electrophoresis once the orange marker reaches the bottom of the gel.
6. From each well cut the gel slice from 250 to 450 bp.

   PAUSE POINT: Gel slices can be stored at −20°C for up to 1 week.
7. Weigh the gel slices and transfer all of them to one 50-ml Falcon tube. Add 3× the weight of the slices volume of Qiagen QC buffer.
8. Solubilize the gel slices at room temperature through vigorous vortexing. To speed up the process, cut each gel slices into 2–3 smaller pieces. Add one volume of 100% isopropanol. Divide the mix between several MinElute columns (one column per 400 mg gel slice). Clean following the standard MinElute gel purification kit. To increase the concentration, elute half of the columns with 10 µl EB each. Use the same eluate to elute the second half of the columns.

   PAUSE POINT: The ready libraries can be stored at −20°C for up to 6 months.

3.8. Final Library Quantification by qPCR

1. Run 1 µl from the final multiplexed library after the gel cleanup step (in duplicate) on Agilent DNA 1000 chips. If the concentration is >15 ng/µl, prepare 1:300 and 1:450 dilutions in EB + 0.1% Tween; if <15 ng/µl, prepare 1:150 and 1:300 dilutions.
2. For each PCR, mix the following reagents (final concentrations are given in parentheses):

10× Platinum Taq buffer	2.5 µl
50 mM $MgCl_2$	0.75 µl (→1.5 mM)
Template DNA	1 µl
10 µM TaqMan probe	0.625 µl (→250 nM)
50× Rox	0.5 µl (→1×)
10 µM c_qPCR_v2.1	0.75 µl (→300 nM)

10 μM PE_qPCR_v2.2	0.75 μl (→300 nM)
2.5 mM dNTPs	2 μl (→200 μM)
5 U/μl Platinum Taq	0.1 μl (→0.02 U/μl)
Water	16.025 μl

3. Conduct the qPCR using the following cycling conditions:

40 cycles of	94°C for 2 min 94°C for 15 s 62°C for 15 s 72°C for 32 s

4. Establish the concentration of the libraries using standards. The standards are libraries with 200 bp insert size that have already been sequenced and for which the cluster number at a given loading concentration is known.
5. The standards have concentrations of 1, 10, and 100 pM, based on the concentrations measured on the Agilent Bioanalyzer 2100. Both the standards and the libraries with unknown concentration are run in triplicate.
6. Prior to sequencing, dilute the libraries to the required concentration (e.g. 4.5 pM) following the Illumina cluster generation protocol. To sequence the tag, use the index-sequencing primer.

3.9. Timing

1. Sample fragmentation: 10 min per sample
2. End-repair, A-tailing, and ligation: 6 h
3. SPRI beads cleanup: 50 min
4. Indexing enrichment PCR:

Setup	0.5 h
PCR	0.5 h
Cleanup	40 min

5. Normalization and pooling:

Reaction setup	0.5 h
qPCR	2 h
Analysis	0.5 h

6. Size-selection step:

Gel casting	1 h
Gel setup and running	3 h
Gel slices purification	5 h

7. Final qPCR quantification:

Reaction setup	0.5 h
qPCR	1.5 h
Analysis	0.5 h

4. Notes

1. Unless specified otherwise, use electronic multichannel pipettes.
2. If the DNA amount is >2.5 μg, increase the duration of shearing to 160 s.
3. Use only MicroAmp optical 96-well type of plates, since they fit well in the SPRIPlate.
4. If the DNA amount is >2.5 μg, elute the samples in 10 μl EB + 7 μl water.
5. The volume of the adapters used depends on the DNA quantity and, consequently, on the volumes in which the samples are eluted.

   As explained in the experimental design, an increased volume of the ligation reaction may be required. In this case, for each sample, the following MM has to be prepared:

	1×
2× Quick DNA ligase buffer	35 μl
A-tailed sample	17 μl
40 μM Adapter oligo mix	1–12 μl
Water	12 μl-adapter oligo mix

6. If the initial DNA concentration is >2.5 μg, elute in 24 μl.
7. It is recommended to work with only half of the samples simultaneously.
8. If the initial DNA amount is >2.5 μg use 1 μl.

9. If the initial DNA concentration is >2.5 μg, elute in 20 μl.
10. If working with >50 PCR fragments or observing Ct values >20, use 1:20 dilution.

References

1. Kasschau, K. D., Fahlgren, N., Chapman, E. J., Sullivan, C. M., Cumbie, J. S., Givan, S. A. et al. (2007) Genome-wide profiling and analysis of Arabidopsis siRNAs. *PLoS Biol.* **5**, e57.
2. Meyer, M., Stenzel, U., Myles, S., Prufer, K. and Hofreiter, M. (2007) Targeted high-throughput sequencing of tagged nucleic acid samples. *Nucleic Acids Res.* **35**, e97.
3. Binladen, J., Gilbert, M. T., Bollback, J. P., Panitz, F., Bendixen, C., Nielsen, R. et al. (2007) The use of coded PCR primers enables high-throughput sequencing of multiple homolog amplification products by 454 parallel sequencing. *PloS One* **2**, e197.
4. Hamady, M., Walker, J. J., Harris, J. K., Gold, N. J. and Knight, R. (2008) Error-correcting barcoded primers for pyrosequencing hundreds of samples in multiplex. *Nature Methods* **5**, 235–237.
5. Craig, D. W., Pearson, J. V., Szelinger, S., Sekar, A., Redman, M., Corneveaux, J. J. et al. (2008) Identification of genetic variants using bar-coded multiplexed sequencing. *Nature Methods* **5**, 887–893.
6. Cronn, R., Liston, A., Parks, M., Gernandt, D. S., Shen, R. and Mockler, T. (2008) Multiplex sequencing of plant chloroplast genomes using Solexa sequencing-by-synthesis technology. *Nucleic Acids Res.* **36**, e122.
7. Hamming, R. (1950) Error Detecting and Error Correcting Codes. *The Bell System Technical Journal* **29**, 147–161.
8. Cox-Foster, D. L., Conlan, S., Holmes, E. C., Palacios, G., Evans, J. D., Moran, N. A. et al. (2007) A metagenomic survey of microbes in honey bee colony collapse disorder. *Science* **318**, 283–287.

Index

Young Min Kwon and Steven C. Ricke (eds.), *High-Throughput Next Generation Sequencing: Methods and Applications*, Methods in Molecular Biology, vol. 733, DOI 10.1007/978-1-61779-089-8, © Springer Science+Business Media, LLC 2011

D

E

F

G

H

M

T

MIX
Papier aus verantwortungsvollen Quellen
Paper from responsible sources
FSC® C105338
FSC
www.fsc.org

If you have any concerns about our products,
you can contact us on
ProductSafety@springernature.com

In case Publisher is established outside the EU,
the EU authorized representative is:
Springer Nature Customer Service Center GmbH
Europaplatz 3, 69115 Heidelberg, Germany

Printed by Libri Plureos GmbH
in Hamburg, Germany